Southern Biography Series

William J. Cooper, Jr., Editor

Joseph LeConte

Joseph LeConte

Gentle Prophet of Evolution

Lester D. Stephens

Louisiana State University Press

Baton Rouge and London

Designer: Marcy Johnston
Typeface: Zapf Book
Typesetter: Graphic Composition
Printer: Thomson-Shore, Inc.
Binder: John Dekker and Sons

Library of Congress Cataloging in Publication Data

Stephens, Lester D.
Joseph LeConte, gentle prophet of evolution.

(Southern biography series)
Bibliography: p.
Includes index.
1. LeConte, Joseph, 1823–1901. 2. Geologists—
United States—Biography. 3. Evolution. I. Title.
II. Series.
QE22.L44S74   550′.92′4 [B]   81–20695
ISBN 0–8071–0995–9   AACR2

For Marie

Contents

Illustrations

Preface

Joseph LeConte was born and reared in Georgia, and he spent over half of his life in the South. The distinctive culture of his native region was thus deeply imbedded in his character. His charming manners, high-minded loyalty, conservative social philosophy, and paternalistic racial attitudes were all constant reminders of his origins and up-bringing. Yet LeConte also differed from many of his fellow southern-ers by being much more cosmopolitan in spirit and less provincial in interests. In large measure his catholicity was a result of the influence of his father Louis, a religious liberal and a man of broad scientific curiosity. Joseph's educational training outside the South likewise helped to shape the young man's intellectual inclinations and to break some of the sectional fetters that bound his southern kinsmen. Moreover, his association with liberal educators and able scientists in southern colleges expanded his views and mitigated his parochial outlook. Still, LeConte's intellectual development and influence would likely have been as confined as that of other able southern scientists had not war and the fortunes of defeat rudely snatched this gentle man from the mold of regionalism and thrust him headlong into the mainstream of American intellectual activity.

As a legatee of a large plantation and four dozen slaves, LeConte enjoyed the benefits of the aristocratic planter class until General Sherman and his Federal forces laid waste to the rebellious South in 1865. After enduring incredible hardships as a result of the war and after losing his material wealth, LeConte endeavored to adjust to a new life, but his powerful allegiance to traditional southern culture, his dismay over Reconstruction policies, and his ardent desire to con-tinue his pursuit of abstract science finally compelled him to leave his native country. Unable to find a better place in the South and shunned by northern academic institutions, the hapless man eagerly seized the opportunity offered him in 1869 by the newly founded University of

California. Thus, he accepted the chair of zoology and geology in the fledgling institution, and there he served for the last thirty-two years of his life. Uprooted from the land of his birth and cast into a markedly different culture on the West Coast, LeConte overcame personal difficulties, readily adapted to a new environment, and ultimately attained international recognition for his scientific writings. That he succeeded in doing all of these after he was forty-six years old is a tribute to the strength of his character.

Although LeConte never achieved the recognition of such men as his contemporaries Louis Agassiz, James Dwight Dana, and Asa Gray, he was nevertheless a very significant intellectual figure, and through his research, teaching, and writing he made an indelible impression upon the history of American thought. Born in an era when intellectuals were actively universal in their range of interests, he published over 190 articles and nine books on a variety of topics—scientific, religious, philosophical, and literary. Indeed, his versatility easily qualified him for the label "Renaissance Man." But such ecumenicity came at a high price, for although LeConte began his career when breadth of knowledge was the hallmark of a learned man, he lived on into the modern age of specialization. As a consequence, he failed to carve out a permanent niche in specialized scientific contributions. Nonetheless, his far-ranging inquiries helped to open vast areas of knowledge, and he earned a lasting place in the annals of American scientific, social, and religious thought. For this reason alone, LeConte is deserving of a biographical study.

But there is yet another justification for writing his biography: the life and work of Joseph LeConte clearly echoed the tenor of thought in nineteenth-century America. Like all prominent individuals in the history of humankind, LeConte was not only a leader but also a product of his own age. His ideas thus typified the paradox of the intellectual whose inquiries inevitably challenge contemporary conventions and at the same time mirror the cultural bias and the spirit of the society in which he lives. Nowhere was this paradox more clearly presented than in the nineteenth-century controversies between evolutionary theory and traditional Christian dogma. Indeed, LeConte devoted much time and thought to reconciling the awesome inconsistencies between Christian teachings and scientific findings, and his publications on the subject earned him the admiration of many of his intellectual contemporaries while also eliciting the opprobrium of evangelical Protestants. LeConte wrestled long and hard before accepting evolutionism, but once converted he became one of its most effective popularizers through his books, articles, public addresses, and classroom teaching.

The talented southerner did not confine his work to that topic alone, however, for in addition to writing many quasi-popular articles, LeConte also made a name for himself as a scientist, most particularly in the fields of physiological optics and geology. His published papers and books on those subjects were widely recognized as authoritative, and they earned for the man a firm position among his peers. Moreover, as a highly esteemed professor of science at the University of Georgia, South Carolina College, and the University of California, respectively, he influenced three generations of students by his teaching. His warmth, humor, and dedication, combined with a sense of devotion to his responsibilities as a teacher, endeared him to hundreds of students over a period of fifty years as a professor.

Joseph LeConte was one of those rare individuals whose place in the social and intellectual history of the United States can be overlooked only at the expense of the larger picture of American thought. To overplay his significance would surely be a mistake, but to underplay his influence would just as certainly be wrong, for it would create a hiatus in our knowledge of American life. This biography attempts to give a full account of the man and his career and to assess his role in the development of social and scientific ideas in the nineteenth century and, in so doing, to give us a better image of ourselves today.

In researching the life of LeConte I had to travel around the country in search of primary source materials, for his papers are scattered among his descendants and various archives. Despite the time-consuming nature of this endeavor, it was well worth the effort. In due course I met many LeConte relatives—all of whom are entitled to my greatest thanks for their cooperation, courtesy, and good will. I am especially indebted to the late Emma Talley Shaw, a great-granddaughter of Joseph LeConte. Mrs. Shaw possessed a wealth of LeConte's personal correspondence and various journals, reminiscences, photographs, and other memorabilia, a bequest of her grandmother, Joseph's oldest daughter, Emma LeConte Furman. All of these rich sources Mrs. Shaw freely placed at my disposal, and to them she added valuable personal knowledge of family traditions as well as warm encouragement to explore fully the life of her famous forebear. I am saddened that she did not live to see the completed work. But the papers she held were bequeathed to her daughter Carolyn McMillan, of St. Simons Island, Georgia, and Mrs. McMillan has been equally cooperative and supportive. Her kindness must not go unrecognized.

Other LeConte descendants also deserve my deep appreciation. Miss Helen Malcolm LeConte, of San Francisco, graciously allowed me to use the few but very important papers in her private collection, while Katherine Adams, of Carrboro, North Carolina, did the same. I am also indebted to the late Nancy Smith, who lent me the LeConte-Furman papers she possessed, and in her memory I have chosen to label that collection with her name, although those papers are currently in possession of her sister, Mrs. Adams. Mr. and Mrs. Joseph LeConte Smith of Macon, Georgia, granted me the privilege of researching their LeConte collection, as did Katherine Brookshire of Athens, Georgia; and I am indebted to them for their generosity.

I am obligated to three other LeConte relatives as well. The late Mr. Richard LeConte Anderson of Macon, Georgia, was unfailingly helpful. Mr. Anderson undoubtedly knew more about the LeConte family than anyone else, and he freely shared his time and knowledge to aid me in this work. Both he and Mrs. Anderson devoted considerable time and energy to proofreading the manuscript, and their meticulous attention to detail places me in their debt. In addition, John McKay Sheftall, also of Macon, provided me with very useful information and even a private collection of LeConte papers. The latter are owned by Mrs. Joseph Nisbet LeConte, of Greenville, North Carolina, and, although I have never met Mrs. LeConte, I owe a debt to her for making those papers available.

As every historian is aware, the help of archivists is essential to thorough research. It is not feasible to list all of the individuals who aided me in locating LeConte materials, so I must be content to offer my general thanks to those able librarians who so cordially facilitated my research at the Bancroft Library, University of California; the American Philosophical Society Library; the Southern Historical Collection, University of North Carolina; Duke University Library; the South Caroliniana Library, University of South Carolina; the Library of Congress Manuscripts Division; the Smithsonian Institution Archives; the Georgia Historical Society; and the University of Georgia Libraries. A special note of thanks is due Miss Lee Swift of the Geological Society of America for her assistance.

In part, this work was made possible by two grants from the Penrose Fund of the American Philosophical Society, the first in support of the biography and the second for special work on LeConte's contributions to the theory of mountain formation. The support of the APS is gratefully acknowledged. I also appreciate the special summer research grants awarded me by my own Department of History at the University of Georgia.

Certain individuals read and criticized specialized portions of this biography, and I am indebted to them for their comments and suggestions. They include Luis M. Proenza and Lelon J. Peacock of the Psychology Department, University of Georgia; Vernon J. Hurst, Research Professor of Geology, University of Georgia; George W. White, Research Professor Emeritus of Geology, University of Illinois at Urbana-Champaign; Stephen G. Brush of the Institute for Physical Science and Technology, University of Maryland; and Clarence L. Mohr, of Tulane University. For his thoughtful criticisms of the entire work, James O. Breeden of Southern Methodist University deserves an expression of my gratitude. Very special thanks are due to my colleague and friend Thomas G. Dyer, whose encouragement, criticisms, and suggestions were enormously helpful throughout the project.

To Joseph C. Hammock I express my appreciation for originally suggesting that I undertake this study, and to Nancy Heaton I pay thanks for extraordinary care in the typing of the manuscript. She was ably assisted by Cathy Cartey, LeAnne Fouche, and, especially, Linda Green, in the final stages of the endeavor.

Finally, but most important, I must acknowledge the constant support and encouragement of my wife, Marie Ellis. In addition, her skills as a reference librarian facilitated my research, and her detailed criticism of the manuscript saved me from many errors. Above all, she cared.

To all of these helpful and generous people goes my warmest appreciation. I can only hope they will be pleased with the work they helped me achieve.

Acknowledgments

Permission to quote from LeConte materials in their possession has been kindly granted by Katherine S. (Mrs. J. E.) Adams, of Carrboro, North Carolina; Miss Helen Malcolm LeConte, of San Francisco, California; Mrs. Joseph N. LeConte, of Greenville, North Carolina; and Carolyn Shaw McMillan, of St. Simons Island, Georgia. Material from the diary of Bertha L. Chapman, in the Yosemite National Park Collections, is quoted by permission of the National Park Service, via courtesy of Mrs. Carol C. Montoya, of Vidalia, Georgia. Quotations from materials in the LeConte Family Papers, University of California, are published by permission of the Bancroft Library.

Portions of Chapter XII were previously published as: "Joseph LeConte on Evolution, Education, and the Structure of Knowledge," in *Journal of the History of the Behavioral Sciences*, Vol. XII, No. 2 (1976), 103–19; and "Joseph LeConte's Evolutional Idealism," in *Journal of the History of Ideas*, XXXIX (July, 1978), 465–80. They are reprinted in revised form here by the permission of each journal. Part of Chapter XIII first appeared as "Evolution and Women's Rights in the 1890s: The Views of Joseph LeConte," in *Historian*, XXXVIII (February, 1976), 239–52. I acknowledge the editor's permission to reproduce this material. Sections included in Chapters VIII and XIV were originally published as "Joseph LeConte and the Development of the Physiology and Psychology of Vision in the United States," in *Annals of Science*, XXXVII (1980), 303–21, and appear here in revised form by permission of Taylor & Francis, Ltd., London, England.

Joseph LeConte

—••❡ I ❡••—

As the Twig Is Bent: The Early Years

Flowing gently through the lowlands of Liberty County, Georgia, the shallow waters of the little South Newport River meander in their crooked channel until they merge with their northern counterpart and then glide imperceptibly into a maze of swamps and tidal creeks, emptying finally into the inlet on the southern fringe of St. Catherines Sound. The swampy terrain of the environs is typical of the entire coastal plain of southeast Georgia—flat, fertile, and studded with pines, bald cypresses, and water oaks. By the late eighteenth century, a number of enterprising Americans had bought up vast acres of this land, cleared large portions of timber, and established rice and cotton plantations. One of the most successful of such endeavors was the plantation called Woodmanston, operated by Louis LeConte in the first third of the nineteenth century. Covering an expanse of 3,356 acres, Woodmanston was situated near the South Newport River and in the heart of Bulltown Swamp, a dozen miles inland from the coast and forty-five miles due southwest of Savannah.

The original tract of land that became Woodmanston was acquired in 1769 by William and John Eatton LeConte, natives of New Jersey. The LeConte brothers also established a plantation called Sans Souci on the Ogeechee River near Savannah. Although they lived together for many years at one or the other plantation, William eventually settled at Sans Souci, and John Eatton at Woodmanston. William became a permanent resident of Georgia, and during the Revolutionary War he served as a member of the state's Council on Safety and later, of the provincial congress. The younger LeConte continued to reside in New Jersey during the winters and in Georgia during the summers. In 1787, after the death of his brother, John Eatton arranged a settlement with William's widow, whereby he relinquished claim

1

to Sans Souci and acquired virtual ownership of Woodmanston.[1]

Married to Jane Sloane of New York, John Eatton LeConte sired three children, William, Louis, and John Eatton, Jr. His son William never married, but the other two brothers did, and it is through them that the LeConte line was continued. The third son, John Eatton, Jr., was born in 1784 and joined the United States Army in 1817. Serving for ten years as a captain and three years as a major in the Corps of Topological Engineers, the younger John Eatton established himself as a leading naturalist, and he continued to collect and study plant and animal specimens virtually until the end of his life in 1860. Although he spent much time in Georgia, John Eatton, Jr., resided officially in New York City until 1852, when he moved to Philadelphia. Widely recognized for his research, Major LeConte was personally acquainted with most of the leading naturalists of his time, and he published numerous papers on a variety of scientific topics. His only son, John Lawrence LeConte, became one of the most prominent entomologists in the United States in the period 1850–1883.[2]

It was the second of the elder John Eatton's sons, however, who came to be most intimately associated with Woodmanston. Born in 1782, Louis was graduated from Columbia University in 1799. He studied medicine briefly thereafter but apparently did not take the M.D. degree. Around 1809 he moved to Georgia and, about a year later, "assumed the management" of Woodmanston, eventually acquiring full possession of the Georgia estate. Louis LeConte was a man of great intellectual curiosity and unorthodox religious inclination. This was to be expected, since he was heir to a legacy of inquiring minds and dissenting religious views. His great grandfather was Guillaume Le-Conte (1659–1710), a Huguenot who had fled France near the time of the revocation of the Edict of Nantes by Louis XIV in 1685. Several

1 Joseph LeConte, *The Autobiography of Joseph LeConte* (New York, 1903), 4; *The Colonial Records of Georgia* (28 vols.; Atlanta and Athens, Ga., 1904–79), VIII, 336, 665, 754; *ibid.*, IX, 178, 227, 458.

2 The information on the younger John Eatton LeConte is derived from numerous sources: the LeConte Family Bible (in possession of Katherine Brookshire, Athens, Ga.); LeConte, *Autobiography*, 1–21; and the LeConte Family Papers, American Philosophical Society Library, Philadelphia (hereinafter cited as LeConte Family Papers, APS). In addition, see Jeannette E. Gaustein, *Thomas Nuttall, Naturalist* (Cambridge, Mass., 1967), 99, 120, 160; Asa Gray, "Some North American Botanists: IV. John Eatton LeConte," *Botanical Gazette*, VIII (April, 1883), 197–99; William Darlington (comp.), *Reliquiae Baldwinianae* (Philadelphia, 1843), *passim*; and Richard Adicks (ed.), *LeConte's Report on East Florida* (Orlando, Fla., 1978), 6–16.

years before the turn of the century, Guillaume migrated to New Rochelle, New York. His son Pierre became a prominent physician in colonial New Jersey. Pierre's second wife, Valeria Eatton, bore several children, including the two sons who acquired the Georgia properties, William and the senior John Eatton.[3]

When Louis LeConte arrived in Liberty County, he found himself "an outsider, an interloper," in the firmly religious community dominated by English-descended Puritans. Although he was never completely accepted by the closely knit clan, he quickly established a reputation as a man of great integrity and responsibility, and in 1812 he married Ann Quarterman, one of the "insider" Puritans. From their union were born four sons and three daughters, one of whom died in infancy. The sixth of their children was Joseph, born on February 26, 1823. Joseph would be strongly influenced by his able father, a man of versatile talent, devotion to scientific study, liberal religious views, and prodigious energy. This influence, coupled with a childhood in a rich natural environment, made an indelible impression upon Joseph in his formative years.[4]

At some point shortly after he took charge of Woodmanston, Louis established a botanical garden on an acre of ground only a few yards from the knoll on which the Woodmanston home lay. In it he planted and cultivated many imported bulbous plants, especially from the Cape of Good Hope, and a great variety of flowering shrubs. The garden became a showpiece that attracted widespread attention and

3 The LeConte genealogy is given in several sources, including LeConte, *Autobiography*, 1–21; Samuel H. Scudder, "Memoir of John Lawrence LeConte, 1825–1883," *National Academy of Sciences Biographical Memoirs*, II (1886), 261–307; and Emma LeConte Furman, "Sketch of the LeConte Family of the Nonant Line in America" (Typescript in R. Means Davis Papers, South Caroliniana Library, University of South Carolina, Columbia). Mrs. Furman also compiled a genealogy book, formerly in the private collection of the late Nancy Smith, Milledgeville, Ga., and now in possession of Katherine Adams, Carrboro, N.C. (hereinafter cited as Nancy Smith Collection). These genealogies are occasionally inaccurate. The most complete and accurate research on the LeConte line is Richard LeConte Anderson, *LeConte History and Genealogy* (2 vols.; Macon, Ga.: Privately printed, 1981). On related branches of the family, John McKay Sheftall, of Macon, Ga., has compiled useful information.

4 LeConte, *Autobiography*, 1–21. Louis LeConte is listed as a graduate of the New York college in 1799 (*Columbia University Officers and Alumni, 1754–1857* [New York, 1936], 118). See also typescript copy of the will of Jane Sloane LeConte, from the Clerk's Office of the Surrogate's Court, Hall of Records, New York, N.Y. (copy made by Richard LeConte Anderson; now on deposit in LeConte Genealogical Collection, University of Georgia Library, Athens).

drew visitors from both the U.S. and abroad. The well-known Scottish horticulturist Alexander Gordon, who visited Woodmanston in 1832, later wrote that LeConte's botanical garden was one of the most notable he had ever seen. Joseph was profoundly impressed by his father's beautiful garden, and he often fondly recalled that the "garden was the joy and delight" of his childhood.[5]

As a devoted cultivator of science, Louis also contributed to the development of botanical science in other ways, particularly through his numerous specimen-collecting excursions in the surrounding region and in the vicinity of the Altamaha River. Although he declined to publish his own findings, the able botanist freely shared his knowledge of Georgia flora with his brother and other eminent naturalists such as William Baldwin and Asa Gray. Intimately acquainted with both the Linnaean and the natural systems of classification, LeConte observed and identified numerous species of southern flora.[6]

Louis was a master of Latin, which he could read "at sight almost as readily as he did English," and he was "thoroughly acquainted" with the classics. Indeed, his knowledge of the latter had a decided impact upon Joseph, who often revealed a marked familiarity with the great literary works of the past. Moreover, Louis was equally adept in mathematics, physics, chemistry, and zoology. Like most of his intellectual contemporaries, he was a universalist, and two of his sons, John and Joseph, emulated him in their broad approach to scientific matters. As Alexander Stephens was to remark long after the death of Louis, the unpretentious naturalist was extraordinarily gifted. Stephens was in a position to know: he was hired to tutor Louis' younger children from 1832 to 1833, and he was in close contact with the elder LeConte during that time.[7]

A paternalistic slave owner, Louis diligently tended to the physical

5 Alexander Gordon, "Principal Nurseries and Private Gardens in the United States of America," in J. C. Loudon (ed.), *Gardener's Magazine*, VIII (1832), 286–89. See also John C. Loudon, *An Encyclopedia of Gardening* (Rev. ed.; London, 1850), 332; LeConte, *Autobiography*, 9–10.

6 LeConte, *Autobiography*, 9–10; Gray, "Some North American Botanists"; Darlington (comp.), *Reliquiae Baldwinianae, passim*.

7 LeConte, *Autobiography*, 11–12; Rudolph von Abele, *Alexander H. Stephens: A Biography* (New York, 1946), 49–51; Myrta Lockett Avery (ed.), *Recollections of Alexander H. Stephens* (New York, 1910), 487–90; James Z. Rabun (ed.), "Alexander H. Stephens's Diary," *Georgia Historical Quarterly*, XXXVI (March, 1952), 79–80. Stephens wrote the article on Louis LeConte in *Johnson's New Universal Encyclopedia* (New York, 1877), 1703. See also

needs of his bondsmen, and he looked after their "moral and spiritual welfare" by encouraging local clergymen to instruct them in Christian doctrine. Among those with whom Louis freely cooperated in the task was the Rev. Dr. Charles Colcock Jones, the well-known Presbyterian minister who served for a time as pastor of the Midway Congregationalist Church. In fact, Louis himself attended the Midway Church, located only a few miles north of Woodmanston, and he required his children to attend services there every Sunday for many years. But he personally refused to become a church member, choosing instead a more naturalistic position on religious matters.[8] This attitude exerted a lasting influence upon his son Joseph.

Relatively little is known of Joseph's mother. Born in 1793, she died in 1826, when Joseph was only three years old. Yet, as Joseph reflected in later years, Ann Quarterman LeConte probably molded and shaped his life more than he really knew. She was a "deeply and genuinely pious" person, and she was "passionately fond of art, and especially of music." Perhaps, as Joseph suggested, it was his mother's influence that directed his taste for art and music, and surely his own deep piety can in large measure be attributed to Ann's early teaching of her infant son.[9]

In his boyhood days Joseph was most intimately associated with his brother Lewis and his sister Ann. This was only natural since his oldest brother, William, and his oldest sister, Jane, were his seniors by eleven and nine years respectively. His brother John was five years older than he, and thus, by the time Joseph had reached the age of ten, John was already away at college. Lewis was only two years older than Joseph, however, and the two boys not only spent much time together at play and in school but also attended the same classes at the University of Georgia. The youngest child, Ann, was born in 1825, only two years after Joseph. She often participated with him and

Louis LeConte to William LeConte, October 29, 1832, in Clifford Anderson Papers, Southern Historical Collection, University of North Carolina, Chapel Hill.

8 LeConte, *Autobiography*, 6–7, 12–13, 16–17. On the missionary work of Dr. C. C. Jones, see Robert Manson Myers (ed.), *The Children of Pride* (New Haven, 1972), *passim*. See also Annie Sabra Ramsey (ed.), "Church Going at Midway, Georgia, as Remembered by Matilda Harden Stevens," *Georgia Historical Quarterly*, XXVIII (December, 1944), 270–80.

9 LeConte, *Autobiography*, 5–6, 15.

Lewis in social activities and accompanied them on outdoor excursions.[10]

Woodmanston was an ideal place for sporting, and Joseph and Lewis frequently hunted and played together in the vast expanse of their father's plantation. Lewis gained a reputation as a superior marksman, and he dearly loved guns, a devotion that ultimately brought on his untimely end. Joseph, although he thoroughly enjoyed specimen-collecting and made his own mark as a "Nimrod," was extremely sensitive to the death of animals, and he never favored hunting as a sport. Instead, he found greater pleasure in canoeing about the shallow waters of Bulltown Swamp and in simply observing the myriad forms of natural life.[11]

In 1832 Joseph's brother William graduated from the University of Georgia. William's roommate at the Athens institution had been Alexander H. Stephens, a slight young man, frail of form but able of mind and scholarly by nature. William encouraged his father and other neighboring plantation owners to employ Stephens as a teacher. Stephens accepted their invitation in December, 1832, and for a year taught the younger children of Louis LeConte and other prominent Liberty County families. For Joseph, this experience constituted his only formal preparatory training for college. Although Stephens taught him for only a year, Joseph always maintained that no one else other than his father had contributed more to his early education.[12]

Meanwhile, Joseph's oldest sister, Jane, had become engaged to John M. B. Harden, a young physician recently moved to the county. In 1833 she and Harden were married, and they settled at Woodmanston. Harden was keenly interested in science, and he published numerous articles dealing with medicine, physiology, botany, and geography. He often worked with his father-in-law in the little chemical laboratory constructed by Louis LeConte in the attic of the Woodmanston home. It is significant that Harden's interests and scientific in-

---

10 *Ibid.*, 23–24. As with the name of his father, the spelling of the name of Joseph's younger brother is recorded variously as "Lewis" or "Louis." For purposes of clarity, the name of the senior LeConte is given herein as Louis and that of the junior as Lewis, although the correct spelling of each is Lewis (information supplied by Richard LeConte Anderson).

11 LeConte, *Autobiography*, 25–29.

12 *Ibid.*

quisitiveness bore fruit in their influence upon Joseph. In fact, it was with Harden that Joseph first discussed the idea of organic evolution, which ultimately became the focal point of the latter's thought and writing.[13]

The third surviving child of Louis LeConte was John. He too attended the University of Georgia, graduating in 1839. Somewhat aloof and formal in his relationships throughout his life, John did not make friends easily, but he was a serious student and distinguished himself in mathematics and physics. After receiving his M.D. degree in 1841 from the College of Physicians and Surgeons in New York, John married Eleanor Josephine Graham of New York and returned to the South to establish a medical practice in Savannah. His interests were primarily scholarly, however, and he devoted considerable time to research and writing—in medicine during the early years of his career and later in physics. Eventually, he and Joseph were to become fellow faculty members at the University of Georgia, the College of South Carolina, and, finally, the University of California.[14]

Like his father and his brother John, Joseph displayed a natural curiosity for learning. He read widely, learned to play the flute, and inquired into many subjects. Before he reached his early teens, he mastered the ability to observe optical phenomena, often sitting alone for hours to study visual illusions and to increase his skill in controlling his sight. He also frequently accompanied his father on specimen-collecting trips, and he acquired extensive knowledge from his father's chemical experiments and mathematical investigations.

By the age of fifteen, Joseph was prepared to enter college, and he and his brother Lewis arranged to set out for Athens to enter the University of Georgia in the new school term beginning in January, 1838. The young boys "had never been farther from the plantation home than Midway Church," some eight miles away, and they had never before worn any formal attire. All seemed well as the nervous and excited youths prepared for the trip to Athens, but on the day before

13 Lester D. Stephens, "Of Mercury, Moses, and Medicine: Views of Dr. John M. B. Harden," *Georgia Historical Quarterly*, LIX (Winter, 1975), 402–15.

14 Joseph LeConte, "Memoir of John LeConte, 1818–1891," *National Academy of Sciences Biographical Memoirs*, III (1895), 369–93; John Samuel Lupold, "From Physician to Physicist: The Scientific Career of John LeConte, 1818–1891" (Ph.D. dissertation, University of South Carolina, 1970).

their scheduled departure, Louis took to his bed with a fatal illness. He died from "blood-poisoning" on January 9, 1838.[15]

The blow to the LeConte children was severe, but it was particularly devastating to young Joseph, who was already in a state of anxiety over leaving home. As he watched his father slipping away in agony, the sensitive youth was struck by a barely perceptible but nonetheless real recollection of his dying mother and the basin of blood let from her by attending physicians. Louis' death was subsequently to play a role in the "conversion" of Joseph to Christianity as his brother William adjured him to avoid the uncertainty of his father's immortality by preparing his own soul for the afterlife.[16]

Still, youth is resilient, and both Joseph and Lewis quickly shook off the numbing effects of this loss and embarked for Athens, some three hundred miles distant. Arriving a week after classes had begun, the homesick but ambitious young men plunged into their studies and quickly proved themselves serious and able students. Chartered in 1785, the University of Georgia had not actually opened its doors until 1801. By the time Joseph and Lewis arrived, the college had grown very little, and its curriculum resembled that of many other small institutions where the classics and religion predominated. In fact, it was not really a university at all but only a small liberal arts college, and the name by which it was commonly called, Franklin College, was fittingly appropriate.[17] Since the sciences were secondary to other subjects in the curriculum, Joseph received little worthwhile training in them, except for mathematics and physics, which were ably taught by Professor Charles F. McCay, a man with whom Joseph was later to become a colleague at the University of Georgia and subsequently at the College of South Carolina. On the other hand, a thorough exposure to the classics intensified his acquaintance with important literary works, which served Joseph well throughout his long life as a universalist in interests and a philosopher at heart.

Slim and lanky, the fifteen-year-old youth (he turned sixteen a month after entering college) "suffered severely from nostalgia" during the first few weeks away from home. "My yearning for the old planta-

---

15 LeConte, *Autobiography*, 30–36. John M. B. Harden gave a detailed report on Louis LeConte's death in his "Cases of Diffuse Cellular Inflamation," *American Journal of the Medical Sciences*, XX (August, 1838), 396–402.

16 LeConte, *Autobiography*, 6, 36.

17 E. Merton Coulter, *College Life in the Old South* (2nd ed.; Athens, Ga., 1951), 1–89.

tion and the beautiful garden was intense," he said. His spiritual malaise was deepened further when his brother William wrote to him and Lewis, describing in detail the horrible deathbed scene of their father. A "thoroughly religious man ... of the old orthodox type," as Joseph later called him, William urged upon the two boys "the absolute necessity of 'fleeing from the wrath to come.'" The letter "distressed me greatly," recalled Joseph in later life, but for several months he merely lived in a state of anxiety regarding his soul's salvation—that is, until he and Lewis both joined the Presbyterian Church during a religious revival in Athens in 1840. Despite the spiritual crisis, however, Joseph weathered the temptation to leave college and return to the "old plantation."[18] After all, it was not the same place he had always known, and it would never be the same again.

Joseph had no disposition toward revelry, and he shunned the activities of the college men who sought a good time in drinking, hazing, and prank-playing. In addition to devoting his time to study, he joined the Phi Kappa Literary Society and actively participated in debates with the rival Demosthenian group. By his own admission, he was "too young and sensitive, too easily embarrassed to make a debater." Nevertheless, he participated in the lengthy Saturday exercises, some of which consumed an entire day and, if "the question was a living one," sometimes "continued until midnight." During his junior and senior years, in spite of his reluctance to speak in public, Joseph was elected as an orator by the faculty.[19] Overall, these efforts paid rich dividends, serving Joseph well in the future when he became a teacher and a much-sought-after speaker, particularly on the controversial subject of the theory of evolution in relation to religion.

The rival Greek societies were a central part of life at Franklin College during Joseph's student days, and the members took their business seriously. Debating was, of course, the focal point of their organization, and Joseph watched the Phi Kappans practice their art almost every Saturday for the duration of his stay at the college. Among the subjects fervently debated were religion, philosophy, and

18 LeConte, *Autobiography*, 40–41; William LeConte to Lewis LeConte, August 23, 1838, and William LeConte to Joseph and Lewis LeConte, February 14, 1838, both in possession of Carolyn McMillan, St. Simons Island, Georgia (hereinafter cited as McMillan Collection); Joseph LeConte to his niece Matilda "Tillie" Harden Stevens, September 27, 1897, in LeConte Family Papers, APS.

19 LeConte, *Autobiography*, 41–56.

a host of current issues. Much vigorous discussion was devoted to such questions as: "Should Students in College visit the Young Ladies?" and "Ought Parents to regulate entirely the choice of their children in the all important subject of matrimony?" Aside from the triviality of such issues, these exercises promoted the skill of oratory, and although Joseph never participated as actively as most of his peers, he was nevertheless influenced. The first vivid effect of this influence appeared in 1856, when Joseph delivered a florid, though graceful and learned, address on the otherwise prosaic topic of the origins of coal.[20]

Although he was slight of stature, Joseph was always athletically inclined. "I was light, active, and fleet of foot," he noted, "and became very expert in gymnastics and as a player of town-ball."[21] Indeed, his superior athletic agility served him ably throughout his active life as a mountain climber and outdoorsman, and, as he observed when he was well over seventy years old, his physical activity probably prolonged his life.

In an all-male college, Joseph was associated with women only through activities outside the campus. The comely but socially reticent young man did not court women during his college days, but he often attended social events where he could admire the young ladies of Athens. Perhaps because he had been deprived of the close companionship of a mother during his childhood and adolescent days, Joseph tended to venerate women, especially those who conformed to his view of the culturally refined and socially graceful lady. "Refined women were . . . a sort of superior beings, belonging to another, higher, and purer sphere of existence. I simply worshiped them," he said.[22] The same romantic inclination prevailed throughout his life, and it was later to play a part in his reaction to the woman's rights movement and in his sociological scheme for the sexes. But it can also be said that it served him well in the choice of a mate several years later and in the duration of a happy fifty-four-year marriage.

As a student, Joseph excelled. He and Lewis were among the best

20 Coulter, *College Life in the Old South*, 103–33; Phi Kappa Literary Society Minutes, 1838–41, University of Georgia Library, Athens; Faculty Minutes, June 9, 1840, and June 28, 1841, University of Georgia Library, Athens; Minutes of the University of Georgia Trustees, August 3, 1841, University of Georgia Library, Athens; Joseph LeConte, "Lectures on Coal," *Annual Report of the Board of Regents of the Smithsonian Institution, 1857* (Washington, D.C., 1858), 119–68.

21 LeConte, *Autobiography*, 46.

22 *Ibid.*, 46–47.

of their class, and Joseph was the third honor graduate in 1841. Yet his achievements came through ability and diligence, not through driving ambition. "I never had any great ambition to excel my fellows in the classes," he claimed. Instead, he studied hard for two reasons: "to please my instructors" and "a real taste in the subject of study." The latter was reflected in his high achievement in "mathematics, mechanics, and physics, and in the last year, in mental and moral philosophy." Thus were laid the grounds of his future research in science and philosophy. As an orator of the junior class, he spoke on "True Greatness," and, of the senior class, on "Love of Truth, the Highest Incentive to Effort." Unfortunately, the content of his addresses is unknown, for he later "burned" them "in disgust at their almost childish crudity and immaturity." Still, their titles suggest that Joseph was something of an idealist, and his lofty views remained very much a part of him through his long life.[23]

During his three years at Athens Joseph dutifully pursued his studies, and he never caused his mentors any problems. Only once did he even border on violation of the college rules. That occurred when his brother Lewis got into a fist fight and was badly pummeled by another student who had, in Lewis' opinion, made a rude remark to a female companion of Lewis. After he learned of the affair, Joseph felt honor-bound to defend his brother, and he proceeded straightaway to the student's room with the intention of accosting his brother's victor. Fortunately for him, the student apologized profusely, and Joseph was spared what, in his own words, would have resulted in a beating by the much larger youth.[24] Significantly, this appears to be the only occasion in his life when the gentle Joseph thought of resorting to personal violence.

Shortly before he graduated, tragedy struck Joseph for a third time when his brother William died in January, 1841, almost exactly three years after his father's death. Following in the footsteps of his father, William decided to become a planter, and in 1839 he had applied the share of money he had received from Louis' estate, along with the proceeds of the sale of some New York property acquired by inheritance, to purchase the Cedar Hill plantation, located only a short dis-

23 *Ibid.*, 47–48; Macon *Georgia Telegraph*, August 17, 1841, p. 2.
24 LeConte, *Autobiography*, 49–51; Sallie LeConte Diary, April 21, [1872] (MS in Davis Papers).

tance from Woodmanston. Married to Sarah Nisbet, William had four children, all of whom were later to become very close to their uncle Joseph after he was eventually appointed their financial guardian. Only twenty-eight when he succumbed to a brief illness, William had served as Joseph's guardian since the death of Louis, and Joseph admired and respected him greatly. Thus, the "news of his death came . . . as a terrible shock" to the young man.[25]

"But youth, absence from the scene of grief, [and] diversion of constant duties of study" soon lessened Joseph's sorrow, and the young man continued on toward his A.B. degree, graduating in August, 1841. Immediately afterward, he and Lewis were joined in Athens by their sister Ann, who had just completed her studies at Georgia Female (later, Wesleyan) College. Possessed of the means from their inheritance to engage in extensive travel, the three set out for a tour of the Northeast in the fall of 1841. The green and woefully inexperienced travelers journeyed first to the nation's capital, where, at Ann's insistence, they "put up at the best hotel," taking their meals in the luxurious quarters of their own parlor. Subsequently, they "gave up this expensive habit," said Joseph, but for several weeks they lived the life of royalty. Awed by the architecture of Washington, the three young people ambled about the capital, delighting in their first taste of life in a large city. For a full week they remained there, taking time to visit the halls of Congress, where they watched the "celebrated trio, Webster, Calhoun, and Clay" at their oratorical best.[26]

From Washington they traveled to Baltimore, Philadelphia, Boston, and Cambridge. Marveling over the wonders of those centers of American culture, they visited "everything that was worth seeing." In October, they went on to New York, boarding near the home of their "Uncle Jack" (John Eaton LeConte, Jr.) and his son, John Lawrence. There they spent a most pleasurable and extended visit, for John

---

25 LeConte, *Autobiography*, 51–52. For additional information on William, see a copy of his letter to Colden R. Ketcham, December 4, 1832, and J. A. LeConte to B. Palmer Axson, January 27, 1941, both in LeConte Genealogical Collection.

26 LeConte, *Autobiography*, 52–55. The three siblings received money for their extended trip from the estate of their father, of which there was "$8,696 cash on hand" at that point (Accounts of Executors, Administrators, and Guardians, 1829–54, in Liberty County Records, Georgia State Archives, Atlanta). On May 16, 1845, Joseph received several shares of stock from his father's estate valued at $3,317.50 and presumably divided equally between him and his sister Ann (Superior Court Deeds, Book M, Pt. 2, 1842–47, pp. 374–75, in Liberty County Records).

LeConte, who had recently received his M.D. degree, and his beautiful wife Josephine, or "Josie," were living temporarily with Uncle Jack. At the same time, John Harden and his wife Jane were also visiting in the home of John Eatton. "All of us spent every evening at 'Uncle Jack's' house, and a very happy six weeks we passed under these delightful conditions."[27]

In November, Joseph and the others returned to the South, but as the three youngest children no longer had their own home, they stayed with the Hardens, who were at the time still living in the old Woodmanston house. Jane, who had served as surrogate mother for them, gladly took them in. During this time Joseph's sister-in-law Sarah, William's widow, returned from a visit to Macon in company with her brother, John T. Nisbet, who was Joseph's age. The two young men soon became fast friends, and they spent the winter of 1841–42 together in "grand times," hunting, horseback-riding, and visiting "the ladies" from nearby plantations. It was the kind of carefree life that Joseph was to enjoy for almost two years.[28]

In the spring of 1842, Joseph, Lewis, and John and his wife went to Athens and met with "four of the Nisbets" to plan a trip to the North Georgia mountains. The Nisbets were, of course, related to the Le-Contes through William's widow, and one of them, Harriet, was soon to become Lewis' wife. Hiring two carriages, the gay party struck out for Gainesville, the gateway to the foothills of the Appalachian Range. From there, they pushed on directly northward to Nachooche Valley and Yonah Mountain. After several days at those scenic spots, the group turned in a northeasterly direction, stopping at Tallulah Falls to view the awesome wonder of the deeply etched gorge, which especially fascinated Joseph. Turning south, they ended their trip at Toccoa Falls, another monument of natural splendor. This was the kind of terrain that ultimately drew Joseph to geological study and a lifetime devotion to the mountains. As the weary but elated travelers set out for Athens, Joseph determined that he would return again to explore this region at his leisure.[29]

27 LeConte, *Autobiography*, 56.
28 *Ibid.*, 56–57.
29 *Ibid.*, 57. Joseph claimed they were gone a full month. He had previously visited the area briefly in the spring of 1841. This information is omitted from the published version of the *Autobiography* but is given in the original manuscript (hereinafter cited as LeConte, Autobiographical MS.). There are a few other instances in which the latter

But such pleasurable sojourns could not last forever; John was ready to embark upon his career as a physician, and Lewis was set to study law at Harvard University. Joseph had meanwhile tentatively decided to pursue a career in medicine, and so, as was customary for the time, he apprenticed himself to a practicing physician. He chose as his preceptor Dr. Charles West of Macon, and for the next several months he lived in that central Georgia city. Dr. West was a capable physician, and Joseph learned much from him. His heart was not fully in the endeavor, however, and he took off considerable time for naturalist excursions, primarily in search of bird specimens, of which he eventually acquired a large collection. By late fall of 1842, he had once more returned to the old plantation.[30]

In the winter of 1843, Joseph "nominally continued" his medical studies under his brother John, who had opened a practice in Savannah. "But precious little study I did that winter," he admitted, for his cousin John Lawrence LeConte "came South and spent several months with me at the old homestead." Again, Joseph had a "delightful winter, riding, boating, duck-shooting, etc." Serious consideration of his future could be put off for yet awhile. When his cousin left and summer arrived, Joseph and "a companion"—most likely his half first cousin, William Louis Jones—set out on an excursion to the Altamaha River, "a distance of thirty miles, to gather the plants and river shells for which the region is so celebrated." There, like his father before him, Joseph delighted in the fragrance of magnolias, viewed the abundant water lilies, and collected specimens of freshwater mollusca.[31] Had it been an earlier era and had not the day of the old-style naturalist all but passed, Joseph might well have followed in the footsteps of his famous uncle and his father or in those of the renowned naturalists who had frequently visited this same region. But things were different now, and he had not yet committed himself to a career, even though he was already nearing his majority.

Meanwhile, his sister Ann had become engaged to Dr. Josiah P. Stevens, a Liberty County physician, and they were married in June, 1843. Although apparently rather successful as a medical doctor, Stevens

provides some useful details excised from the printed version. The Autobiographical MS. is in the Elizabeth Furman Talley Papers, Southern Historical Collection, University of North Carolina, Chapel Hill.

30 *Ibid.*, 57–58.
31 *Ibid.*, 58–59.

never achieved the fame of his famous in-laws. He did serve as physician to the prominent C. C. Jones family of Liberty County, and, with the advice and assistance of his brothers-in-law John Harden and John LeConte, he subsequently published an article in a medical journal. Eventually, he gave up medicine for the life of plantation owner, taking his wife's inheritance, including slaves, to build a moderately successful enterprise of his own. Joseph and Stevens were never especially close, but Stevens' son Walter LeConte Stevens ultimately became like a son to Joseph.[32]

Shortly after Ann's wedding, Lewis returned from Harvard and married Harriet Nisbet of Athens. By this time Joseph had definitely decided to study medicine at the College of Physicians and Surgeons in New York, and he determined that he would enroll when the next term began. Meanwhile, the young man idled away the remaining months of 1843 by visiting in Macon, then in Liberty County, and finally at Uncle Jack's in New York.[33] His decision had at last been made, but it represented less a commitment to medicine than an avenue to the pursuit of abstract science; and if the ultimate goal proved unattainable, the life of a practicing physician at least offered security and status.

32 LeConte, Autobiographical MS.
33 LeConte, Autobiography, 60–61. See also LeConte, Autobiographical MS.

—➤❧ II ❧◆—
A Winding Route to Medical Practice, 1844–1850

Among the foremost medical schools of the mid-nineteenth century, New York's College of Physicians and Surgeons in the 1840s boasted a roster of prominent professors. That was one of the reasons Joseph LeConte decided to pursue his medical training at the reputable institution. Another was because both his father and his brother John had gone there before him. In choosing to go outside the South for his training, he passed up the opportunity to enroll in the medical college established in Augusta, Georgia, in 1832, and he appears not to have entertained any serious thoughts of attending medical school in Charleston, South Carolina, where his brother-in-law John Harden had studied. But LeConte's decision was typical of many young southerners of prominent families at the time. Southern medical schools enjoyed neither the reputation nor the resources of those in New York, Philadelphia, and other northern cities, although it is true that by the 1840s several able men had joined the faculties of the southern institutions.[1]

In January, 1844, the almost twenty-one-year-old LeConte enrolled at the College of Physicians and Surgeons. "It was a constant grind, grind of lectures, six lectures every day for six days in the week," complained the diligent young man. After a series of general lectures on medicine, students took two months of specialized lecture courses. But the curriculum provided for little laboratory work. Joseph recog-

1 Frederick P. Keppel, *Columbia* (New York, 1914); Charles Stephen Gurr, "Social Leadership and the Medical Profession in Antebellum Georgia" (Ph.D. dissertation, University of Georgia, 1973); Joseph I. Waring, *A History of Medicine in South Carolina, 1825–1900* (Charleston, 1967); Richard H. Shryock, *Medicine and Society in America, 1660–1860* (New York, 1960).

nized the deficiencies of a purely theoretical program of studies, and he therefore took it upon himself to observe surgeons at work in local hospitals. He also participated voluntarily in quiz classes and employed a "coach" to aid him in acquiring a sound knowledge of medicine. In the latter instance, he chose young Dr. Lewis Sayre, who subsequently gained recognition for his skills as an orthopedic surgeon. The friendship that developed from this association was to last for many decades. Joseph fondly recalled in his autobiography that the hard work he did under Sayre in the evenings was often relieved by the latter's jolly good humor. "We sometimes had a 'high old time,'" reported LeConte. But even the work under his coach was demanding, and it required much tedious memorization.[2]

Not overly impressed by the quality of instruction at the medical school, Joseph claimed upon graduation that he hardly felt qualified to practice medicine. Actually, he was right; but, given the nature of contemporary medical education, he was no less qualified than any other inexperienced M.D. of the time. Theoretically, the medical schools required a three-year apprenticeship prior to admission, and Joseph had more or less fulfilled that requirement, although he had not been completely diligent in his efforts. Nevertheless, in spite of its inadequate system of educating physicians, the College of Physicians and Surgeons was as good as any other contemporary medical school, and it was far better than most. Yet LeConte recognized the deficiencies of the program of training, and a few years later, in his first published paper, he would address himself to the subject of medical education.[3]

What did attract LeConte was the study of subjects that he called "the more scientific part of the curriculum." He especially enjoyed "physiology, anatomy, pathology and chemistry." This is revealing, of course, because it clearly indicated his preference for abstract science over practical medical work. He also found dissection "strangely fascinating, the very horror of the thing adding greatly to the fascination."[4] Better prepared than many of his fellow students, who did not even hold the baccalaureate degree, Joseph profited from the study of these subjects; and they would eventually serve him well in his research and scientific hypothesizing. His subsequent studies in phys-

2 LeConte, *Autobiography*, 62–63.
3 *Ibid.*, 104.
4 *Ibid.*, 63.

iological optics and comparative anatomy owed much to his professional medical training.

But the evidence that medicine was not his real love manifested itself after the first two terms had ended. As soon as the second course of lectures was over in mid-May, 1844, Joseph seized the opportunity to take a long, leisurely excursion. With his cousin John Lawrence LeConte, or "John L.," as he usually called him, Joseph planned a "westward" trip. Only a little over one year younger than Joseph, John Lawrence LeConte would enter the College of Physicians and Surgeons the following year. A meticulous and highly industrious student, the nineteen-year-old John Lawrence had already established something of a reputation as an authority on insects. In due course, he would become one of America's leading entomologists and a world authority on coleoptera. But the primary purpose of their trip was pleasure and simple exploration, and they made few plans to gather scientific data—although John Lawrence's indefatigable pursuit of "bugs" resulted in the collection of numerous specimens. Joseph was to "poke fun" at his intensely curious cousin over his insect-collecting obsession throughout the journey, but he helped him nevertheless—that is, whenever it did not interfere with the "higher pleasures" of the exploration.[5]

Their journey carried them first to Niagara Falls and Buffalo, then on to the small settlement of Detroit, where they remained for a week. In a side-trip they visited the University of Michigan, which was "then a very small affair." In Joseph's words, "the best thing they [the university] had was a rather fine collection of minerals." After a few days in Detroit, the two young men decided to strike out for a visit into the rugged country around Lake Superior. Traveling by steamer through Lake Huron, they arrived at Fort Mackinac, on a small island lying in the straits that join Lakes Huron and Michigan. During their stay in Fort Mackinac, the young men explored Arched Rock and Sugar Loaf Rock, and Joseph observed the "drift cobbles" that completely covered the island. He claimed that "this was the first time I was interested in

---

5 *Ibid.*, 63–100; Joseph LeConte, "An Early Geological Excursion," typescript of a paper presented before the Cordilleran Section of the Geological Society of America, December 30, 1899, in Minnesota Historical Society, St. Paul. The original journal of the trip was apparently destroyed by General Sherman's troops in February, 1865. For a brief sketch of the life and work of John L. LeConte, see Scudder, "Memoir of John Lawrence LeConte."

geological phenomena." But there would be other sights to whet his curiosity along the way, although it would be many years before he began to study geology seriously.[6]

Hiring a canoe and two men, the companions struck out up the St. Mary's River for Sault Sainte Marie. They remained in the village for two or three days and then caught a steamer to Keweenaw Point, on the shores of Lake Superior. For three weeks the two young students camped on the beach of the landlocked bay at Eagle Harbor in the northernmost part of the Keweenaw Peninsula, idling away the hours collecting agates, hunting game, and swimming. The beach was also ideal for insect collecting, and "John was in ecstasies over this place." But the experience here was likewise invaluable to Joseph, and he later used illustrations regarding the process of fossilization based upon phenomena he observed at the time.[7]

"On the 3d of July, we regretfully left our delightful camp," wrote Joseph, and went on to LaPointe in the Apostle Islands just off the Wisconsin coast. Several days later, they arranged a canoe trip to the headwaters of the Mississippi River, and on July 8 the pair embarked with two hired Canadian guides, neither of whom could speak good English. As they passed the other Apostle Islands, Joseph observed their "level red sandstone, with bold shores, crowned with heavy forests." He was especially fascinated by the erosive power of the waves upon the sandstone, and he recorded his observations in his journal. As he later claimed, his observation of the geological phenomena may have been the first record on the subject, but it was never published, and the journal itself was subsequently destroyed by General Sherman's troops in 1865. Joseph recorded many other such observations in his journal; his interest in geology was growing.[8]

Reaching the St. Louis River a few days later, the party paddled their canoe up the wide and winding river. At one point, as they portaged around a dangerous falls, Joseph could not resist the temptation to plunge into the "roaring cataract" for a "glorious swim," much to the awe of a group of Indian men and boys, who could not be induced to join the intrepid swimmer in spite of his bantering challenge. Impressed by the ability of his guides to conduct the portage while car-

6 LeConte, *Autobiography*, 63–68; LeConte, *Elements of Geology* (Rev. ed.; New York, 1882), 510–11.
7 LeConte, *Autobiography*, 68–72.
8 *Ibid.*, 72–79.

rying extremely heavy loads, Joseph often recalled this as an excellent example of the superiority of the "white race." It was a sign of "good blood," he was later to remark—though it is interesting to observe that one of the guides was a half-breed Indian. But consistency is not always a virtue, and certainly contemporary anthropological distinctions of race were at best murky and unscientific. For Joseph, the comparison was primarily between blacks, with whom he was so familiar from his plantation experience, and whites, who, in his view, constituted the other group of humans.[9]

The party continued on up the rapid river until they reached Sandy Lake, where they camped for a few days to rest, wash clothes, and repair the canoe. Always an avid swimmer, Joseph enjoyed "the clear, warm water of the lake" while his athletically unskilled cousin scurried about on the shore in search of insect specimens. But they were soon off again and before long "rushed with a hurrah into the swift current of the Mississippi." For ten days and five hundred miles, the little party paddled with the current of the huge river, through Little Falls down to the Falls of St. Anthony. While making the half-mile portage around the great falls, Joseph was struck by the geological phenomena of the river gorge. "I at once saw, or suspected, the mode of origin of this gorge, by the recession of the fall," he said. "My conclusions were completely confirmed by the existence ... of an escarpment at the mouth of the gorge eight miles below the fall." But, although his observations were "duly recorded" in his little journal, Joseph was then "too young to appreciate the importance of the observations" he had made, and he was "too little acquainted with geological literature to know that there was anything new in them." Yet, despite his poor knowledge of geology, Joseph displayed keen insight and a remarkable ability to formulate explanatory hypotheses regarding diastrophic processes. Again, he later drew upon his experience here for illustrations in his geology lectures and textbook.[10]

Shortly after passing the falls, the group turned into the St. Peter's (now Minnesota) River and paddled on to the "little straggling village of St. Peter's." Joseph and John L. intended to go on, traveling northwestward up the river into Sioux country, but their guides balked at

---

9 *Ibid.,* 79–80; LeConte, "Human Beings as Pack-Animals," *Science,* XI (June, 1888), 290.

10 LeConte, *Autobiography,* 80–99; LeConte, *Elements of Geology* (1882 ed.), 13–14.

the prospect of venturing into that dangerous territory. Arguing that they had fulfilled their obligation, the two guides departed for their home at LaPointe, leaving the intrepid but unwise cousins without means to continue their journey. Afterward, the two LeContes took passage on a boat upriver to Fort Snelling, where they remained for a week awaiting the arrival of a steamer.[11]

It was now early August, but Joseph and John L. were not anxious to return to New York. After they caught the steamer to St. Louis, they decided to disembark at Galena, Illinois, so that Joseph could "examine the lead-mines there." The pair also rented a boat and rowed to nearby Dubuque, Iowa, for the same purpose. "To the early interest thus excited," stated Joseph in later life, "I attribute the fact that this [*i.e.*, the nature of ore deposits and metalliferous veins] has been a favorite subject of investigation with me." Indeed, it was; he subsequently became a recognized authority on the subject, winning the esteem of mining geologists for his contributions to their field.[12]

The cousins traveled on to St. Louis and then up the Ohio River to Pittsburgh, where they caught a train back to New York. Joseph's interest in geology had been aroused, but the active pursuit of the subject must wait. Now it was time to return to medical school.

Although he was not required to complete additional study, LeConte apparently felt that he needed more training. Thus, for two more terms he subjected himself to the "old grind." The tedium was relieved mainly by gymnastic exercise and visits with leading naturalists, including the ornithologists John Bell, J. G. Giraud, Spencer F. Baird, and the renowned John James Audubon, who lived only a short distance from New York City. The aging Audubon was a gracious host, and Joseph and his brother John, who had come to see relatives in New York, visited the famous old artist and naturalist on several occasions. In fact, these acquaintances, especially the last, aroused Joseph's old ornithological interests, and for the next four or five years after he returned to Georgia, the aspiring naturalist collected numerous species of avifauna and became something of an authority on the habits and distribution of southeastern birds.[13]

11 LeConte, *Autobiography*, 99–100.
12 *Ibid.*
13 *Ibid.*, 100–103; Lester D. Stephens, "Joseph LeConte's Contribution to American Ornithology," *Georgia Journal of Science*, XXXV (June, 1977), 170–81, and corrigendum, *ibid.*, XXXVI (January, 1978), 48.

The grind of lectures ended in April, 1845, when Joseph completed a "creditable thesis," passed his examinations, and graduated as a doctor of medicine. In his words, he was "utterly unfit . . . to assume the terrible responsibilities of medical practice," but that did not especially bother him, for he had "studied medicine mainly as the best preparation for science." His plans were not firmly set, however, and at this point he was still unsure of the actual career he wished to pursue. That mattered little now, for he had inherited enough money and capital from his father's estate that he did not have to worry immediately about employment. For the time being, he would simply do whatever pleased him most.[14]

After bidding farewell to Uncle Jack and John L., the carefree young doctor set out for Savannah to visit his brothers John and Lewis, the latter of whom had established a law practice in the same city where his older brother was practicing medicine. Now legally old enough to assume possession of his share of the Liberty County property and slaves, Joseph made arrangements for the formal transfer of his property from John's care to his own. Once this was complete, Joseph went on to Liberty County, visiting relatives and simply enjoying himself. Now he had time to read for pleasure, search for specimens, and pursue a free social life—and all of these Joseph did in earnest. In fact, for over two years, he lived without any sense of responsibility regarding his future.[15]

One of the books he read during these idle years was the anonymously authored *Vestiges of the Natural History of Creation*, published in 1844. The book aroused Joseph's interest in the subject of organic evolution, a topic that had been debated for some time—and long before Charles Darwin's famous *Origin of Species* appeared in 1859. In

14 LeConte, *Autobiography*, 103–104; *Columbia University Officers and Alumni, 1754–1857*, 213. On August 6, 1845, Joseph was awarded an M.A. degree from the University of Georgia (Athens [Ga.] *Southern Banner*, August 7, 1845, p. 2). The degree was automatically "conferred upon all graduates of the University of three years standing." The practice was discontinued in 1871 (*Catalogue of the Trustees, Officers, Alumni, and Matriculates of the University of Georgia from 1785 to 1906* [Athens, 1906], 17).

15 LeConte, *Autobiography*, 104–19. In the "Inventory and Appraisement of the Estate of Louis LeConte, 12 July 1838," the value of over 230 slaves was set at $87,980.01, and "goods, etc." at $2,425.00. The 3,356 acres of land were divided among the LeConte children; Joseph eventually received 484 acres. Ultimately, he also acquired approximately 40 slaves. See Wills and Appraisements, Vols. A–B, 1789–1850, pp. 234–37; Superior Court Deeds, Book M, Pt. 1—1842–47, pp. 186–88; and Tax Reports, 15th Militia District, 1847 and 1848, all in Liberty County Records.

spite of many errors of fact and fallacious arguments, *Vestiges* was a provocative little book, and Joseph was fascinated by its thesis of the transformation of species. He "fervently discussed" the work with his brother-in-law, John Harden. "This was my first introduction to the doctrine of evolution," he said.[16] But it would be a long time before Joseph would fully accept the idea of organic evolution and become one of its leading advocates.

In Liberty County one of Joseph's oldest friends was his cousin William Louis Jones. Louis, as he was known to Joseph and other close acquaintances, was the son of William Jones, who was married to a half-sister of Joseph's mother. The elder Jones owned a large plantation close to Woodmanston, and Joseph was a frequent visitor in his Uncle William's home. Louis Jones shared many interests in common with Joseph. Like his cousin, Louis studied medicine (graduating from the same medical school in 1848), and he was intensely interested in natural history.[17] Later, he and Joseph would enroll together at Harvard University to study under Louis Agassiz.

During 1845–1846 Joseph accompanied Louis Jones on several specimen-collecting excursions in the vicinity of Liberty County, and both together and independently they built sizable cabinets of bird specimens. In fact, Joseph himself collected and stuffed over 120 different species of avifauna. In addition, he recorded in his copy of Audubon's *Synopsis of North American Birds* much useful information on the distribution, habits, and morphology of Georgia birds. His notations on some of the rare species and the appearance of other avifauna in Georgia represent some of the first such records for the state. John LeConte also collected birds for awhile, and Joseph collaborated with him in drawing up the first list of avifauna in Georgia, published in 1849. A few years later, after Joseph had committed himself to a career in abstract science, he donated his excellent collection of specimens to the Smithsonian Institution. A number of them were displayed in the United States Museum for many years, and several

16 LeConte, *Autobiography*, 105. In the Autobiographical MS., Joseph says, "I read it [*Vestiges*] with intense avidity." He had previously said the same thing in a letter to R. K. Charles on September 20, 1888 (in Charles Family Papers, South Caroliniana Library, University of South Carolina, Columbia). Although fraught with errors and misconceptions, *Vestiges* was nevertheless ahead of the times in its views on organic evolution. In general, "the most intemperate" criticisms of the noted book "came from scientists" (George H. Daniels, *American Science in the Age of Jackson* [New York, 1968], 57).

17 LeConte, *Autobiography*, 105.

were given to schools and museums both in America and abroad. Spencer Baird later utilized Joseph's collection in his monumental studies of North American birds.[18]

Meanwhile, Joseph gave little thought to settling down for the practice of his profession. In July, 1845, he went to Athens for a month, during which time he and "a party of ladies and gentlemen—nine in all" took a long trip to the North Georgia mountains. Then he went to Decatur to visit his old friend John T. Nisbet. The pair spent a week "tramping around the country and swimming daily in the mill-pond," after which they decided to visit Tallulah Falls. During their excursion the two young men "explored the Tallulah chasm from end to end, doing all sorts of foolhardy things." The daredevil Joseph even swam daily in the treacherous Hawthorn Pool. Less able swimmers had drowned while braving these dangerous waters, but Joseph knew his ability, and he "swam all over" the swirling pool with confidence.[19]

At the end of September, the two friends returned to Decatur, and soon thereafter Joseph went to Macon. By November he was back in Liberty County. He did not at this time take his responsibilities as a plantation owner very seriously, leaving them basically in charge of John and Jane Harden, who in 1843 had built a new home on Jane's property, which was called Halifax. In fact, both of Joseph's surviving brothers, John and Lewis, had chosen not to follow in the footsteps of their father and oldest brother, and they supervised their estates from afar. The particular tract of land bequeathed to Joseph consisted of nearly five hundred acres, and it was called Syfax. Ultimately, Joseph would have to decide what he would do about his large estate, but now was not the time for such serious matters.[20]

In late December, Joseph's sister-in-law Sarah came from Macon to her Cedar Hill plantation for the winter, bringing with her a nephew, a niece, and a cousin. The niece was Caroline Elizabeth Nisbet, the eighteen-year-old daughter of Alfred M. Nisbet of the Midway community near Milledgeville, Georgia. Joseph met "Miss Bessie" early in January, 1846, and, in his own words, it was "literally love at first sight."

18 Stephens, "Joseph LeConte's Contribution to American Ornithology."
19 LeConte, *Autobiography*, 105–108.
20 *Ibid.*, 108. The name is variously spelled Syfax, Scyfax, and Syphax. Its location in relation to Woodmanston and the other divisions of Louis LeConte's estate is given in Liberty County, Georgia, Land Records, Book A–L, 1840, p. 342, and Superior Court Deeds, Book M, Pt. 1—1842–47, pp. 186–88, both in Liberty County Records.

Thenceforth, Joseph's goal was to win Bessie's love, and the ardent suitor began to court her earnestly.[21]

But Joseph wanted to be sure that it was real love and not mere infatuation. Besides, he did not know if Bessie loved him. So he waited. In April Bessie left to visit her aunt in Savannah, and two weeks later Joseph decided it was time to drop in on his two brothers in that city. There he saw Bessie once more. The young woman soon returned to her home in Midway, and Joseph did not see her again for four months. But Joseph now knew for certain that he really loved Bessie. So profound was his adoration for her that he felt a "deep and permanent change had ... taken place in [his] whole nature."[22]

As summer arrived, Joseph decided to visit John T. Nisbet, who was now in Macon. While there, the two companions received an invitation from John T.'s cousin Joseph Nisbet to visit in Midway. Bessie in turn invited some of her close friends, and "we again had a merry party," Joseph reported. Two weeks later, seven of the group decided upon a trip to the falls and mountains of North Georgia. Serving as chaperons were Sarah Nisbet LeConte, William's widow, and the Rev. Thomas Conrad Porter, a Presbyterian minister from Macon, who later became a recognized professor of botany at Lafayette College in Pennsylvania. Their ultimate destination was Joseph's favorite spot, Tallulah Falls. For over a week, the delighted young doctor guided the group through the great gorge. A sermon on Elijah at Mount Horeb and a recitation of Coleridge's "Hymn to Mount Blanc" by the Rev. Porter capped the party's delightful sojourn in that natural wonder. "Such experiences of course fed the flames of love in me," confessed Joseph, but he was yet unsure of his beloved Bessie's feelings toward him. He therefore "determined to settle matters" when he was back in Midway.[23]

On the return trip, the group stayed for several days in Athens,

21 For a brief time (1832–36), Alfred Nisbet was coeditor of the Athens *Southern Banner*. He moved to the state capital in Milledgeville later and became a cashier at the Central Bank of Georgia (LeConte, *Autobiography*, 108). Surviving letters of Joseph to Bessie include four dated November 1, December 5, and December 8, 1846, and January 9, 1847, in McMillan Collection. In the Autobiographical MS., Joseph contrasts his love letters with those of his brother John. His own were "so buoyant, so gay, joyous, humorous, playful, so abounding in life," while John's were "so sentimental, lovesick, almost melancholy."

22 LeConte, *Autobiography*, 110–12.

23 *Ibid.*, 112–18.

where John LeConte was now a professor of physics and chemistry at the University of Georgia. By mid-September all were back in Midway, and the "fateful day" came for Joseph. Anxious and uncertain, he prevailed upon his cousin, Rosa Jones, to arrange for Bessie to walk with Joseph to church on Sunday evening, September 20. His heart filled with love, he blundered out words of devotion to Bessie, who readily confessed her love for him. They decided to marry and set the wedding date for January 14, 1847. In the interim Joseph idled away the tedious hours in Liberty County by thinking of his love and writing her letters almost daily. As the date drew near, he went to Sarah's home in Macon "and there impatiently awaited the appointed time." The wedding took place as scheduled, in Bessie's home.[24]

Still possessed of sufficient means, Joseph had no immediate concern about money and a career. The young couple thus took a leisurely and extended honeymoon, remaining for a few days in Midway, then going on to Macon and thence to "the old homestead in Liberty" for the entire winter and half of spring. Joseph offered Bessie no exotic honeymoon, although he entertained the possibility of a European trip. But, he stated, "we were too happy in each other to care for much else and at the time did not appreciate the importance of foreign travel." By late spring he and Bessie set out to spend the summer at the Central Georgia resort areas of Indian Springs and Rowland Springs. "Here," Joseph later recalled, "we spent several months in the delightful place, riding, hunting, swimming, etc."[25]

But Joseph could not prolong any further the question of his future. By now he was twenty-four years old, and Bessie was pregnant. He must soon decide what he would do to earn a living. There was his plantation, which, though not lucrative, would provide an adequate income. He could not abide the thought of such a sedentary life, however, and he had been exposed to too much of the other world to be happy in such intellectually isolated surroundings. Moreover, science was what he loved, and since the old days of leisurely, self-supporting naturalists were gone forever, he could not follow in the footsteps of his father. Yet, no opportunities for employment as a professor of sci-

24 *Ibid.*, 119; LeConte, Autobiographical MS.; Cornelia Jones Pond, *Life on a Liberty County Plantation*, ed. Josephine Bacon Martin (Darien, Ga., 1974); Joseph to Bessie, December 5, 1846, in McMillan Collection.

25 LeConte, *Autobiography*, 121–22; Emma LeConte Furman, "Personal Reminiscences of Childhood & Youth" (MS in McMillan Collection).

ence appeared on the horizon. Besides, Joseph felt that he was inadequately prepared to take such a position even if it were offered. His brother John had worked hard at scientific writing for several years before he landed his post at the University of Georgia. Thus, the choices were limited, and only the practice of medicine seemed to offer both the promise of an adequate income and the social status that the somewhat aristocratic LeConte could accept. Joseph therefore reluctantly made his decision to practice medicine, and Macon seemed the best city in which to begin.[26]

First, however, Joseph must set his affairs in order. In October he took Bessie to her mother's home to await the birth of their first child. On December 10, 1847, the baby arrived and was duly christened Emma Florence. "Oh, the joy and yet the strangeness of fatherhood," cried Joseph. Of his five children, Emma was always to hold a very dear spot in his heart. Shortly after Emma's birth, Joseph went on alone to Liberty County, where he spent one evening with his brother Lewis, who was just recuperating from a bout with the measles. During the time of his medical training, Joseph had often visited others who had the disease, and he therefore thought it perfectly safe to visit with Lewis. But by the time he had returned to Macon, he had a high fever and took to his bed. He could not go on to Midway, and for ten days he lay miserably ill with a severe case of measles. Impatient to see his wife and baby, Joseph left his sick bed before he had fully convalesced. It took him a long time to regain his "former vigor." In fact, he insisted that perhaps he never did; but his strenuous physical activities even past age seventy-five effectively contradict his claim.[27]

Macon was a thriving little city in 1848; its location in the middle of the state and its importance as a railway, agricultural, and commercial center made it a potentially promising place for a young physician to open a practice. But first the young doctor must establish a reputation, and the city already abounded with physicians. Joseph did have contacts there, however, including his old preceptor, Dr. Charles West,

26 LeConte, *Autobiography*, 123–24. John's earliest publications dealt primarily with medical topics, although one of them (in which Joseph participated) was an experimental study of "the seat of volition in the alligator" (Lupold, "From Physician to Physicist," 24–63). By the time he was appointed to the faculty of the university, John had already published a dozen articles.

27 LeConte, *Autobiography*, 122–23; Bessie LeConte to Ann LeConte Stevens, January 30, 1848, in McMillan Collection.

and his wife's famous uncle, Eugenius A. Nisbet, a prominent jurist and statesman. Thus, when Joseph and his wife and daughter took up their abode in the central Georgia city, they were not total strangers. By mid-winter Joseph had rented an office in the Old Commercial Bank Building, and he soon began to receive patients.[28]

Trying hard to make a go of medical practice and establish a permanent residence in Macon, Joseph decided to contract for a new house. This would require careful planning since his cash resources were limited. His father had left him a large plantation, but he had not bequeathed him much money. Moreover, much of what Joseph did receive had already been spent. He had no income during the meantime, of course, except from his plantation—and that represented a modest amount. It was sufficient, however, to keep him going and to help him build a savings. He had now become frugal, particularly since his medical practice produced little revenue during the first year. Yet he had enough money to build a house, and by the summer of 1849, construction was under way. In late July, the LeContes moved into the partially completed house with two of the slaves Joseph had brought up from his plantation to act as house servants. Since the carpenters were still at work, Joseph thought it best to send Bessie and the baby to Athens for a visit with his brother John. Bessie seemed eager to go; her social life had been restricted since the move to Macon. It was supposed to be a short visit, but it turned out to be an extended stay of over two months.[29]

Joseph's correspondence with Bessie during this time reveals much about the man and his disenchantment with medical practice. Bessie was, of course, still a young woman, and she was used to living without financial worries. Like most contemporary males, Joseph held a paternalistic attitude toward his young wife; in fact, he always maintained a father-child relationship with Bessie, which the adoring wife willingly accepted. The nature of this relationship was fully revealed shortly after Bessie had arrived in Athens. Failing to receive a letter

---

28 Macon *Georgia Telegraph*, June 26, 1849, p. 1; Macon *Georgia Journal and Messenger*, April 11, 1849. West was active in founding the Georgia Medical Society, while Nisbet was a well-known Georgia statesman who served as a U.S. congressman for many years and in 1845 was one of the three judges appointed to the newly established Supreme Court of Georgia. See John McKay Sheftall, "Eugenius A. Nisbet" (A.B. honors thesis, University of Georgia, 1980).

29 LeConte, *Autobiography*, 124; LeConte, Autobiographical MS.; Joseph to Bessie (in Athens), July 27 [?], 1849, in McMillan Collection.

from her, the impatient young husband reproached her gently but firmly for failing to write to him. He also reproved her for being irresponsible about her physical condition. "Don't be imprudent," he said. "You have already made one mistake viz leaving your iron pills." Then he urged her to take "moderate *regular* exercise." Within a few days, he was writing to her again, this time because he had received no letter "for four mornings in succession." Then he reproached her further: "If I don't receive a letter from you tomorrow I shall write to John or Lewis Jones to know what is the matter." Since he received no letter the next day, he kept his promise by writing to John to inquire about Bessie.[30]

But poor Bessie had been ill, and when Joseph finally received a letter from her, she cried that he had "distrusted her attachment." Joseph should have been shamefaced over the matter, but instead he replied that she had misinterpreted his intentions. He did apologize indirectly, and he assured his spouse of his undying devotion. Still, his future letters continued to be filled with fatherly admonition, instructing Bessie, for example, on how to discipline Emma and scolding her for spending too much money for a cloak and bonnet.[31] But his views merely represented the expectations of a devoted husband who was playing out a normal role for the times. He loved his young wife deeply, and their marriage was never threatened.

In part, of course, Bessie had not yet come to realize that her husband did not have unlimited resources and that they could no longer be as free as they had been during their halcyon honeymoon. After a few weeks in Athens, she began to urge Joseph to come up for a visit before she returned. Bessie obviously enjoyed Athens more than Macon. But Joseph replied that he could not close his office; he had already been away for "a month this year" (just where he had been is unknown, but probably to take care of plantation business). Then, he noted, he would have to be away for six weeks during the coming winter (again, perhaps for business but probably also for the extended vacation they expected to take in Liberty County). This, he said, "is fully sufficient unless I expect to become a man of *elegant leisure.*" As the month of August drew to a close, Joseph finally agreed to come to Athens for a short stay and to bring Bessie home, but his plan was

30 Joseph to Bessie, July 31 and August 2, 1849, and Joseph to John LeConte, August 3, 1849, in McMillan Collection.

31 Joseph to Bessie, August 7, 10, and 13, 1849, in *ibid.*

thwarted by the severe illness of the infant daughter of his wife's Uncle
Eugenius. For several weeks Joseph was confined to Macon, trying to
save the child from a case of "chronic diarrhea." He had already lost
"a little patient" from the same illness only shortly before. The struggle
was in vain, however, for Joseph knew neither the cause of the malady
nor a suitable remedy, and the child finally died in late September.
The young doctor was struck hard by the death of children under his
care, and this seems to be one of the reasons he finally decided to give
up medicine.[32]

By October the new house was completed, and Bessie had returned. But Joseph was not happy, and he sought an outlet from his
intellectually stifling practice in various ways, one of which was to go
fishing with Eugenius Nisbet. He was attracted to the influential jurist
because Nisbet loved to discuss literature and philosophy with him.
In fact, Nisbet recognized that Joseph had both the potential and the
interest to engage in scientific and literary pursuits, and he encouraged him to do so. At the same time, Joseph began to read the works
of the anatomist Richard Owen, and he became particularly fond of
the Englishman's *The Archetype and Homologies of the Vertebrate System*. Indeed, as Joseph claimed, it was this work that influenced him
to study under the famous Swiss-American scientist Louis Agassiz,
and, later, when he wrote a thesis, LeConte took as his subject "The
Homologies of the Radiati." Medical practice was simply too prosaic
for the aspiring scientist; he wanted to pursue "abstract science."
Thus, when his cousin Louis Jones paid him a visit in the spring of
1850, the die was cast. Jones had already decided to enroll in the
newly formed Lawrence Scientific School of Harvard University, and
he encouraged Joseph to go with him.[33]

The decision was somewhat difficult for the twenty-seven-year-old
LeConte because, after all, he now had a successful medical practice,
and he had just been invited to enter into a partnership with Dr. C. B.
Nottingham, "an old and distinguished physician" who had recently
come to Macon. In addition, LeConte owned a new home, and all of

32 Joseph to Bessie, August 16, 21, 25, 28, 30, 31, and September 3, 5, and 6, 1849, in
ibid. The Nisbet child must have been Corinne Alexander Nisbet, b. January 21, 1848, d.
September 24, 1849.

33 LeConte, *Autobiography*, 127; E. A. Nisbet to Joseph LeConte, March 29, 1855, in
McMillan Collection; William Louis Jones, "Joseph LeConte," *Transactions of the Medical
Association of Georgia* (1902), 42.

his family relationships were in Georgia. There was, moreover, the responsibility for his Syfax plantation, which, although operated by an overseer, required supervision from time to time. But in Joseph's mind the benefits outweighed the costs. The reluctant physician was disenchanted with medicine, he preferred a life in abstract science, and he desired to become a teacher. In fact, as he admitted, he had enjoyed the instruction of a few medical students for whom he had served as preceptor much more than he had his work with patients. Theoretical science beckoned, and the call was too strong to ignore when the opportunity arose. "It was now or never," he said, and in August, 1850, he "broke up, sold out, left Macon, and went to Cambridge."[34]

As he departed, LeConte could look back upon his brief experience in medicine with a degree of pride. He had succeeded reasonably well as a physician, and he had helped in a small way to advance the profession. In the latter instance, he had participated actively in the organization of both the Macon Medical Society and the Georgia State Medical Society. But, like other state and national medical societies, including the American Medical Association, these organizations expressed little interest in the improvement of medical education. For LeConte, however, the inadequacies of current programs of medical training were a matter of great concern, and when an opportunity to address the Macon Medical Society arose in 1849, he chose to speak on "The Science of Medicine." In this address, LeConte charged that people had lost their confidence in "the so-called orthodox system of medicine" because it was making no significant discoveries of laws pertaining to medical pathology. The backwardness of the system would prevail, argued LeConte, until the profession recognized the importance of training physicians in the inductive sciences and statistical observations. Intuitive impressions and deductive inquiries could never lead to the advancement of medicine, he concluded.[35]

LeConte's address was hardly profound, of course. It added no new knowledge to medicine, but it was a pertinent exhortation that reflected the poor state of medicine at the time. Over and beyond that,

34 LeConte, *Autobiography*, 126.

35 *Ibid.*, 125; Macon *Georgia Telegraph*, February 19, 1849; Ida Young, *et al., History of Macon, Georgia* (Macon, 1950), 139–42; William G. Rothstein, *American Physicians in the Nineteenth Century* (Baltimore, 1972); Joseph LeConte, "On the Science of Medicine and the Causes Which Have Retarded Its Progress," *Southern Medical and Surgical Journal*, n.s., VI (August, 1850), 456–74.

the timely address represented LeConte's own feeling that scientific research was necessary to the advancement of medicine. LeConte would soon have his chance to contribute to scientific progress, and he would succeed beyond his own expectations, although it would be largely in fields other than medicine.

—◆◆❮ III ❭◆◆—

"Abstract Science": With Agassiz and Back to Georgia

Filled with youthful enthusiasm, LeConte arrived in the quiet little town of Cambridge, Massachusetts, in August, 1850, almost two months before the fall term was scheduled to begin. No sooner had he settled his family than he and Jones hastened to the campus to find Agassiz and get under way with their study. Joseph could not wait until October; he had come to learn, not merely to observe the formalities of a university program. Having already met Agassiz a few days earlier at the meeting of the American Association for the Advancement of Science in New Haven, Connecticut, which he "attended on the way to Cambridge," LeConte would need no formal introduction.[1]

Louis Agassiz deserved the esteem in which he was held by Le-Conte and a host of American and European scientists and other intellectuals, for he had clearly established himself as a foremost authority in natural history. Ambitious, dedicated, and brilliant, the renowned Swiss scientist had come to the United States in 1846 and shortly thereafter accepted an appointment to the faculty of Harvard University. In the summer of 1850, Agassiz was offered a professorship in Harvard's newly formed Lawrence Scientific School to teach courses in geology and zoology.[2] Quite naturally, as a student and great admirer of Agassiz, LeConte chose the same fields in which to pursue his work.

As LeConte and Jones eagerly approached the great man, they ob-

---

1 LeConte, *Autobiography*, 128. LeConte is first listed as a member of AAAS in 1850 (*Proceedings of the American Association for the Advancement of Science*, III [1850], xiii).
2 Edward Lurie, *Louis Agassiz: A Life in Science* (Chicago, 1960).

viously expected him to begin their instruction in some formal manner, but little did they know of Agassiz's approach to teaching. After pondering for a few moments, Agassiz "pulled out a drawer containing from five hundred to a thousand separated valves of Unios, of from fifty to a hundred different species, all mixed together," and instructed the expectant men to "pair these valves and classify into species." With those terse instructions, recalled a bewildered LeConte, "he left us alone, very severely alone." For an entire week, LeConte and Jones labored at the difficult assignment, and although Agassiz occasionally examined their work, he made not a single remark. Less able men might have wilted under such circumstances, but neither LeConte nor Jones succumbed to the pressures of uncertainty and anxiety. Instead, they arranged, rearranged, and rearranged again, until they felt they "had done the best we could." Then they informed Agassiz that they had completed the task to the best of their ability. Just as Agassiz was looking over their work, a friend of his entered the laboratory and, after introducing him to LeConte and Jones, Agassiz paid the now-anxious young men a long-awaited compliment, saying "these pupils . . . had just amended correctly the classification" of Isaac Lea, the great authority on conchology.[3]

This was the beginning of LeConté's work under Agassiz. It was the commencement of his undeviating regard for the Swiss scientist, and, in later years, LeConte would often comment on the profound impact Agassiz had made upon him. Indeed, he eventually claimed that, "more than any other" person, Agassiz had "inspired" his "subsequent life." His work with the brilliant naturalist for "eight hours every day" during a fifteen-month period resulted in his "second and higher intellectual birth." Over two decades after he had completed his study at Harvard and upon the occasion of a memorial tribute to the deceased intellectual giant, LeConte called Agassiz the "great *master-mason*" in the "construction of the temple of science." Agassiz, he said, "was the originator of new ideas, and the introducer, or at least the perfecter, of new methods in science." Although LeConte praised Agassiz for his work in glacial geology, he was much more impressed with Agassiz as "the perfecter of the great method of organic science." By this, LeConte meant that Agassiz had refined the "method of comparison" in zoology, and that method, in LeConte's view, had become

3 LeConte, *Autobiography*, 128–29.

the key not only to all biological study but also to every form of social science.[4]

Agassiz stressed the necessity for field work, and LeConte accompanied him on brief geological excursions into the Massachusetts mountains and the Catskills of New York. It was on the latter trip that LeConte first met the famous geologist James Hall of Albany. As a direct result of these field experiences, LeConte developed a strong interest in the subject of orography, or mountain-formation. But, whether in the field or in the laboratory, LeConte was Agassiz's companion day in and day out for several months, and he dearly loved every moment of the association.[5]

Late in 1850, Alexander Dallas Bache, head of the U.S. Coast Survey, arranged for Agassiz to travel to the Florida reefs to gather information on the origin and nature of coral formation and to collect marine specimens. Agassiz was authorized to pay the expenses of two student assistants, and he chose LeConte and Jones. Elated over this unique opportunity to make firsthand scientific investigations, Joseph eagerly informed Bessie of the offer. But Bessie did not receive the news so joyfully, for now she had a new baby, two-month-old Sarah Elizabeth (usually called "Sallie"). Yet, realizing the significance of the excursion to Joseph's career and to his own personal satisfaction, she offered no protest. The dutiful wife would often have to endure such separations in the interest of her husband's scientific pursuits, and this was only the first of many occasions when the devoted couple would be apart. Despite his enthusiasm for the project, LeConte was reluctant to leave Bessie and the two girls. "Oh, the pain, the distress, the pity of it!" he moaned. But Bessie "urged" him to go, and he departed on her birthday, January 1, 1851.[6]

LeConte and Jones arrived in Charleston a few days later and there awaited the steamer that would carry them on to Key West. Within a

---

4 *Ibid.*, 144–53; LeConte, "Agassiz Memorial Address," *Proceedings of the California Academy of Sciences* (1874), 230–36. Later tributes to Agassiz by LeConte include "Review of *Louis Agassiz: His Life and Correspondence*," *Overland Monthly*, 2nd ser., VII (January, 1886), 103–105; and "Agassiz and Evolution," *Popular Science Monthly*, XXXII (November, 1887), 17–26.

5 LeConte, *Autobiography*, 130, 141. On several occasions during the Catskills excursion, LeConte was invited to dine with the prominent geologists James Hall, James Dwight Dana, and Josiah D. Whitney (John M. Clarke, *James Hall of Albany: Geologist and Palaeontologist* [Albany, 1923], 225).

6 LeConte, *Autobiography*, 130–31; Lurie, *Louis Agassiz*, *passim*.

few days, they set foot upon the "exquisitely tropical" island, and Joseph quickly found himself in the highest of spirits. But there was little time for relaxation; Agassiz expected his assistants to be "incessantly at work." He soon had them exploring the reefs, boating into the Everglades, and collecting specimens. Before the expedition ended, said LeConte, the "collections were enormous." Constantly "observing, noting, and gathering specimens," the scientists would spend entire days in "waist-deep" water examining the coral formations. These activities were interspersed with work in the temporary laboratory set up on the wharf, where the scientists made microscopic examinations and drawings and then carefully packed away the specimens or preserved them in alcohol. LeConte was an industrious worker possessed of virtually inexhaustible energy, but he could not match the indefatigable stamina of Agassiz, who was consumed by a passion to spend virtually every moment of life in his scientific endeavors.[7]

During the course of the expedition, Agassiz gave special assignments to his two assistants, one of which came during their visit to the Dry Tortugas. Agassiz sent LeConte and Jones out on a small sailing vessel "to examine a little island about 8 or 10 miles to the northwest." The pair quickly finished their investigations, but they were prevented from returning by a becalmed sea. After three days of waiting, LeConte and Jones began to worry that Agassiz would need them, so the captain ordered two of his men to row them back in a small boat. Midway on their return, "the boat suddenly grounded on the close-set prongs of an extensive grove of madrepores." Struck by the peculiar nature of these corals, LeConte insisted that they investigate them further. He made a fortuitous discovery that led, over twenty years later, to the publication of a brief scientific paper. Actually, LeConte wrote the paper as a confirmation of a statement made by James Dwight Dana in his authoritative work, *Corals and Coral Islands*, published in 1872. What LeConte discovered was a correlation between tide-level and the growth of the tree-pronged grove of madrepores. His graphic description of the phenomenon provided an excellent, if minor, contribution to knowledge of the growth of this particular species of coral.[8]

Most of LeConte's discoveries on the Keys expedition were incor-

7 LeConte, *Autobiography*, 131; Pond, *Life on a Liberty County Plantation*, 25–26.
8 LeConte, *Autobiography*, 132–37; Lurie, *Louis Agassiz, passim*; LeConte, "Rate of Growth of Corals," *American Journal of Science*, 3rd ser., X (July, 1875), 34–36.

porated into Agassiz's *U.S. Coast Survey Report* for 1851, as they rightly should have been, since Agassiz was the man assigned to the task and LeConte was only an assistant. But LeConte did make an important discovery on his own. Since he was so busy for the next five years, however, and because the hypothesis had to be mulled over for a long time, LeConte did not offer his finding until he addressed the American Association for the Advancement of Science in 1856. "This was," said LeConte, "my first really scientific paper." It also proved to be one of the earliest scientific explanations of the geological development of the peninsula of Florida.[9]

In this paper, LeConte argued that "coral agency alone is not sufficient to account for the phenomena" of the geological formation of the Florida peninsula and the Keys. He accepted Agassiz's hypothesis that the initial foundation of the recently formed peninsula and the Key islands had been laid by reef-building corals, but, as he had already observed, corals cease to grow when they reach the surface of the water. LeConte contended that Agassiz's hypothesis did not go far enough because it failed to explain the continual buildup of the peninsula. He also refuted Michael Tuomey's theory that the peninsula had developed through subsidence and elevation by pointing up the recency of the formation of the peninsula. Instead, LeConte suggested that the phenomenon occurred through "the agency of the Gulf Stream." That powerful current, he argued, had carried great quantities of sediment from the Mississippi River to the coral reefs. As the Gulf Stream flowed parallel to the land mass, it swerved in a southward direction, hence constantly depositing sediment among the reefs. Over a long period of time, a land base was thus formed. Since the stream then had to flow around the obstructions it had created, the aggrading process must end as the powerful Gulf Stream moved closer to the island of Cuba, leaving only a relatively narrow channel for its course. LeConte concluded, then, that the geological process of building the Florida peninsula was now complete.[10] Although this ex-

9 *Annual Report of the Superintendent of the Coast Survey During the Year Ending November, 1851*, Senate Executive Documents, No. III, 32nd Cong., 1st Sess., Appendix No. X, 145–60; LeConte, *Autobiography*, 138. The paper was published as "On the Agency of the Gulf-Stream in the Formation of the Peninsula of Florida," *Proceedings of the AAAS*, X, Pt. 2 (August, 1856), 103–19; reprinted in *American Journal of Science*, 2nd ser., XXIII (May, 1857), 46–60.

10 LeConte, "On the Agency of the Gulf Stream," 46–60. On February 4, 1856, Agassiz

planation was eventually modified, it certainly represented at the time a sign of true ingenuity.

After six weeks in the sunny climate of Key West, the Agassiz expedition ended, and the party returned to the "arctic winter" in Cambridge. LeConte plunged back into his studies, adding to his work in geology and zoology under Agassiz a course in botany under Asa Gray. It was excellent training, and although he never became deeply involved in the study of plants, LeConte did acquire enough working knowledge in that field to offer small contributions later in his career. By spring, LeConte felt that he had gotten what he had come to Harvard for, and he began to think of returning to Georgia. But Agassiz prevailed upon him to write a thesis and graduate from the Lawrence Scientific School. LeConte had not originally intended to earn a degree, since he already held an A.B. and the M.D. He did not object to Agassiz's suggestion, however, and he consented to stay on, completing by mid-summer a thesis on the common structural features of echinoderms and coelenterates.[11]

After this, LeConte remained for several more weeks. "I continued to study right along . . . sometimes in Agassiz laboratory, sometimes by myself," he said. On one occasion, he took Bessie and the children on an excursion to the Atlantic Coast, near the little town of Cohasset, for a brief vacation. While he was there, LeConte "made a careful study under the microscope of the development of Bryozoa," and he produced many drawings from his studies, which "greatly delighted" Agassiz. In fact, Agassiz encouraged him to draft a paper on his research and present it to the next AAAS meeting, but LeConte demurred because he felt he was "not sufficiently acquainted with the literature of the subject." Still, the suggestion represented Agassiz's high opinion of his student, and a year later when LeConte requested a letter of recommendation from his mentor, Agassiz wrote that he considered LeConte "as the most prominent among all the young men who have lately entered their career in science." Indeed, he went on to say, "I would not hesitate to nominate him as the ablest candidate to fill any Professorship . . . [in his fields of training] in any of our colleges North or South. I have never known a better student."[12]

wrote to LeConte to tell him that "your suggestion about the reef is ingenious & quite new to me" (in Talley Papers).

11 LeConte, *Autobiography*, 140–42.

12 *Ibid.*, 142–43; Agassiz to James H. Couper, October 26, 1852, in Talley Papers. On

By October, 1851, LeConte had decided to return to Georgia. He was loath to leave Cambridge and its "galaxy of stars" and the "intellectual atmosphere" that was so "stimulating . . . to thought," but the time had come for him to cultivate new fields and tend to affairs at home. Stopping first in New York, the LeContes visited Joseph's "dear old Uncle Jack," and ten days later they embarked in high spirits by steamer for Savannah. The mood of excitement and joy over returning to their native region was met with sad news, however, for no sooner had they stepped off the ship than they were greeted with the depressing message that Joseph's brother Lewis had been killed by his own shotgun only two days before the steamer docked at the Savannah wharf. The tragic news was shocking to LeConte, for he had already lost his mother and father, his oldest brother, and his brother-in-law John Harden (who died in 1848 at age thirty-eight, from tuberculosis). Now, he had lost the brother who had been closest in age and temperament to him. The shock was compounded by the nature of the tragedy. Some people believed that Lewis had been gunned down by two men whom he had previously prosecuted for stealing his cattle and against whom he had still another complaint for the same offense. The truth would never be known, however, for the bodies of the two men were found two weeks later, washed up on a large sandbar in the Altamaha River, some forty miles from the scene of the tragedy. The jury of inquest could find no strong evidence to support the murder theory, and they therefore ruled the death accidental.[13]

Still grieving over the death of Lewis, Joseph and his family went on to Liberty County, where they lodged with his widowed sister Jane. He could not stay at his Syfax plantation because he had never built a home there, and he could not live in the old home place because it had been greatly neglected since the Hardens had built a new home on their own Halifax plantation. Things were changing drastically at

---

the same date, Agassiz wrote to LeConte: "It is with no little pride I state that I have always considered you as my best pupil" (in Talley Papers).

13 LeConte, *Autobiography*, 143–54; William Louis LeConte, "Some Events of My Life" (MS in LeConte Family Papers, University of Georgia Library, Athens [hereinafter cited as LeConte Family Papers, UG]); Savannah (Ga.) *Republican*, October 23, 1851, p. 2. The article on the jury's findings was written by Thomas H. Harden, a relative of Joseph's sister Jane. The belief that the tragedy was an accident was confirmed by an old friend of Lewis many years later (Lollie Belle Wylie [ed.], *Memoirs of Judge Richard H. Clark* [Atlanta, 1898], 250). Of his brother's death, Joseph says only that Lewis "accidentally shot himself" (LeConte, *Autobiography*, 154).

the once-famous Woodmanston estate. Now overgrown with fennel and broom grass, little remained to reflect the former splendor of the marvelous garden. William was dead, and his wife Sarah had remarried and lived mainly in Macon; Lewis, who had chosen not to live on his Beech Hill plantation but to operate it from afar, was dead; Ann and her husband J. P. Stevens spent most of their time in Baker County, where Stevens owned another plantation; John was a professor at the University of Georgia and a genuinely absentee plantation owner; and Joseph was only an infrequent visitor to the land of his birth. Only the widowed Jane remained, and she would not reside permanently at her Halifax estate as her children matured and left home. Yet, it was in a sense still home for Joseph, and until the end of the Civil War some fourteen years later, he would continue to come back once a year to check on the welfare of his slaves.[14]

In December, 1851, Joseph received word from the Oglethorpe University trustees that they had accepted his application for a position on their faculty, beginning in January, 1852. This was not the job LeConte preferred, but it was acceptable. Long before he left Cambridge, he had learned of a vacancy in the chair of geology and natural history at his alma mater, the University of Georgia. He really wanted this position: the chair suited his interests ideally, the university was the best school in the state, and he desired to be associated with his brother John, who had already been at the institution for six years. Unfortunately for him, Louis Jones had learned of the vacant position first and told Joseph he intended to apply.[15] Joseph would not compete with his old friend and thus resigned himself to the appointment at Oglethorpe.

Founded in 1835 in Midway, near Milledgeville, Oglethorpe University was a small Presbyterian college. It paid Joseph a salary of $1,000 a year, which was only half as much as Jones received at the state

---

14 LeConte, *Autobiography*, 154. The description of the Woodmanston garden was reported to Laura E. Maxwell in a letter from Mary Sharpe Jones, March 13, 1856 (Myers [ed.], *Children of Pride*, 195). Despite such changes, however, Joseph foresaw no decline in the prosperity of the region, for, in May, 1852, he purchased two tracts of land (500 and 80 acres respectively) near his own plantation, at a cost of $1,000 (Superior Court Deeds, Book N, Pt. 2, 1847–54, pp. 553–54, in Liberty County Records). The visits to Syfax can be reconstructed fairly accurately from the *Autobiography* and various personal letters, although LeConte's recollections in the former appear in a few instances to be off by a year.

15 LeConte, *Autobiography*, 154.

university. But since LeConte had "other sources of income," he could afford to take the position. At least it was a job, and it did offer him a chance to break into college teaching. The responsibilities were heavy at the small school, and LeConte had to teach "mechanics, physics, chemistry, geology, and botany." Such large and diverse loads were, unlike those in many contemporary northern colleges, quite common in the South, allowing professors of science little time to specialize or to conduct research. LeConte taught no course in zoology, his special field of preparation; but he was glad for that because the school had no laboratory, and "there was not even a text-book on the subject" at the time. He did especially enjoy teaching the botany course, however, for Gray's "excellent text-book" was available, and the college had managed to purchase "a first-rate microscope." In later years, LeConte viewed these varied teaching responsibilities as "excellent training" that enhanced his "interest in all departments of science." This broad background helped him become a better geologist, he claimed.[16]

His tenure at Oglethorpe lasted only one year, however, for Jones "got on badly with the president" of the University of Georgia and re-signed after two academic terms. Joseph decided immediately to ap-ply for the vacated post, and he wrote to his brother John to tell him the news. Perhaps Joseph should have been forewarned by his cousin's failure to get along with the "bigoted, dogmatic, and imperi-ous old man" Alonzo Church, who had been president of the Athens institution since 1829, but Joseph seemed to think he could succeed where Jones had failed. He also had ample warning from John, who for some time had been unhappy with Church. Nevertheless, Joseph was eager to move, and he gladly accepted when the post was offered in November, 1852. His principal concern was over the assignment itself; zoology did not go with the chair, but French did! The classics-oriented school had appended the teaching of the modern language to whatever chair it could, and just now that task resided with the geology and botany post. LeConte "read French with ease," but, in spite of his ancestry, he could not speak the language. Thus, in prepa-

16 *Ibid.*, 155–56; Allen P. Tankersly, *College Life at Old Oglethorpe* (Athens, 1951), 1–82; Milledgeville (Ga.) *Southern Recorder*, November 30, 1852, p. 3; Augusta (Ga.) *Chronicle*, January 29, 1853, p. 2; Stanley M. Guralnick, *Science and the Ante-Bellum American College* (Philadelphia, 1975), *passim*; Guralnick, "The American Scientist in Higher Educa-tion, 1820–1910," in Nathan Reingold (ed.), *The Sciences in the American Context: New Perspectives* (Washington, D.C., 1979), 99–141.

ration for this necessary evil, he "immediately began taking lessons from an excellent native French teacher."[17]

Arriving in Athens just before the new term began in January, Le-Conte and his family moved in with John. The university campus was situated on the south side of Broad Street, less than a block away from the town. The rustic setting of the oak-shaded campus with its small handful of buildings gave it the peculiar dignity that characterizes so many of the southern university and college campuses even today. Athens itself existed in a sort of symbiotic relationship with the college, and the prosperity of the town was closely linked to the success of the institution. Founded in 1801 when the university opened its doors, Athens grew spasmodically with the fortunes of the university. By the time LeConte arrived there in 1853, the town had begun to flourish with the coming of cotton factories and railroads, and it could boast a population of 3,000 souls, of whom approximately one-half were slaves.[18]

The shortage of suitable housing made it necessary for Joseph to live with his brother for a year. John and Josephine LeConte, who had two sons and a daughter, were somewhat crowded by the presence of the additions to their household, but the strong filial ties between the two men overcame the difficulties. John was a respected professor, though he was never considered so able a teacher as his brother. Aloof, formal, and aristocratic in bearing, the thirty-five-year-old John tended to remain isolated from his peers at the university, but his wife Josephine was the center of attention. By all accounts, Josephine was an extremely beautiful woman who reveled in the attention her comeliness attracted. "She had constant adorers from all classes," recalled Joseph's daughter, Emma, and "the callow college boys fluttered around her like moths," while John's acquaintances "bowed at her shrine." Josephine gloried in her coterie of admirers, and she set high cultural standards for her husband and children. Into this situation came the more humble and gregarious Joseph and his somewhat self-effacing spouse Bessie, who was herself a pretty but not a beautiful

17 LeConte, *Autobiography*, 156; John LeConte to the editor of the Athens *Southern Banner*, January 24, 1856. Joseph was elected on November 17, 1852 (in Minutes of the University of Georgia Trustees). The post was to carry a salary of $1,700.00 per annum (Athens *Southern Banner*, September 23, 1852, p. 2; Joseph LeConte to E. A. Nisbet, September 10, 1852, in the University of Georgia Library Special Collections).

18 LeConte, *Autobiography*, 156; Ernest C. Hynds, *Antebellum Athens and Clarke County, Georgia* (Athens, 1974), 1–78.

woman. Fortunately, such differences created no friction, and Joseph's and John's children got along like brothers and sisters.[19] The ties between the two brothers became very strong, and henceforth they would always be close companions.

Enrolling approximately 150 students at the time Joseph joined the faculty, the university had changed little since he had attended it from 1838 to 1841. Its curriculum was still basically classical, and its president, Alonzo Church, hoped to keep things that way. Church was also the professor of political economy and of moral and mental philosophy, while James P. Waddell was the professor of ancient languages. Charles F. McCay served as professor of mathematics, astronomy, and civil engineering, and the Reverend W. T. Brantly taught belles lettres, oratory, and history. John LeConte filled the chair of physics and natural philosophy, and Joseph, of course, held the professorship of geology and natural history. In addition, the university employed a professor of law and several adjunct instructors. It was not a distinguished faculty overall, but it did claim several very able men.[20]

The first year passed without any serious hint of the trouble that was to come to LeConte and the university. He struggled through his French courses for six months until he was relieved of that onerous burden when the university hired Emmanuel V. Scherb as a French teacher. Particularly excited over his geology course, Joseph showed no displeasure because he could not teach zoology. In fact, he claimed that he was "glad" he did not have to teach his favorite subject because "Agassiz had introduced an entirely new mode of studying and teaching zoology, and my preparation was entirely ahead of the times; the colleges were not yet ready for the new method." The significance of Joseph's assignment was that the young scientist would relegate zoology to a second love and elevate geology to the front of his research and writing; but the dividends did not come until much later. In addition to his duties in geology and botany, Joseph was responsible for the university's botanical garden, and he taught a Monday morning class in natural theology, which he "greatly enjoyed."[21] Indeed, his

19 Emma LeConte Furman, "Personal Reminiscences."

20 Coulter, *College Life in the Old South*, 33–58. The LeContes were acquainted with Crawford W. Long, discoverer of ether (Francis Long Taylor, *Crawford W. Long and the Discovery of Ether Anesthesia* [New York, 1928], 98).

21 LeConte, *Autobiography*, 156–57. On the general condition of science in contemporary southern universities, see Thomas Cary Johnson, Jr., *Scientific Interests in the Old South* (New York, 1936), 11–45.

work in that field of study would also pay dividends later when he became a leader in the effort to reconcile religion and evolutionary theory.

During his three years at the university, LeConte wrote four minor articles. These popular pieces he "regarded mainly as a practice in the art of exposition." Published in the students' magazine, the first two dealt with minor scientific subjects: "Salt Beds, and Their Connection with Climate" and "Sun Drawing Water."[22] These were merely explanations of natural phenomena given in laymen's language. The other two bear closer scrutiny because they represented LeConte's views on contemporary education. In his "Utilitarian Spirit of the Age," the young professor deplored the trend toward an excessive emphasis upon practical education. "The utilitarian spirit, like a dreadful vampire, is sucking the blood of our spiritual life," he declared. Metaphorical license aside, LeConte correctly assessed the contemporary situation in America. "The tendency of the age," he observed, "is to make education entirely subservient to material success in life—to the accumulation of wealth . . . [and] to the utter neglect of all other higher objects." This tendency was partly a result of the course of affairs in the country, he said, but it was also due to the failure of "the present system of University education." Although he recognized the need for changes, LeConte knew that liberal education must not be compromised; it must "make" the individual "the complete representation of *modern* civilization in its widest sense."[23]

Following this theme in his next essay, "Classics vs. Mathematics," LeConte called for curriculum revisions that would provide a better balance between the study of ancient languages and of mathematics. He argued that, whereas the ancient languages were once useful, modern man had superseded the ancients in every way; therefore, although the classics furnish "*exercise* to the mind," they are inferior to the study of mathematics, which increases "the power of intense thought and close logical reasoning." Neither the classics nor mathematics, he added, are sufficient by themselves or even together; they must be complemented by the study of science. Given the context of contemporary education, these views indicate that LeConte had become a progressive in that field. He wished to conserve what he

22 *Georgia University Magazine*, V (April, 1853), 8–14; *ibid.*, VII (August, 1854), 225–29.
23 *Ibid.*, VII (March, 1855), 353–62.

thought was good in the traditional curriculum, but he likewise wanted to give greater emphasis to modern studies. He desired, in his words, "conservation" without "stagnation."[24] Nowhere did this change seem more necessary than at his own alma mater, but nowhere could it have met stronger resistance. President Alonzo Church had already taken note of LeConte's reformist philosophy.

President Church was a forbidding man of authoritarian and dogmatic views on the way a college should be run. Trained in the classics, this stubborn Presbyterian minister had imbibed deeply of Calvinist doctrine, and he would brook no disloyalty from his subordinates. He administered the University of Georgia with an iron hand, certain that truth and right were his peculiar possessions. For many years he had suffered no unusually adverse reaction to his policies and procedures, but things began to change in 1850. Within a space of five years, several capable professors left the university because of Church.[25]

In every instance, these men resented the authoritarian treatment they received at Church's hands, but, although several newspapers played up each of these resignations, the college trustees simply ignored the matter. A thunderstorm was brewing, however, and the first flash of lightning slashed across the campus sky in early November, 1855, when John LeConte suddenly resigned to take a post at the College of Physicians and Surgeons in New York. Actually, John tried to leave quietly, making no charges against Church or even hinting at a disagreement with the imperious president. But several newspapers quickly expressed their regret over John's resignation. "There is no man in Georgia, perhaps none in the Southern States, so highly competent . . . as Dr. John LeConte," declared the Milledgeville *Federal Union*. "To lose him would be a state calamity." Other papers questioned the reason for LeConte's resignation and inferred that Church

---

24 *Ibid.*, X (June, 1856), 69–75.

25 In his *Autobiography*, LeConte barely mentions Church and only briefly refers to the controversy. Joseph's nephew, a student at the university during the height of the controversy, later wrote that Church was "a man of iron will and when crossed, of *tyrannical* disposition." He claimed that students nicknamed the president "Dionysius the Tyrant" (William Louis LeConte, "Some Events of My Life"). The entire controversy that ensued is discussed rather fully in E. Merton Coulter, "Why John and Joseph LeConte Left the University of Georgia, 1855–1856," *Georgia Historical Quarterly*, LIII (September, 1969), 16–40.

was responsible for driving out LeConte and many other able professors.[26]

Stung by this mild slap, the irascible president struck back with fury. In a letter to the Savannah *Republican* on October 12, 1855, he deplored the attacks on his administration and defended his policies, but he carefully avoided any name-calling and made no direct accusations against any of those who had resigned. To counter the charge that LeConte had left for a higher salary, Church noted that the university paid its professors almost as well as any other college in the nation. LeConte had left, he said, because his faculty load would be lighter at the New York school, and such "a distinguished man of science" deserved to be relieved of "the drudgery of constantly teaching the elements of knowledge" in order to "pursue his favorite studies." Church then capped his lengthy letter with a thinly veiled reproach of LeConte. Although he spoke in general terms and did not mention LeConte in his concluding statement, the vengeful president declared that some professors ought to leave because they become bored with teaching after a lengthy period of service. To justify his own tenure of thirty-six years at the university in contrast to LeConte's nine years, Church said that some professors retain their enthusiasm for teaching; indeed, it "increases with age and practice" among such dedicated men. Moreover, he averred, the university would not suffer from the loss of LeConte, for he would be readily replaced.[27]

Privately, however, Church expressed animosity toward LeConte, and in the November meeting of the Board of Trustees he criticized the physics professor for resigning without prior notice of intention. Actually, LeConte had resigned only a few days before the new term was to begin, and Church was thus compelled to apportion LeConte's duties among the other faculty members. The exact reason for the sudden resignation is unknown, but it appears that the New York offer

26 Milledgeville *Federal Union*, October 9, 1855, pp. 3–4. The university students also lamented the resignation of John LeConte, passing seven resolutions deploring his "loss" and urging the trustees "to retain him in some capacity" (*Georgia University Magazine*, IX [November, 1855], 28, 61).

27 Church probably chose the Savannah newspaper because it had previously published a letter by "Amicus" criticizing his administration. The anonymous critic was castigated by the Athens *Southern Banner* on June 29, 1854, and Church was defended by the local newspaper both in this instance and throughout the controversy. On October 18, 1855, the *Southern Banner* took the *Federal Union* to task for criticizing President Church and maintained that John LeConte quit solely for a better position.

came so suddenly that it left John no time for lengthy pondering, and since he was already very dissatisfied with Church, he welcomed the offer as a timely opportunity. But, for Church, such a decision represented selfishness. "The interests of the Institution," he told the trustees, "ought to be paramount to those of an individual." The old curmudgeon was further piqued because Joseph refused to accept a heavy portion of his brother's vacated duties.[28] By now, Joseph had discovered firsthand how Church operated, and he showed a tenacity equal to that of the president.

Reactions to Church's newspaper statement came quickly, but a response from John LeConte did not appear until November 20. Proud, sensitive, and ever-mindful of his reputation, John viewed the matter as a personal attack upon him by the president, and he could not let the matter pass. He therefore sent a letter to several leading Georgia newspapers in which he charged that Church had poisoned the thinking of the public "by insinuation and implication." Denying that he had ever acted discourteously to Church, LeConte freely admitted that "no great degree of cordiality existed" between the two, but cordiality was "impossible," he said, for they differed drastically in their "tastes and sympathies." Then, assuming that all of Church's general criticisms were directed toward him, LeConte responded like an adversary. The charge that his studies would be lighter in the College of Physicians and Surgeons was true, he said, but that was not the principal reason for his resignation. He had previously declined to publicize his criticisms of Church because "it would have been in bad taste to thrust them forward." To the charge of loss of enthusiasm for teaching, LeConte said simply that his students and colleagues would testify otherwise, as indeed they did. Then, unable to let pass an opportunity to give the backhand to Church, he suggested that the charge should be made against the man who had probably grown weary of teaching after thirty-six years.[29]

John LeConte then began to swing with great verbal blows. First, he raised a matter that Church had neglected to include in his letter but one that was commonly known among the faculty to be a source of serious contention between him and Church. It had to do with the

28 Minutes of the University of Georgia Trustees, November 7, 1855.
29 LeConte's letter was sent to several newspapers, including the Athens *Southern Banner*, the Savannah *Republican*, and the Augusta *Weekly Chronicle & Sentinel*.

university rule that required faculty to police students in the dormitory. The rule stipulated that every university instructor must "daily visit the rooms of the students at such hours as may be assigned to them by the President." The faculty were instructed to make these visits at their own discretion, and they were specifically permitted to enter the rooms without prior warning. For quite awhile, John had neglected to carry out this assignment because, he claimed, his laboratory duties required so much time that he could not get around to the spying duties. Moreover, he declared that it was common knowledge among the faculty and the trustees that Church had not enforced the rule in John's case. But the matter ran deeper than mere neglect of duty: both of the LeContes objected to the paternalistic policy, and they firmly believed that spying deterred the maturation of students into responsible adults.[30]

LeConte swung his second big blow against Church on another source of ill-feeling, which Church had also ignored in his letter. Neither John nor Joseph was an unbeliever; indeed, they were devout Christians. But their liberal religious views were manifested in some of their lectures. It is highly doubtful that either of the LeContes held "*infidel sentiments*," as the truculent president privately confided, although both of them considered it essential to the interests of science and a liberal education to expose their students to new ideas.[31] To the firmly orthodox Church, however, that exposure was tantamount to the heretical corruption of youth, and it was certainly one of the big bones of contention between the old-fashioned classicist and the new-fangled scientists.

The verbal imbroglio that ensued for the next nine months ended only after the situation had reached intolerable limits, and it thrust Joseph into the thick of the fray. When Church responded to John's letter on December 13, he brought Joseph into the matter. In his response, Church carefully avoided answering John's criticisms, focusing instead upon two minor points. First, he expressed astonishment that John could have remained so long at the university if conditions were as bad as he claimed, and he vehemently denied that his actions had caused any of the resignations. Then, to put John on the defen-

---

30 Minutes of the University of Georgia Trustees, November 7, 1855; Coulter, "Why John and Joseph LeConte Left the University of Georgia," 26–29.

31 Coulter, "Why John and Joseph LeConte Left the University of Georgia," 28.

sive, he pointed out that the latter had sought to remain as a part-time faculty member in Athens. Why, queried Church, would LeConte have sought to keep his association with the university if the situation were so deplorable. In addition, he asked why Joseph came to the university and was still there if he, Church, had been a poor administrator. He insisted that John actively encouraged Joseph to take the post, and he suggested that the university was becoming a LeConte family affair.[32]

Both of the LeContes shot back replies through the newspapers. Joseph asserted that his brother had not helped him secure his position at the university, and he implied that Church was playing with the truth regarding a part-time appointment for John when the latter resigned. His brother had been asked by local civic leaders to remain on the faculty in contemplation of a reorganization of the university in which John would head a separate school that reported directly to the trustees and not through Church. In fact, the chairman of a local committee of the trustees had asked John to remain, promising a good salary if he did so, and the older LeConte duly noted this fact in his letter of December 29. To this, John added statements that he had solicited from some of the departed professors, all of which supported his contention that they had left because of Church. He concluded his already damning letter by saying that, in spite of his wishes to the contrary, he felt duty-bound "to expose the lamentable frailties of one who presides over my Alma Mater."[33]

That final statement was tantamount to sticking a pin in the only soft spot on Church's otherwise tough hide, for it questioned the old president's competence. Since he could no longer get at John except through public letters—and those were proving not only ineffective but also injurious to his own cause—Church turned his guns on Joseph. This was a bad decision, for Joseph was even more popular than John among the students and his colleagues. But Church was not unpopular with students either, and he did enjoy the esteem of local townspeople as well as the support of several influential state leaders.

32 Augusta *Weekly Chronicle & Sentinel*, December 12, 1855, p. 4. According to Joseph LeConte, Scherb was "a genius, but a queer, highstrung, excitable, erratic, irresponsible creature," and therefore "could not long avoid violent collision with Dr. Church" (LeConte, Autobiographical MS.). Again, Church sent his letter to several newspapers; see, for example, the Augusta *Weekly Chronicle & Sentinel*, December 19, 1855, pp. 5–6.

33 Coulter, "Why John and Joseph LeConte Left the University of Georgia," 30.

Joseph had followed in the footsteps of his brother by neglecting to carry out his police duties, and now Church began to insist that Joseph follow orders. Joseph replied that university regulations did not require him to do so. But he was wrong on that score, and when Church carried the matter to the trustees, they not only upheld the president but added that he had complete authority to assign specific duties and times for faculty visitation of students' rooms. Unmoved by the decision, Joseph carried the quarrel to the Prudential Committee, a local group of trustees who had extraordinary powers regarding the governance of the university. Confronting Church at the stormy session, LeConte charged that the president "was oppressive in his official conduct toward him," but the committee denied the validity of his complaint and instructed him to carry out the assigned duties. Reluctantly, he agreed to comply, but he concluded that it would be "an oppression" to him. Actually, the defeat went beyond LeConte to the other faculty members who also objected in particular to Church's enforcement of their police duties.[34]

The anti-Church newspapers continued to criticize the president, and eventually the controversy began to wear upon the beleaguered man. He submitted his resignation in mid-August, to take effect at the end of the term in October. The pro-Church faction raised a great outcry, and a number of them protested openly over the president's resignation. In his report to the trustees in October, Church made an acerbic attack upon his critics and castigated former professors Jones, Charles McCay, and John LeConte, as well as the current faculty. Among the latter, he, of course, criticized Joseph LeConte for his "open determined hostility to the reputation and success" of the university. The hotheaded and high-handed old president would never admit defeat nor would he ever entertain the slightest doubt about his administrative ability. Faced with a dilemma they could not resolve, the trustees requested the resignation of the entire faculty on October 15, to take effect at the end of the term in late October. Thus, the university now had neither a president nor a faculty, and the trustees would not make new appointments until a month after the term had ended.[35]

34 University of Georgia Prudential Committee Minutes, September 19 and December 6, 1855, and February 23, 1856, in University of Georgia Library, Athens.

35 Athens *Southern Watchman*, December 4, 1856, p. 2; *Georgia University Magazine*, XI (November, 1856), 92. Notice of the vacancies was published in the Athens *Southern*

Meanwhile, students waited to learn who would be appointed to teach them in January, while the faculty suffered through the anxiety of waiting to see if they would be reappointed. In their meeting on December 10, the trustees reelected Church as president by a narrow margin. The "old rascal," as Professor Charles Venable called him, had triumphed. Among the former faculty only one, the Reverend Brantly, was reappointed. All the others lost their bid, but some of them did not wish to return anyway if Church was rehired. It made no difference to Joseph LeConte, for, in the meantime, he had sought and secured an appointment as professor of geology and chemistry at South Carolina College, where his brother John had already taken a post. As one of Professor LeRoy Broun's friends said, "I think you may feel well rid of such a poor set of fools as govern this college." In the opinion of Alonzo Church's daughter, however, the president "has at last a Faculty, some of whom he knows to be possessed of common sense as well as Book learning—he anticipates happy association and more effective Faculty than he has had for many years." It may well be that during his final three years as president, Church had a happy situation; at least he was rid of those he viewed as troublemakers. But he was likewise rid of some very capable men, not the least of whom were the LeConte brothers, whose move to South Carolina College helped to push that institution to the forefront of science in the antebellum Deep South.[36]

Georgia's loss was South Carolina's gain, but Georgians as a whole

---

Banner, November 13, 1856, p. 4; the newspaper reported that the trustees would reorganize the university at their meeting on December 10. U.S. Congressman and University of Georgia Trustee Howell Cobb (of Athens) and his wife took a special interest in the controversy; they intimated that the LeContes were primarily to blame for the difficulty (letters dated from January 16 to August 14, 1856, in Howell Cobb Papers, University of Georgia Library, Athens; used by permission of Howell C. Erwin, Jr., Athens, Ga.).

36 Howell Cobb to Mrs. Cobb, December 11, 1856, and Elizabeth Church Craig to Mrs. Cobb, December 15, 1856, both in Cobb Papers; Charles Venable to William LeRoy Broun, December 9, 1856, J. M. Stevens to Broun, December 14, 1856, and H. Hull, Jr., to Broun, December 11, 1856, all in William LeRoy Broun Correspondence, Auburn University Library, Auburn, Alabama; Joseph LeConte to Bessie LeConte, December 16 and 29, 1856, both in McMillan Collection; *Georgia University Magazine*, XI (February, 1857), 124; Athens *Southern Banner*, December 4 and 18, 1856, p. 2. In his report to the Senatus Academicus for 1858, President Church said: "During the thirty-nine years of my connection with it [the University of Georgia], I have never witnessed as much good order and as strict attention to all duties . . . as I have seen during the year just closed." *Report of Alonzo Church, President of Franklin College, University of Georgia, to the Senatus Academicus* (Milledgeville, Ga.: Boughton, Nisbet, and Barnes, State Printers, 1858).

failed to comprehend fully the significance of their loss and especially of the LeContes' departure. Twenty years later, the university attempted to lure the brothers back, but failed. In 1879, the trustees bestowed honorary degrees upon them and many decades later named a building in their honor. These feeble gestures hardly invalidate the ancient maxim that a prophet receives no honor in his own country. Although the LeContes would maintain their ties in Georgia for a long time, they could not easily forget the rough treatment they had received from their beloved alma mater.

—◦◦✿ IV ✿◦◦—
Before the Storm Burst: South Carolina Years, 1857–1861

The University of Georgia affair may have bruised the ego of Joseph LeConte, but it certainly did not mar his reputation. Indeed, while the verbal fracas was reaching its peak in September, 1856, LeConte was in Albany, New York, reading to the AAAS his paper on the agency of the Gulf Stream and gaining the esteem of prominent American scientists. Among those who took special notice was Joseph Henry, Secretary of the Smithsonian Institution. Soon after LeConte presented his paper, Henry invited him to deliver a series of lectures under the auspices of the Smithsonian.[1]

In spite of the difficulties that he faced in Athens, Joseph began immediately to prepare three lectures on coral formation and three on the origins of coal. In mid-December he traveled to the nation's capital to deliver the lectures at the new Smithsonian building. Shortly after arriving, he wrote to Bessie that he was nervous: "I have just been looking at the lecture room. The size of it rather alarms me—but I hope I will get through it safely. If I can only succeed the first evening all will be well." But he had no reason to be anxious, for the lectures came off successfully. After the first, he reported to his wife that he had done "as well as I expected—in fact better.... Several persons complimented me on my effort last night." Then, upon completing the entire series, he wrote home again, saying, "My lectures in Washington

---

1 LeConte, *Autobiography*, 162–63. Henry had already asked Joseph and John LeConte to serve as directors of a project to develop "a series of tables on the Constants of Nature" (*Annual Report of the Board of Regents of the Smithsonian Institution, 1873* [Washington, D.C., 1874], 23–24). On January 6, 1858, he asked Joseph LeConte to prepare a set of geological maps (letter in LeConte Family Papers, Manuscripts Division, Bancroft Library, University of California, Berkeley [hereinafter cited as LeConte Family Papers, UCB]).

succeeded beyond my expectations. . . . Prof. Henry wants me to write
them out and publish them." He was obviously justified in making a
confident statement, and the comment of the highly respected geologist J. Peter Lesley serves as testimony to his success. Lesley said in a
letter to his wife, "Last evening Prof. LeConte of the south gave us his
eloquent lecture on coal." In addition, LeConte received the praise of
former President Millard Fillmore and several French scientists who
later had the coal lectures translated into their own language.[2]

As a lasting contribution to scientific knowledge the lectures on
coal were limited, much of their content being later superseded by
more exact knowledge and other important geological discoveries. In
contemporary context, however, they represented an important synthesis of information on the origin and nature of coal deposits. Perhaps one of the most significant of LeConte's points dealt with "the
affinities of gymnosperms"—that is, the relationship between the
structure of those plants that produce seeds not encased or enclosed
in an ovary, and the angiosperms, in which the seeds are so enclosed.
LeConte's speculation "anticipated by thirty years similar views" espoused by the great American paleobotanist Lester Frank Ward and
the accomplished German botanist Adolf Engler.[3] But LeConte's lectures on coal also provide additional insight into the man himself and
the nature of his thought at that stage of his professional career. Any
consideration of LeConte's intellectual development toward evolutionism must begin with these lectures.

In a rather lengthy introduction bearing upon science and religion
but having little to do directly with the topic of coal deposits, LeConte
eloquently expounded upon the subject of scientific progress. "Nature
is a book in which are revealed the divine character and mind," he
began. "Science is the human interpretation of this divine book, hu-

2 LeConte, *Autobiography*, 164; Joseph LeConte, "Lectures on Coal," *Annual Report of the Board of Regents of the Smithsonian Institution, 1857* (Washington, D.C., 1858), 119–68. In the *Annual Report, 1856*, Joseph Henry indicated that LeConte had delivered three lectures on "Coal" and three on "Coral" (p. 46). See also Joseph LeConte to Bessie Le-Conte, December 13, 16, and 29, 1856, all in McMillan Collection; J. Peter Lesley to Susan Lesley, December 16, 1856, in *Life and Letters of Peter and Susan Lesley*, ed. Mary Lesley Ames (2 vols.; New York, 1909), I, 341; Leo Lesquereaux to J. Peter Lesley, in J. Peter Lesley Papers, American Philosophical Society Library, Philadelphia. The French extract of the lectures was published as "Sur les plantes des terrains carbonifères," *Mémoires de l'Académie de Dijon* (1862).

3 LeConte, *Autobiography*, 165.

man attempts to understand the thoughts and plans of Deity." But, he continued, some areas of scientific inquiry are by their very nature more significant than others, foremost of which are astronomy and geology. The first deals with space, the other with time. "As astronomy measures her distances by billions of miles, or millions of earth radii, so geology her epochs by millions of years, *i.e.*, earth revolutions." The former, he averred, "carries us upwards by the relations of geometry," while the latter "carries us backwards by the relations of cause and effect," or, in other words, "astronomy steps from point to point of the universe by a chain of triangles," while geology "steps from epoch to epoch of the earth's history by a chain of mechanical and organical laws." The great purpose of science is "to establish the universality of law . . . unity in the midst of diversity." Thus, concluded LeConte, astronomy and especially geology serve religion, and soon the "revelations of geology" would carry mankind into the "promised land" of knowledge about nature, hence about mankind itself.[4]

Although in no wise either profound or new, LeConte's statement is a revelation of the course of his thought. Religion was obviously important to him, but it was something to be understood through rational means rather than through conventional authority. Arguing from what was essentially a uniformitarian view of the history of the earth—although his position was wracked occasionally by inconsistent insertions of the catastrophic theory—LeConte had begun to divest himself of the traditional biblical explanation of an incessantly intervening divine power. This was further evidenced in his rejection of the idea that the earth was a unique phenomenon in the vast universe. Although he avoided any mention of the origins of the heavenly bodies, LeConte was clearly straying from biblical to natural theology, and he was declaring that the study of the past as revealed in the book of geology provided the key to understanding the work of God. He was now speaking of an earth that evolved over millions of years according to universal laws and alluding to a "tree" of development.[5]

Although LeConte did not realize it, and though in these addresses he denied the possibility of transmutation of species, he was laying

4 LeConte, "Lectures on Coal," 119–25.

5 LeConte's ideas on this subject are essentially representative of a contemporary school of thought. For an excellent discussion of the context into which LeConte's ideas fitted, see Charles Coulston Gillispie, *Genesis and Geology* (Cambridge, Mass., 1951), especially Chapters IV–VI.

the ground that would ultimately lead him to adopt the theory of organic evolution. The Agassiz-trained scientist was certainly tampering with traditional faith, and he was unwittingly beginning to step from the mold of his mentor, who would always hold fast to the idea that the fossil record was but an account of special divine creations, not an indication of the transmutation of species over the vast expanse of time. The seed had been planted, but it would not flower until much later when LeConte encountered the Darwinian theory and, after a long intellectual struggle, finally abandoned the views of Agassiz.

His lectures delivered, LeConte returned to Georgia to finish packing for the move to Columbia, South Carolina. By early January, 1857, he was ready to take up his new duties, which included teaching courses in geology, chemistry, and freshman mathematics. The first presented no problem, of course, for LeConte had now taught the subject for four years. But he had taught chemistry only once before, at Oglethorpe University. He liked the subject, however, and a few years later he decided to coauthor a chemistry textbook with his brother John. Eventually, he produced a lengthy manuscript of the textbook, but, unfortunately, it was destroyed along with his other papers by the marauders who sacked and burned his wagon in 1865. To his great credit, LeConte rewrote the entire manuscript (over four hundred tightly written pages in his tiny scrawl). His brother never redid his section of the work, however, and the project was eventually dropped. To Joseph's responsibilities for instruction in geology and chemistry was added the duty of helping Charles McCay teach the freshman mathematics course. The burden was so heavy, "ten exercises a week in all," that LeConte found it "impossible to do any original work."[6]

After only six months in South Carolina a frightening incident occurred, reminiscent of the University of Georgia affair and once again involving Charles McCay and the LeContes. Founded in 1801, South Carolina College first opened its doors in 1805. From the outset, the college had experienced problems because of religious factions that desired control of the institution. In December, 1851, James H. Thorn-

---

6 Joseph LeConte to Bessie LeConte, from Athens, Ga., December 29, 1856, in McMillan Collection; LeConte, *Autobiography*, 165–66. The manuscript of the proposed chemistry textbook is in LeConte Family Papers, UCB.

well, an orthodox Presbyterian theologian and a long-time professor in the college, was appointed to the presidency of the institution. An able scholar, Thornwell followed other distinguished men who had brought considerable prestige to the school, but poor health and dissatisfaction with administrative routine led to his resignation in December, 1855. A number of candidates were put forward, including Francis Lieber, a native German who had joined the institution as a political scientist in 1835. Despite the fact that the fifty-seven-year-old Lieber was a nationalist in sympathy, an erstwhile critic of slavery, and an outspoken opponent of evangelical Christianity, he was held in high esteem as a scholar. Yet his candid views did not go unnoticed in the restive political situation that characterized South Carolina in the 1850s. One of his chief opponents was the influential Thornwell, who helped to block his bid by strongly supporting McCay, a good Presbyterian and a man with no enemies yet.[7]

McCay was eventually elected, but he soon found himself in hot water. Lieber resigned in disgust after his defeat, and his supporters blamed McCay not only for the defeat of their candidate but also for his resignation. In addition, McCay enjoyed no real standing in South Carolina; he had been at the institution for only two years before his unexpected appointment to the presidency, and he possessed no special qualifications for the post. By February, 1856, one professor had died; another was too ill to teach; and Thornwell's and Lieber's chairs had not been filled. To cover the vacant chairs, McCay did the only possible thing he could do: he assigned the duties to other professors, who then blamed McCay for their added burden.[8]

It would have taken a man of extraordinary skill to cope with the problems facing McCay, but, unfortunately, the hapless fellow possessed little administrative ability. Students quickly took advantage of the tenuous situation to create trouble for the new president. First, they called upon the board of trustees to reconsider their appointment of McCay, and, when the petition was ignored, they began to engage in disorderly and disruptive activities, which merely increased as McCay turned his head. Already upset by the increase in their load, the faculty lifted not a hand to quell the disturbances. Soon the rau-

7 Daniel Walker Hollis, *University of South Carolina* (2 vols.; Columbia, S.C., 1951), I, 160–211.
8 *Ibid.*, I, 194–203.

cous students precipitated a series of ugly events that led to a serious confrontation between armed camps of town citizens and students. Thornwell, now at the nearby Presbyterian seminary, saved the day, but the college was subsequently closed for three weeks.[9]

McCay offered to resign, but, to save face, the trustees urged him to remain in office. Many students left, and others were suspended. Meanwhile, the trustees hired John LeConte to replace the ill professor and William J. Rivers to fill the chair of Greek literature, which had been vacant since the death of its former occupant a few weeks earlier. These circumstances all played a role in quieting the hostilities of faculty and students, but McCay's troubles were not yet over. By early December, two additional faculty members had been hired, one of whom was Joseph LeConte. McCay also shuffled the faculty around in a reasonable effort at reorganization, but the classicists cried that in so doing he had depreciated the role of their subjects while elevating the sciences. The beleaguered president simply could not win; any way he turned, he was pilloried by his critics.[10]

The LeContes, especially Joseph, were most certainly aware of the vexatious situation into which they were entering, but they may not have understood the political ramifications involved and the depth of the rampant recriminations. Moreover, they both had high regard for their old colleague McCay, although their knowledge was limited to his scholarly attainments, and they knew nothing of his administrative abilities. At any rate, John was eager to return to his native South, and Joseph, of course, could not conceive of a more deplorable situation than the one in which he was involved at the University of Georgia. So to South Carolina they came. Although they must have sensed a certain tension in the air in the early months of 1857, they could not foresee the trouble that followed in April and May.

When a group of students disrupted chapel services in late April, McCay and the faculty dismissed three of them, whereupon the junior class petitioned the president to reconsider on the ground that the students had been suspended without sufficient evidence of their guilt. McCay ignored the petition, and when he would not cooperate with a special committee of students commissioned to work out an agreement with him, the junior class ceased to attend his lectures.

9 *Ibid.*, I, 196–203; LeConte, *Autobiography*, 166–78.
10 Hollis, *University of South Carolina*, I, 201–203.

Unhappily for McCay, some of the disgruntled faculty tacitly supported the students, and the volatile situation reached the point of ignition by the end of May. At last McCay had used up his large reservoir of patience, and "he accused the faculty of bad faith and heatedly resigned." John LeConte thereupon proposed the adoption of the following resolution by the faculty: "That the Professors have unabated confidence in the honor, integrity & purity of motives of the President." The resolution represented something of a gamble, given the sour mood of the faculty and the fact that the LeContes were new to the college. Only John and Joseph voted in favor of the resolution; all the others abstained.[11]

It is likely that a favorable vote would have mattered little, for the trustees were now less concerned about saving face over their appointment of McCay than they were about saving the college from any further squabbling. On June 11, they approved a motion calling for the resignation of the entire faculty. The results differed drastically from the same situation in Athens some seven months before, however, for the trustees met again on the same day and unanimously reappointed the LeConte brothers. They also reappointed W. J. Rivers, but they replaced two other faculty members and left unfilled the positions of the remaining four, including Maximilian LaBorde, a longtime, successful professor of logic, rhetoric, and belles lettres. The institution was then closed for the summer, with no president and only three faculty members, the two replacements having declined their appointments.[12]

Under a withering round of criticism, the trustees met again in September, when they attempted to induce Thornwell to take up his old job. But the former president adamantly refused, and the board simply dropped the matter for awhile. Finally, they got together a faculty

11 LeConte, *Autobiography*, 48–49, 167; Hollis, *University of South Carolina*, I, 203; South Carolina College Faculty Minute Books (South Caroliniana Library, University of South Carolina, Columbia), June 1, 1857.

12 L. L. Fraser, Secretary, Board of Trustees, South Carolina College, to John LeConte, Joseph LeConte, and William J. Rivers, June 11, 1857, in William J. Rivers Papers, South Caroliniana Library, University of South Carolina, Columbia. The three reelected professors met on the same date and chose Rivers as acting president (South Carolina College Faculty Minute Books, June 11, 1857). Joseph's letter of acceptance of reappointment was written to Fraser on June 12, 1857 (in Joseph LeConte Papers, South Caroliniana Library, University of South Carolina, Columbia; hereinafter cited as LeConte Papers, SCL/USC). See also Hollis, *University of South Carolina*, I, 204.

of seven, including LaBorde, and the school opened on schedule in October—still without a president. It was not until late November that the trustees finally agreed upon a man to fill the post. He was Augustus Baldwin Longstreet, an experienced and proven administrator who had served successfully as president of Emory and Centenary Colleges and of the University of Mississippi. Although already sixty-seven years old, the new president was still full of vigor and had lost none of his verve. In addition, the trustees hired Charles F. Venable, the former colleague of Joseph LeConte who had been involved in the quarrel with Alonzo Church.[13]

At long last, the difficulties of South Carolina College seemed at an end, but in fact they were not, for the students felt it necessary to test their new president. In Joseph LeConte's words, Longstreet was "an able lawyer and judge and a distinguished humorist ... but he was utterly unfit for the presidency of a college. He was not in any sense a cultured man, and could not inspire the highest respect of either the students or the faculty." Although Joseph's description bears a mark of bias against a man who did not fit LeConte's image of a true scholar and gentleman, it did accurately portray the view of both students and faculty. Longstreet "was not, however, wanting in firmness." Indeed, the old man was a stern and harsh disciplinarian of the old school. He quickly had a chance to prove his ability to maintain order shortly after he took the reins of office in January, 1858.[14]

Clamping down upon disorderly conduct, Longstreet aroused the ire of the high-spirited students. When they subsequently demanded suspension of classes in commemoration of John Calhoun's birthday, Longstreet stoutly denied their request, whereupon the students retaliated by tarring the classroom benches. Longstreet simply decreed that classes would be held in other available rooms and promptly suspended over one hundred students. His decision stuck faster than the tar the students had so freely smeared upon the benches, and when the fall term began in October only about fifty of the suspended students were readmitted, and then only after a grueling examination. In January, 1859, another twenty-five or so were readmitted, but some

13 Hollis, *University of South Carolina*, I, 204–208.
14 LeConte, *Autobiography*, 169–70; John Donald Wade, *Augustus Baldwin Longstreet: A Study of the Development of Culture in the South* (New York, 1924), 317 and *passim*.

were never allowed to return.[15] Respected or not, Longstreet was in full control, and after that unpleasant experience the college faced only a few minor disturbances.

In spite of the difficulties at the college in 1857, Joseph was happy in Columbia, and he kept himself occupied with his classes, some occasional writing, and social activities. After vacationing with his family at Virginia Springs early in the summer, he left in August for Montreal to attend the annual AAAS meeting. It was his first trip outside the United States, and although he did not present a paper, he thoroughly enjoyed the renewal of association with prominent scientists, and he was especially intrigued by James Hall's address on the subsidence-sedimentation theory of mountain-formation. Hall's idea later played a role in the development of LeConte's own theory of orogeny. When it was first presented, however, neither LeConte nor other geologists were "ready for the truth" presented by the great geologist. In fact, LeConte told Arnold Guyot at the time that he understood "not a word" of Hall's explanation.[16] That would change, however, and LeConte would not only subsequently comprehend it but also show its deficiencies and its strengths, fitting it properly into a more comprehensive theory.

Back in Columbia for the unsettled opening of the college in the fall of 1857, LeConte began to draft his inaugural address, a customary expectation of all new faculty members at the time. The paper, completed and delivered on December 8, 1857, was published as simply "Inaugural Address," but later LeConte wrote in the title as "On the Function of Geology in a Course of Education." Like the introduction to his lectures on coal, the paper represented a mixture of reflections on science and religion, but it also contained numerous allusions to literature. Ever prone to form a hierarchy of disciplines on the "order of their increasing complexity," LeConte maintained that "mathematics is the simplest of all sciences, since it includes only ideas of *number and quantity*." It was followed by the development of mechanics, astronomy, and physics, which, "in addition to ideas of number and quantity, include those peculiar to themselves, viz: ideas of *force*."

15 LeConte, *Autobiography*, 170; Hollis, *University of South Carolina*, I, 209–10; South Carolina College Faculty Minute Books, March 25, 1858.
16 LeConte, *Autobiography*, 167–69. For some reason LeConte judged Canadians to be "miserable snobs" (see his letter to his wife, August 16, 1857, in McMillan Collection).

These "departments of science" were then followed respectively by chemistry, biology, and geology, each of which included preceding ideas but added their own—that is, successively, "chemical affinity," "life," and "historic development." As each discipline "matured," it formed "the foundation upon which the next was raised." Thus, taken together, they constitute "a magnificent temple of which mathematics forms the solid foundation and Geology the heaven-pointing spire— the most wonderful and perfect work which human genius has yet erected in honor of Deity."[17]

Upon this base LeConte argued that a liberal education must go through all of the sciences, including geology. "I think it cannot be denied," he said, "that any course of education which commences with Mathematics is radically bad, distorting and contracting the mind unless it passes upwards through all the sciences and particularly unless all equal attention is paid to the more complex departments of Biology and Geology." He then argued that a "complete course of education is naturally divisible into three subordinate courses," the first of which begins with mathematics and moves progressively through the sciences, ending with geology. The second begins with a study of the ancient languages and moves "upward through all literature and art." The final division begins with metaphysics and ends in the study of "political and social philosophy."

After sketching the historical development of thought, LeConte tried to show how the role of the sciences in education strengthened religious belief. "I know," he said, "that Geology is looked upon by many with dread and suspicion as the enemy of faith," and he confessed that the "collision" of geology with theology "has seemed to be opposed to Scripture." But, he concluded, "the history of Science proves that this collision is only apparent, and the evil therefore transitory, while the service is real and lasting. Thirty years more and Geology will be universally acknowledged as the chief hand-maid of religion among the sciences."[18] LeConte thus not only anticipated the impending conflict between theology and science but also understood the need for serious reform of the staid old southern college

17 "Inaugural Address," delivered in the State House, December 8, 1857; reprinted at Columbia, S.C., 1858. The modified paper is in the possession of Helen Malcolm LeConte, San Francisco, Calif. LeConte wrote a six-page "Addendum" to this paper in 1871, which reflects his thoughts on the subject in light of his leanings toward evolutionism.
18 LeConte, "Inaugural Address."

curriculum. The full significance of his statement was not immediately apprehended by the stalwarts of orthodoxy, but neither did LeConte then understand the direction in which his liberal interpretation was carrying his thought.

Although the situation at South Carolina College was not conducive to original scientific research, it did not dampen LeConte's growing enthusiasm for writing. If the heavy burden of preparing classroom lectures retarded original scientific inquiry, it nevertheless increased the popular teacher's philosophical and theoretical interests. Such interests were sharpened by LeConte's association with Thornwell, William C. Preston, former president of South Carolina College, and Wade Hampton, a prominent statesman and later a Confederate military leader. The latter two men encouraged LeConte to write out his thoughts on various philosophical and literary topics. They also helped to shape his views on states' rights, but since LeConte was not a political animal, he shunned opportunities to express opinions on governmental affairs. Of much greater interest to him were philosophy, education, literature, and religion. These interests coincided with those of Thornwell, who inspired LeConte to write a series of essays on the fine arts and education and their correlation with religion, all of which were published in the *Southern Presbyterian Review*. LeConte praised Thornwell, Preston, and Hampton as gifted friends who "powerfully stimulated" his "intellectual activity."[19]

In the summer of 1858, Joseph, John, and their families vacationed at Flat Rock, North Carolina, a splendid summer resort that attracted "many of the most cultured families of Charleston and the low countries generally." Among those whom Joseph met on this occasion was Langdon Cheves, son of a well-known statesman of the same name. Cheves, a prominent planter and ardent proponent of secession, was a well-read intellectual who believed in the evolution of species by transmutation. Since both he and LeConte were intimately familiar with *Vestiges of the Natural History of Creation*, they had grounds for a "hot and heavy" debate. Cheves argued for the transformation of species, while LeConte took the side of his beloved teacher Agassiz, who believed that separate species constituted specific creations by divine intelligence or, in other words, that the Supreme Power had created new species in separate stages of a preordained design of life-

19 LeConte, *Autobiography*, 172–73.

forms. Quite naturally, then, LeConte maintained that, although fossil species found at differing levels of geological strata appeared to be similar, they had not evolved by transmutation but were created as they were by the Supreme Intelligence. Nevertheless, LeConte admitted that he was "struck . . . very forcibly" by Cheves's argument.[20]

Toward the end of his stay in Flat Rock, LeConte agreed to man a barometer station near Waynesville for his old friend Arnold Guyot, the Swiss-American geographer from Princeton University who compiled massive climatological records for the eastern section of the United States. One of Guyot's large undertakings was the compilation of accurate altitude measurements of the Appalachian Mountains. He asked Joseph to observe a stationary barometer situated among the highest peaks of the Smoky Mountains. LeConte did so for several days, and it appears that his service was rewarded when Guyot named a mountain in his honor. Formerly known as Central Peak, by 1860 the mountain was already appearing in the literature as Mount LeConte, and the name was officially recognized many years later. It was the first of two mountains, as well as numerous other landmarks, that came to bear his name.[21]

As the months passed in South Carolina, LeConte kept busy with his "philosophical" writings. Several of these dealt with education and were given as addresses to various groups. In one of them, titled "The Principles of a Liberal Education," LeConte reiterated his disagreement with the utilitarian approach to college education. "I fearlessly assert," he began, "that it is not the object of liberal Education to prepare one for any business, trade, profession, or pursuit in life; but, on

20 *Ibid.*, 174–76. See also Clement Eaton, *Freedom of Thought in the Old South* (Durham, N.C., 1940), 309; and W. J. Cash, *The Mind of the South* (New York, 1941), 91–94.

21 James B. Buckley, "Mountains of North Carolina and Tennessee," *American Journal of Science*, 2nd ser., XXVII (May, 1859), 287; Arnold Guyot, "On the Appalachian Mountain System," *ibid.*, 2nd ser., XXXI (March, 1861), 187; Paul M. Fink, "Smoky Mountains History as Told in Place Names," *East Tennessee Historical Society's Publications*, VI (1934), 7; Paul M. Fink and Myron H. Avery, "The Nomenclature of the Great Smoky Mountains," *ibid.*, IX (1937), 63. In a letter to the author from Donald J. Orth, U.S. Board of Geographic Names, Washington, D.C., November 29, 1976, the naming of the mountain is confirmed. The evidence seems to refute the statement of Thomas L. Connelly, *Discovering the Appalachians* (Harrisburg, Pa., 1968), 180, that the mountain was named for LeConte in honor of his discovery of "saltpeter deposits . . . at the present Alum Cave Bluffs." During the Civil War, LeConte did visit "all the niter caves in northern Georgia, Alabama, and Tennessee" (LeConte, *Autobiography*, 184), but this was long after the mountain was named.

the contrary, that the educational life is the *complement*, and, in some sense, even the *antagonist* of business life." Comparing human society to an organism, he maintained that, as civilization advanced and became more complex, individuals ceased to carry out broad functions and became more and more victims of a "division of labor" into specialized activities. Independence of action was therefore curtailed by the progress of civilization. If "material success" were the only goal, this specialization would work to man's advantage. But since man is also a spiritual being, he needs the benefits of cultural uplift as well. A liberal education, said LeConte, can combat the "evil tendencies" of materialism, aided by the church and the family, which together constitute "a glorious *Trinity*" overseeing "man's highest good."[22]

Once more LeConte stressed the importance of a curriculum consisting of three hierarchically arranged departments of knowledge. To this he added extensive comments on the role of the inductive method in developing rational and truth-seeking minds. Critical of ancient and medieval systems and unsympathetic toward modern European approaches to education, he commended the general program followed in the United States, but he feared it was in danger of succumbing to the utilitarian drive. "The object of liberal Education," he concluded, "is not to make specialists or experts in any department, but to communicate general principles" and methods which aid the individual to make "the proper discrimination of the essential from the non-essential."[23] It was a philosophy that would please many modern critics of undergraduate specialization, despite its somewhat rigid structural organization. LeConte later expanded upon his "ideal" curriculum in numerous other papers.

The education of females was likewise important to LeConte, and he elaborated at length upon the subject. Like most southern gentlemen, he placed women on a pedestal, giving them a special place in society. Women, he fervently maintained, have a "moulding influence . . . on the plastic mind of childhood," and they are thus more effective than men in preserving civilization. As a being possessed of a "distinctive nature and sphere," the female must be educated in such a way that her distinctive features and social role are specially cultivated. At

---

22 "The Principles of a Liberal Education," address delivered on April 20, 1859, and published by R. W. Gibbes in Columbia, S.C.; also published in the *Southern Presbyterian Review*, XII (July, 1859), 310–35.

23 *Ibid.*

the time he wrote the paper, LeConte was terribly disturbed by the "woman's rights movement," which, although primarily aimed at gaining the franchise, he viewed, like many of his southern contemporaries, as an attempt "to assimilate the male and female character, the male and female spheres of activity, and the male and female education." He joyfully exclaimed that the "madness, the frenzy, the absurdity of this spirit has not touched us here at the South. . . . Woman has not unsexed herself here by invading the sphere and donning the habiliments of man." He agreed, however, that the movement had succeeded in "elevating the standard of female education, and in making it more solid and thorough." Yet he expressed fear that overall the movement had "tended to degrade" women.[24]

LeConte steadfastly claimed to believe in the equality of the sexes, but just as steadfastly he believed in a dual standard. The woman's rights advocates, he averred, operate from a "palpably false premise," that is, "that there is but one standard of human excellence—one ideal of humanity, and that is the male standard." It is upon "this Procrustes bed her tender limbs must be stretched," he lamented. The tendency to "assimilate" is dangerous, for it leads to the mistake made by the French in their great revolution, that is, by trying to abolish "distinctions of every kind," leading ultimately to the destruction of "every relation of society." Thus, argued LeConte, man must define the spheres of activity for each sex and develop appropriate educational programs accordingly. First, female education must be more informal than that for males because woman's mind is "more susceptible of indirect teaching, and less susceptible of didatic [sic] teaching." "Man's sphere is the great public world"; woman's is the domestic world. Secondly, "woman's education may be made more practical than man's," for the female inherently possesses many of those cultural qualities that the male can gain only through liberal education. Finally, falling back upon his tripartite division of liberal education, LeConte insisted that "the Art course must forever constitute the main building" of female education. Other areas must not be neglected; but woman by nature is governed by "the *intuitive* reason," whereas "the *formal* reason" predominates in man. LeConte concluded that language, literature, composition, and art are open avenues to the female character.[25]

24 "Female Education," *Southern Presbyterian Review*, XIII (April, 1861), 60–91.
25 *Ibid.*

Although his interest in the defense of the special role of women in society would soon subside, it never died, and when the woman's rights movement reached a peak of fervor in California in the mid-1890s, LeConte would return to the subject with renewed vigor.

Although LeConte devoted most of his writing during this period to such ruminations, he did not totally neglect scientific subjects. Indeed, he prepared two papers for presentation before the AAAS meeting at Springfield, Massachusetts, in early August, 1859. The first of these was a defense of the theory of a molten or liquid earth-interior. Representing LeConte's first foray into the field of diastrophic geology, the paper was mainly a defense of the views of Professor George B. Airy, an English astronomer, and therefore constituted no truly original thought. The actual contents of the paper are unknown, for it too was among the manuscripts destroyed in 1865. A synopsis of the address was made by one of the geologists in attendance, however, and that summary was subsequently published in the *Canadian Naturalist*. From that summary and later comments by LeConte, it is clear that at the time he subscribed to the theory that the earth's interior region was fluid. His "elaborate explanation" was "illustrated by diagrams of different bodies floating upon water," and it utilized laws of flotation to show how a solidified crust would, according to its uneven thickness, either rise into continents and mountains or sink into ocean beds and depressions. In the opinion of T. Sterry Hunt, the geologist who summarized LeConte's paper, the argument "as a whole was remarkably clear, logical, and conclusive, and presented many points worthy of study."[26] Although he would subsequently abandon his theory, LeConte was able at this point to demonstrate his remarkable ability for explaining and illustrating complex scientific phenomena. This was an ability that ultimately distinguished him as one of the most capable writers of his kind.

LeConte's second paper generated even more interest than his first. Later recast to conform to his altered views on the theory of evolution, his "The Correlation of Physical, Chemical and Vital Force, and the Conservation of Force in Vital Phenomena" was a concise synthesis of findings in the field of physiology and a thoughtful treatise on the

26 LeConte, *Autobiography*, 173–74; Joseph LeConte to his wife, August 11, 1859, in McMillan Collection; "Theory of the Formation of Continents," *Canadian Naturalist and Geologist*, IV (August, 1859), 291–94.

transformation of energy. Contrary to his oft-repeated emphasis on the importance of the inductive method, this elaborate discussion of changes in the forms of matter was based largely upon "deductive reasoning," and hence, because it depended upon certain premises that were later invalidated, it contained some fallacies. Nevertheless, it was an excellent summary of the relationships between and among various forms of energy and the processes of their transformation, and, more than any of his early papers, it enhanced LeConte's reputation as a scientist. It was, as the title suggests, an inquiry into one of his favorite, if heretofore neglected, subjects (*i.e.*, physiology), and although the paper did not stand the ultimate test of scientific scrutiny, it stimulated further thought in the field.[27] Likewise, the paper indicated LeConte's broad grasp of several scientific disciplines. Such universalism certainly summoned recognition at this point in LeConte's career, but in the long run it hampered specialization. Thus, during the last quarter of the nineteenth century, when others were staking out their specialties, LeConte continued to play the role of Renaissance man.

By "vital forces" LeConte seems to have meant the physiological phenomena of *human* life, although he also referred to animal life as well. In fact, he identified "four planes of material existence," of which the highest was the animal, followed in descending order by the vegetable, the mineral, and the elemental. As material things could be hierarchically arranged by structure, so could "the forces of nature" be graded by function, from physical to chemical to vital. LeConte's lengthy discussion of the processes by which physical forces—that is, matter—were transformed into chemical forces and thence into life forces ranged over a large number of metabolic activities, each illustrating the phenomena of composition and decomposition in plants, animals, and humans. Although LeConte spoke of a "pyramid of na-

27 The paper on correlation of forces appeared as "The Correlation of Physical, Chemical, and Vital Force, and the Conservation of Force in Vital Phenomena" in several leading publications, including the *Proceedings of the AAAS*, XIII (August, 1859), 187–203, and the *American Journal of Science*, 2nd ser., XXVIII (November, 1859), 305–19. It was revised in the *Popular Science Monthly*, IV (December, 1873), 156–70, and published a year later in Balfour Stewart (ed.), *The Conservation of Energy* (New York, 1874), 169–201. The revised essay reflects LeConte's conversion to the theory of evolution. LeConte was especially proud of this paper (LeConte, *Autobiography*, 173–74 and 333). Merle Curti, *The Growth of American Thought* (2nd ed.; New York, 1951), 438, cites it as being very influential both in the United States and abroad.

ture" whereby the "organic" or vital forces worked upward to their highest form in man, he stoutly denied that *new* forms of life, or different species, could evolve from such processes. The "question of the origination of species," he said, is "utterly beyond the limits of human science." The transformation of physical, chemical, and vital forces takes place within an already existing organism.

Still very much the Agassizian, LeConte was yet a long way from accepting the transmutational theory that was at that very moment appearing in Darwin's revolutionary book. When LeConte revised the paper for publication in 1873, he was well-nigh converted to evolutionism, but not totally; thus, he skirted the issue, concluding only that his views were not to be taken as a defense of "pure materialism."[28] As he would so often argue, any conflicts between science and religion, or between competing theories of any kind, could be reconciled whenever sufficient rational thought had been brought to bear upon them.

By 1859 LeConte was well established in his career; he had become "the idol of the college," and he had made a place in the professional world. The future looked bright, except for the ominous clouds of secession that rolled across the South Carolina sky, darkening more and more as the political crisis worsened. Joseph tried to ignore the matter as much as possible. Yet he was a slave owner, and he believed he had a moral responsibility to care for the blacks he had inherited from his father's estate. Each year he managed to visit his plantation, and he apparently also looked in on the plantation affairs of John and of his sister Jane, who had moved to Orangeburg, South Carolina, for the education of her two youngest daughters. Like other plantation owners who had a surplus of slaves, Joseph sometimes hired a few of them out, but he would not sell any, for he claimed that he had a personal obligation to take care of his bondsmen.[29]

28 LeConte, "The Correlation of Physical, Chemical and Vital Force," *Proceedings of the AAAS,* XIII (August, 1859), 187–203, and *Popular Science Monthly,* IV (December, 1873), 156–70. Contrasts in the final paragraphs of the original and the revised versions are especially notable, and in a letter to E. L. Youmans of the *Popular Science Monthly,* April 8, 1873, LeConte said that the revised article was "a restatement with some addition of my own views on the subject. It contains a good deal that is new" (E. L. Youmans Letter, Historical Society of Pennsylvania, Philadelphia).

29 Mary E. Robarts to Mary Jones, December 9, 1858, in Myers (ed.), *Children of Pride,* 462; LeConte, *Autobiography,* 231–32; LeConte, "The South Revisited," *Overland Monthly,*

Joseph's letters to Bessie during his visits to Syfax in December of
1858 and 1859 reveal much about LeConte's attitude toward his
charges. On one occasion he wrote: "Dear Bessie it really touches and
warms my heart to see their [the slaves'] evident attachment & apparent devotion to myself & you." Such "attachment" and "devotion" became painful, however, as Joseph's sister Ann and her husband J. P.
Stevens began to carry out their plans to move their Negroes to their
new estate in Baker County, thereby splitting familial ties among some
of the LeConte slaves. Recounting the event, Joseph told Bessie:

You should have seen how the whole of them, men, women & children, all
without exception gathered around me. And such a shaking of hands and
such a sending of howdy to *Misses* to the children & to Mary Ann [a house
servant in Columbia]. Such expressions of devotion to my interest you never
saw the like in your life. Dear Bessie we must never neglect coming down to
see them every year. A little trouble, a little kindness on our part gives them so
much happiness.

He continued:

At first one of my men & one of my women had some notion [of] going with
husband and wife but when it came to the pinch they have both backed out,
preferring to give up husband and wife [rather] than their master. I am very
glad it is so. I look upon my negroes as a sort of second family of mine. I
dislike to interfere with the dignity of the family. . . . I confess I was quite
touched.

Such was LeConte's feeling toward the more than sixty slaves he now
owned.[30] After emancipation, he strongly affirmed that he had wanted
to be rid of the burden of slaveowning but that he could not stand the
thought of selling his slaves. Nowhere does it appear that he ever entertained the notion of manumission.

LeConte's paternalism, his belief that he *must* care for his "second
family," was genuine, however misguided. The plantation was not a
very profitable enterprise, and he could, as he claimed, have made
more money by selling out and investing the money for a higher return. An obvious sense of guilt plagued Joseph, for he also believed
that a slave owner should "personally" oversee his charges, as his
father had done, but he confessed he could not do that "without sac-

2nd ser., XIV (July, 1889), 22; LeConte, "The Race Problem in the South," in *Man and the
State* (New York, 1892), 349–50.

    30 Joseph LeConte to Bessie LeConte, December 23, 1858, and December 24, 1859,
both in McMillan Collection; Tax Reports, 15th Militia District, 1861, in Liberty County
Records.

rificing all my ambition in life and the health of my family." He denied, however, that he "felt any conscientious scruples" about owning slaves, although he admitted that he was constantly distressed over his responsibility.[31] The dilemma was settled only when the Civil War tore it asunder and freed LeConte of his awesome burden. The impact of his plantation life would always shape his view toward black people, however, and it would also later make it easier for him to work out a racial scheme to fit his evolutionary theory.

As the political turmoil heightened in 1859 and 1860, LeConte faced his own personal crisis. His northern relative and close friend John L. LeConte was unsympathetic toward the southern cause. Just how his Uncle Jack felt about the matter is not at all clear. John Eatton LeConte was, of course, the son and brother of slave owners, yet he chose not to take up residence in the South or to get involved in the operation of the plantation. This may have been simply because of his desire to be associated with northern-based naturalists, or it may indeed have reflected his distaste for slaveowning. At any rate, his son John Lawrence was, if not antislavery in attitude, at least strongly opposed to the secessionist movement. From his home in Philadelphia, late in 1860, John L. LeConte wrote to his second cousin Matilda ("Tillie") Harden: "I hope the disunion sentiment will soon subside. If it is persisted in, it means civil war, servile insurrection, and all the attendant horrors." To this he added: "The North will suffer greatly, but the South will be utterly ruined. I trust the demagogues of either side may pause before leading the people to such enormities." Although the highly esteemed entomologist was critical of secessionism and would join the Union Army as a surgeon during the war, he never allowed his sympathies to degenerate into ill-feelings toward his southern cousins; and, although he remained adamantly loyal to the Union cause, he readily lent money to his desperately impoverished relatives after the war had ended and even brought his influence to bear in securing the quick release of one of those who had been captured by Federal troops.[32]

The strong filial relations between the northern and southern Le-Contes were especially evident just before the war began. By early

31 LeConte, *Autobiography*, 232.
32 John Lawrence LeConte to Matilda "Tillie" Harden, December 26, 1860, in LeConte Family Papers, APS.

1860, Uncle Jack's health had deteriorated badly. Joseph wrote to his uncle of the "affectionate reverence" that his family held for the elderly man, and a few days later he told John Lawrence that he considered Uncle Jack "almost . . . as my father also." In fact, so devoted was Joseph to his dear uncle that he made it a point to pay him an extended visit on his return from the AAAS convention in August, 1860, and when the old soldier and naturalist died in early November, Joseph sincerely lamented his loss. In spite of the strain of differences with his cousin over the issue of secession, Joseph maintained communication with John Lawrence until the war broke out, and he apparently felt free to express his opinions on the crisis. In early January, 1861, for example, he wrote a long, friendly letter to his cousin. "The state of the country," he said, "is truly melancholy in the extreme," and he observed that the secession "of all the cotton states is absolutely inevitable." Although he deplored recent actions in the South that had upset the "status quo," Joseph optimistically informed his cousin that the southern economy was functioning well, and he assured him that he had never seen "the negroes more loyal all over the South than now." Yet he expressed fear that the "dictate of passion and not reason" would win out and war would ensue. "I wish I could say God preserve our much loved Union," declared Joseph, but "I can only now say God save the whole country from the dire calamity of civil war."[33]

Joseph's political sentiments were growing in spite of his desire to remain aloof from such mundane matters. In fact, while he was in attendance at the AAAS meeting in Newport, Rhode Island, in August, 1860, he declined to attend a picnic to which the association members had been invited because it was to take place at Julia Ward Howe's estate in nearby Lawton's Valley. Although it would have been proper for LeConte, as the newly elected general secretary of the AAAS, to attend, he could not force himself "to go to Mrs. Howe's or enjoy her hospitality in any way after her abuse of southern hospitality." Evi-

33 Joseph LeConte to John Eatton LeConte, April 14, September 18, and October 20, 1860, Joseph LeConte to John Lawrence LeConte, April 14, 1860, and January 10, 1861, Matilda "Tillie" Harden to John Lawrence LeConte, April 28, 1859, and to John Eatton LeConte, July 31, 1859, and February 12, 1860, Josephine LeConte to John Eatton LeConte, September 24 and November 12, 1860, Joseph LeConte to Matilda "Tillie" Harden, November 21, 1860, Ann LeConte Stevens to John Lawrence LeConte, December 7, 1860, John L. LeConte to Ann LeC. Stevens, December 15, 1860, Jane LeConte Harden to John L. LeConte, March 26, 1859, and May 4, 1860, all in *ibid.* Joseph LeConte to Bessie LeConte, July 28, July 30, and August 9, 1860, all in McMillan Collection.

dently, the activities and criticisms of the antisouthern, prosuffragist woman were more than LeConte could bear. "Our next meeting will be at Nashville, Tenn.," and the association members "will experience something of southern hospitality there," he said with an air of disdain for northern critics.[34]

When Abraham Lincoln won the presidential election of November, 1860, the "storm burst" as South Carolina seceded from the Union, followed a few months later by "the dreadful war between the North and South," as LeConte expressed it in his autobiography. "At first," he said, "I was extremely reluctant to join in, and was even opposed to the secession movement; I doubted its necessity and dreaded the impending conflict and its result." But, caught up in the surging sentiment of loyalty to his native South, LeConte was swept along into the spirit of "enthusiastic unanimity," and he entered wholeheartedly into the great cause. So certain was he of the justification of the cause that he condemned most of the dissenters as "untrue men in every way." He did admit to exceptions, however, and he thus accepted the position of the Union sympathizer James L. Petigru because he had confidence in Petigru's "integrity." Yet LeConte was himself convinced that the ensuing conflict was neither an "insurrection" nor a "rebellion"; it was a war between two legitimate governments or, indeed, "a war between two nations." As a loyal citizen, he must therefore defend his nation, and he did so, both in moral support and, subsequently, in service to his government.[35]

The die was cast, and when Fort Sumter was bombarded by Confederate troops on April 12, 1861, LeConte rejoiced as much as any other South Carolinian. Although he surely realized that blood would be shed and his own career disrupted, LeConte was swept along in the great tide of patriotism. Like his sectional kinsmen, the sentimental southerner thought that war was always something that touched elsewhere, not in one's own front yard. Before the conflict ended, LeConte and his family would be thrust headlong into suffering, and, indeed, the war would quite literally come to their own front yard.

34 Joseph LeConte to Bessie LeConte, August 6, December 31, 1860, and January 2, 1861, all in McMillan Collection; *Proceedings of the AAAS, 1860*, 235.

35 LeConte, *Autobiography*, 178–81. In his notebook, "Misc[ellaneous] 1894," LeConte summarized (in notes) his views on the origins of the war, viewing the conflict as a step in the evolutionary process of society (MS in LeConte Family Papers, UCB). His summary followed the interpretation of George Lunt, *The Origin of the Late War* (New York, 1866), which he cited in "The Race Problem in the South," 355.

The Dreadful Ordeal of War

The surge of enthusiasm for war had an intoxicating effect upon the students of South Carolina College. With visions of heroism kindling their fervor, these high-spirited young men vowed to sacrifice life or limb in defense of their cause. As early as November, 1860, in spite of his appeal to the college students to temper their patriotism with prudence and to stick to their books, President Longstreet had fulminated against the Unionists in such heady oratory that the students were aroused to action. Thus indirectly encouraged, the zealous patriots had quickly organized a new company of cadets. They would be prepared to take up arms in defense of slavery and secession.[1]

The college faculty wanted to keep tighter reins on their charges than they had held during the town-and-gown rivalry of 1856, however, and so, although they recognized the cadet unit, they also drafted a list of firm rules whereby the corps could be abolished at their whim. Moreover, they strictly forbade the unit to take any significant action without the college president's express approval. But the latter provision soon proved unenforceable. As the crisis worsened in Charleston in early April, the cadet corps ignored the president's denial of their request to go to the defense of the city and quickly moved by train to the scene of the conflict. Although disappointed over their failure to receive battle orders, the unit happily accepted assignments of guard duty. Many of them returned to campus with great fanfare a few weeks after Fort Sumter had fallen. Some of the aspiring young soldiers returned to their studies, but others began to leave the college for active duty, and by June, 1861, only 108 students were still enrolled.[2]

The fall term commenced on schedule in early October, but only 75

1 Hollis, *University of South Carolina*, I, 212–13.
2 *Ibid.*, I, 213–15.

74

students showed up for classes. Moreover, there were now only six faculty members, since Professors Venable and Barnwell had left for active duty in Virginia. Under these trying circumstances the college continued to function as best it could. Then, when Federal troops captured Port Royal, near the southern tip of the state, many students from that region requested permission to go home. Longstreet denied their request, but he could not keep some of them from leaving anyway. The situation was growing desperate. Longstreet resigned at the end of the fall term, as he had planned to do for some time, and Professor LaBorde was appointed chairman of the faculty. The trustees strove to keep the college going, but they could not prevent the steady loss of young men to individual militia units. Thus, when the spring term began on March 10, 1862, not a single student showed up for classes. As acting chairman, LaBorde summoned the other five faculty members to discuss the matter. The faculty decided to continue, however, and they published a notice that classes would resume on March 17. A total of 9 students appeared as a result of their appeal.[3]

The college continued to operate through June, but obviously it could not go on with a student body only slightly larger than the faculty. Things grew worse when Confederate officials issued requests to use several college buildings as medical facilities. Both the faculty and the trustees resisted, but in the end they lost control of many of their buildings. The faculty tried to regain possession of the buildings for the October term, but the governor, noting the virtual abandonment of college exercises, ruled against them. LaBorde and the trustees continued their struggle, but their efforts were futile. In the face of increasing military reversals, South Carolinians registered even less concern about the operation of the college. No classes were held after June, 1862, but it was not until December 2, 1863, that the trustees bowed to the inevitable and officially suspended operations at South Carolina College. Technically, the faculty were still employed by the state. They were to be paid from rental proceeds by the Confederate government for use of college buildings, but the treasury was soon too depleted to meet the obligation. In fact, the professors never collected all of their back pay.[4]

3 *Ibid.*, I, 215–22; South Carolina College Faculty Minute Books, April, 1861, sessions. In his *Autobiography*, LeConte chose not to "speak in detail" on the matter, and, strangely, he claimed that the "College went on quietly during 1860 and 1861" (p. 181).
4 Hollis, *University of South Carolina*, I, 219–24; Maximilian LaBorde, *History of the*

Meanwhile, Joseph LeConte suffered not only from the closing of the college but also from personal difficulties as well. The first problem involved his sisters Ann and Jane and the latter's eldest daughter, Matilda. In 1861, "Tillie" Harden informed her mother that she planned to marry Thomas Sumner Stevens, the younger brother of Ann's husband J. P. Stevens. For some reason, this greatly displeased her mother. In fact, Jane had objected to Ann's marriage to J. P. Stevens back in 1843, apparently because Dr. Stevens was not from a prominent family. Her disapproval had created a strain between the two sisters, and when the twenty-four-year-old Tillie persisted in her intention to marry the forty-one-year-old bachelor Sumner, Jane was incensed. But finally, she recognized the futility of her protests and reluctantly consented. The stubborn woman refused to give her blessing to the union, however, and she told Tillie that, although she would allow the ceremony in the Halifax home, she would not attend. Under such humiliating circumstances, Tillie rejected the offer, choosing instead to be married in a church in Walthourville, where Ann now lived. Naturally, this caused a further rift between the LeConte sisters, and it created a virtually unbridgeable gap between Tillie and her mother.[5]

Poor Joseph was dragged into the squabble against his will. As the most amiable of the LeConte siblings and the recognized family mediator, all parties appealed to him to arrange a settlement. This dilemma caused him great pain, coming as it did just at the time when the college problems were increasing and when he was distressed over the health of his baby daughter Josephine. Joseph loved his sisters: Jane had been like a mother to him during his boyhood, and Ann had been his close childhood companion. Moreover, he loved Tillie almost as a daughter, and he had great respect for Sumner, who was an old college acquaintance. Joseph therefore did the best he could, counseling moderation by all parties and begging everyone to look to the happiness of each other. To Tillie he addressed a long letter, filled with tender advice; but he refused to criticize her engagement to Sumner. To Jane he penned "a long affectionate and very sorrowful letter,"

South Carolina College from Its Incorporation, Dec. 19, 1801, to Dec. 19, 1865 (Charleston, 1874), 471–507; South Carolina College Faculty Minute Books, May 11, 1863, to December 8, 1863.

5 Emma LeConte Furman, "Recollections of Emma Florence LeConte's Youth," [1931?] (MS in McMillan Collection). Contrary to Emma's belief, Tillie and Sumner were not married in Ann LeConte Stevens' home but in a Walthourville church.

urging her "not to allow anger to govern her conduct." Tillie and Sumner were married in June, 1861, thus invoking the undying animosity of her intractable mother. Jane never fully forgave either her daughter or her sister for defying her will.[6]

While thrust into this unhappy affair, Joseph was faced with an even more serious concern, the poor health of his baby daughter Josephine Eloise. Ever since her birth on September 29, 1859, the little girl had been frail and sickly. Joseph and Bessie had long wanted another child, and they were delighted when the baby arrived some nine years after the birth of Sallie. Little "Dodie" or "Josie," as she was affectionately called, was, in Joseph's words, "the most beautiful child we ever had," and Bessie said of her, "I don't think I ever had just such feelings for any child before[;] she seems to be so holy and pure—as a gift fresh from the hand of God." Not long after Josie was baptized in the Columbia Presbyterian Church on April 15, 1860, Joseph expressed "very great anxiety for her," and in the summer of 1860 he wrote that she was in a state of "precarious health." During the following months, however, the baby began to pick up weight and improve in health. For the first time, the LeContes felt Josie would survive. They believed she was now healthy enough to accompany them on a vacation, and they decided to take a trip to Lookout Mountain in Tennessee.[7]

The war was far removed from the thoughts of Joseph and Bessie, and Josie seemed to be full of vigor and life. After several weeks of delightful relaxation, they went to Raccoon Mountain to visit Joseph's nephew James Nisbet LeConte. But a few days later, Josie came down with whooping cough. Her frail constitution could not withstand the dreaded childhood disease, and within a day after the symptoms appeared, she was at the point of death. Holding the little girl in his arms

---

6 Emma LeConte Furman, "Recollections"; Joseph LeConte to Matilda "Tillie" Harden, April 18, 1861, in LeConte Family Papers, APS; Mary E. Robarts to Mary Jones, May 31, 1861, in Myers (ed.), *Children of Pride*, 689. Trained in engineering at the U.S. Military Academy in West Point, N.Y., Stevens was injured while working on the construction of a Georgia railway. Afterwards, he returned to teaching, which had been his profession prior to his training in engineering.

7 During much of this time, Emma was away at school in Orangeburg, S.C., living with her aunt Jane Harden. Joseph, Bessie, and Sallie wrote to her often, telling of little Josephine's condition; see, for example, Bessie to Emma, January 8 and 17, 1860, Sallie to Emma, January 17 and April 15, 1860, Joseph to Emma, January 31, February 20, March 25, April 20, May 20, and June 13, 1860, Joseph to Bessie, July 28 and December 31, 1860, all in McMillan Collection.

throughout most of the terrible ordeal, Joseph agonized as Josie clutched his neck during her spasms of coughing. Unable to bear the impending fate after Josie fell unconscious, Bessie took Emma out "to sit on the bluff in the quiet solitude of nature." "As the sun sank below the sky line," Joseph "came walking down the path" toward Bessie and Emma, and they "knew it was over." Sadly, the grief-stricken family set out for home with the remains of the little girl who had, in Joseph's words, "a halo of glory about her sunny curls." For the distraught father, "the shock was so great" that for several days afterward he "was prostrated with fever." Josie was buried on the shaded slope of a little rise in the Columbia cemetery in early September, 1861, just a few days before her second birthday. The loss remained vividly in Joseph's mind until the end of his days.[8]

Life could not stand in suspension, however, and Joseph went back to his duties at the college, which was now becoming a terminal for young men headed into the bloody conflict that was proving so costly to the Confederacy. Occupied with teaching duties, beset by financial problems, concerned over the fate of his plantation—and, above all, grieving over the loss of his baby daughter—Joseph found it difficult to turn his mind to writing. By the summer of 1862, another personal crisis arose when Bessie's brother Edwin Nisbet caught typhoid fever while on duty with C. S. A. forces in Richmond, Virginia. Realizing the inadequacies of hospital facilities, Joseph and Bessie left Emma and Sallie in the care of John and Josephine and went to Richmond to nurse the "desperately sick" Edwin. Bessie wrote to Emma on July 14 that "the whole air is laden with oppression, and I am afraid the amount of sickness will be appalling from the number of typhoid cases, the wounded affecting the atmosphere." Joseph later spoke of the wretched circumstances in Richmond, and he recorded that there were 25,000 sick and wounded men in the local and camp hospitals. "The swarm of flies was simply incredible," he said. "I never saw the like. They crawled over the faces into the mouths and over the eyeballs of the dying. It was horrible."[9] Never before this had he conceived that war could be so cruel.

    8 Emma LeConte Furman, "Personal Reminiscences of Childhood & Youth"; LeConte, *Autobiography*, 177. Greater detail is given in LeConte, Autobiographical MS. The nephew James Nisbet LeConte was the eldest son of Joseph's brother William.
    9 LeConte, *Autobiography*, 182–83; LeConte, Autobiographical MS.; Bessie LeConte to Emma LeConte, July 14, 21, 26, 30, and August 14 [1862], all in McMillan Collection.

When Joseph and Bessie were finally able to leave the convalescing Edwin in August, they returned to Columbia. By now they were beginning to face serious financial difficulties. Before the war, Joseph possessed real estate and personal property valued at $75,000. Most of this represented capital, of course, and not actual cash in reserve. Although LeConte recorded bank deposits and savings in one of his notebooks, the actual amount of money he possessed is uncertain. Nevertheless, it appears that between 1861 and 1863 he held bonds worth $7,000 and had a bank account averaging about $1,500. Although he received approximately $1,500 per year from the produce of his plantation, LeConte actually realized little net profit from this source. Still, until 1863 he was better off than many of his compatriots. But with the soaring cost of food and clothing, the enormous inflation of the Confederate economy, and the loss of his faculty salary, LeConte soon found it necessary to supplement his income. This he did by taking on duties for the Confederate government, the first of which was to serve as an arbitrator between the government and the owner of a nitre cave in Bartow County, Georgia. For his services LeConte was paid the sum of $1,125.[10] This certainly helped, but it was only a stopgap; he must find a steady source of additional income.

Early in 1863, LeConte was hired as a chemist by "a large manufactory of medicines for the army," located just outside the city of Columbia. For a year and a half, he supervised the production of "alcohol, nitrate of silver, chloroform, sulfuric ether, nitric ether," and other medicines to supply the "whole army." Eventually, his services were needed more desperately by the Nitre and Mining Corps, which was established by an act of the Confederate Congress on June 9, 1864. His specific position was "Chemist and Professional Assistant," and his function was "to test all nitrous earth, whether from caves or nitre beds." Part of the time he spent in the college laboratory and the rest

---

10 In the Census of 1860, under "Free Inhabitants of Columbia, South Carolina," Joseph LeConte's assets were given at $5,000 in real estate and $70,000 in personal property. The figures on Joseph LeConte's personal finances are given in his "Notebook, 1860–1881" (MS in LeConte Family Papers, SCL/USC). See also Joseph LeConte to W. D. Simpson, in Lawrenceville, Ga., December 9, 1863, in W. D. Simpson Collection, Duke University Library, Durham, N.C.; Bessie LeConte to Emma LeConte [December, 1863?], in McMillan Collection. References to the nitre-caves controversy are found in LeConte, *Autobiography*, 183, and a letter from Joseph to Bessie, from Atlanta, October 22, 1862, in McMillan Collection. See also Ralph W. Donnelly, "The Bartow County Confederate Saltpetre Works," *Georgia Historical Quarterly*, LIV (Fall, 1970), 310.

in the field. During the summer of 1864, LeConte "visited all the nitre caves in northern Georgia, Alabama, and Tennessee, all the nitre beds in South Carolina, Georgia, and Alabama, and the iron mines and blast furnace at Shelbyville, Alabama." John LeConte was also employed in the service of the Nitre and Mining Bureau, serving as superintendent of the works in Columbia.[11]

In his autobiography, Joseph noted that he was "appointed . . . with the rank and pay of major," leaving the impression that he was actually in the Confederate army. The letter of appointment from Secretary of War James A. Sedden merely stated, however, that he would receive compensation "equivalent to that of a major in the commissioned corps." Moreover, official records list Joseph LeConte under the "Scientific Arm of [the] Corps" with an assignment as "Consulting Chemist." Thus, LeConte was not considered to be an army officer, although he did report to Colonel Isaac M. St. John, a regular commissioned officer. Many years later, one of Joseph's daughters reported that "my father has always said that he and his brother were officers in the army," but even she doubted the validity of the claim. At any rate, Joseph was directly aiding the war effort by his service, and he believed that if he were captured by Federal troops, he would be treated not as a civilian but as a prisoner of war.[12]

During the war, LeConte found it increasingly difficult to look after affairs on his plantation. He had inherited over forty slaves after his father's death, and by "natural increase" the number had grown in excess of sixty by 1864. At one point, Joseph had more than a dozen slaves with him in Columbia, and toward the end of the war he and John brought up even more, including some of Jane's, and hired them

11 LeConte, *Autobiography*, 184–85; Bessie LeConte to Emma LeConte, November 28 [1863], December 20 [1863], and [December, 1863?], all in McMillan Collection. In a letter to Matilda "Tillie" Harden Stevens, September 30, 1864, Joseph called the position "an excellent one—suited to my taste and good pay" and said, "the pay is equivalent (including allowances for rations, wood, rooms, clothing, etc.) to something like $6,000 a year" (in LeConte Family Papers, APS). See also Mary Boykin Chesnut, *A Diary from Dixie*, ed. Ben Ames Williams (Boston, 1949), 199, 252, 278; and Lupold, "From Physician to Physicist," 171–77. Joseph drafted a paper on "Instruction for the Manufacture of Saltpetre" in 1862 (MS in LeConte Papers, SCL/USC).

12 James A. Sedden to John LeConte, August 22, 1864, in LeConte Family Papers, UCB; U.S. War Department, *The War of the Rebellion*, Series 4, Vol. III (Washington, D.C., 1900), 702; Caroline Eaton LeConte to Harold Small, September 23 and November 17, 1936, both in Caroline E. LeConte Letters, Files of the University of California Press, Berkeley.

out, mainly in the nitre works. In fact, Joseph and John were quite uncertain about the total number of slaves they did own, and at one point Joseph asked John to write to Liberty County and "get the exact number of negroes we have—also the exact number of male hands and of women." In one of his pocket notebooks Joseph seems to have tried to construct a list of his slaves, but in many cases he placed a question mark afterward and in some instances he did not even know the names of recently born children. Yet he was not indifferent to the needs of these people and he expressed great concern over being able to clothe them properly as goods became increasingly scarce.[13]

At one point during the war, LeConte sought to engage in some writing, but because he was cut off from scientific publications emanating from the North and received only a few that managed to get in from Europe, he could not keep abreast of the current literature. After publishing another essay on education in June, 1861, he drafted only one other article during the entire course of the war. This lengthy, two-part essay, titled "On the Nature and Uses of Art," was published in 1863 in the *Southern Presbyterian Review*, one of the few journals open to him at the time.[14] The article was really an extension of an earlier paper on "Morphology and Its Connection with Fine Art."

In his article on "Morphology," LeConte identified two great branches of science: physical and natural. The former, he claimed, had already become well established, whereas the latter had barely moved into a second stage of development. The article reflected once more Agassiz's basic view regarding the special creation of species by Divine Intelligence, and it rejected yet again the transmutational theory. Perhaps it was a response to the Darwinian theory, published in the same

13 The "Slave Schedules" of the federal census for 1860, in the City of Columbia, South Carolina (p. 38), list twelve slaves for Joseph LeConte. In his Autobiographical MS., LeConte claimed he had seventy slaves on his plantation and seventeen in Columbia at the end of the war. Evidence of his hiring out slaves can be found in his "Notebook, 1860–1881," and in an agreement between John D. Watkins and Clifford Anderson (husband of LeConte's niece Anne, second child of William LeConte), December 24, 1858, in Anderson Papers. Other comments regarding his slaves are in: Joseph LeConte to Bessie LeConte, October 22, 1862, and August 9, 1863, Bessie LeConte to Emma LeConte, November 28 [1863?], December 20 [1863?], and [December, 1863?], all in McMillan Collection; and Joseph LeConte to Eugenius A. Nisbet, February 8, 1864, in E. A. Nisbet Manuscripts, Duke University Library, Durham, N.C.

14 "On the Nature and Uses of Art," *Southern Presbyterian Review*, XV (January, 1863), 311–48, and *ibid.*, XV (April, 1863), 515–48.

year, but nowhere did LeConte refer to *The Origin of Species*. Yet the paper was an off-handed refutation of the theory of the organic trans-formation of species. Said LeConte, "The transmutation of species is as impossible as the transmutation of metals. And the development theorists are engaged in as bootless a search as the old Alchemists of the middle ages." Other references to "the Divine architecture" and to the special creation of new and higher forms of species at successive levels of geological time clearly indicated LeConte's strict adherence to the view of his mentor, but the overall nature of the discussion reveals that the pupil was already thinking more flexibly on the topic than did his former teacher.[15]

The chief purpose of the article was to show that in the ascending scale of organic forms "the diversity of functions [of species] imposes the necessity of diversity of form." Underneath all of these complex forms, however, exists an "identity of typical structure" that is the in-dividual cells that comprise the essence of every organism. These in-dividual cells represent the work of "divine thought or conception," and the function of the morphologist is therefore to "remove one com-plication after another" until he has discovered the "naked simplicity" of the Divine plan. Morphology and fine art are connected in that the latter is "the material expression" of the former. Thus, science and art go hand in hand, and, although art always precedes science, it is per-fected by scientific discoveries and advancement. The day would come, said LeConte, when art "will be no longer the *unconscious, in-tuitive*, but the *conscious* and *rational* embodiment of morphological laws."[16]

In his second paper on the nature and uses of art, LeConte dis-played a wide range of learning and informed criticism. He was es-pecially proud of this paper, and in his autobiography he called it "one of my best." Although it reflected an occasional bias and narrow-ness growing out of LeConte's own peculiar experience and some-what isolated exposure to scholarly observations, the essay clearly re-veals that its author grasped the theoretical framework of cultural criticism and possessed detailed knowledge of many works of art, mu-sic, and literature. As avocational pursuits, these were the interests

15 "Morphology and Its Connection with Fine Art," *Southern Presbyterian Review*, XII (April, 1859), 83–114. Handwritten modifications and addenda to this and the articles on art are in the possession of Helen Malcolm LeConte, San Francisco.

16 "Morphology and Its Connection with Fine Art," 83–114.

that occupied much of his time, and he would continue to offer insightful reflections upon the arts throughout his life. To him they were all a part of the grand harmony of thought. LeConte was thus one of the last individuals who endeavored to span the chasm between the older universalists and the modern specialists.[17]

Keenly aware of the criticisms of modern art and the old argument that decadence in that sphere leads to cultural corruption and national decline, LeConte blamed this attitude upon the continuing Puritan influence in America. He believed that "the Puritan spirit was evidently a healthy, natural, and necessary reaction against the abuses of the times; an indignant uprising of the spirit of man ... against the miserable sensuousness and frivolities of the age." But he argued that the ascetic spirit of the Puritans had gone too far, while at the same time modern reactions to that spirit had exceeded the bounds of moderation. He thus sought to find a middle ground between the extreme views on art—that is, between the religious and irreligious, the ascetic and the frivolous. To accomplish his end, LeConte felt that he must first "define the *true nature of art*," and then "point out its *true uses*, and its *abuses*." Fine art, he said, can be divided into two categories, the "imitative" and the "non-imitative." To the first belong sculpture, painting, drama, and the novel, while to the second, music, poetry, and architecture. It is the function of both kinds of art to overcome man's fallen, depraved condition and "restore ... the pristine harmony of the soul." True art, he claimed, can "release us" from the "bondage" of nature-worship or "fetishism" by appealing to our "higher faculties." "The true mission of art," he continued, is to "subdue the sensuous and emotional" and strengthen "the imagination, the sense of beauty." The most worthy art—that is, "high art"—impresses not only the "senses, passions, or emotions" but even more "the intellect."[18]

LeConte thus proposed a kind of graded evolutionary development in art that would lead ultimately to the triumph of intellectualization. His ideas reflected the influence of Hegel's scheme of mental (and spiritual) progression, and they also provided a framework into which LeConte could later fit the theory of organic evolution. Although he would always remain a passionate lover of art, he would continue to strive to place it beyond the pale of pure emotion and lift it to the

17 "On the Nature and Uses of Art," 311–48; LeConte, *Autobiography*, 183.
18 "On the Nature and Uses of Art," 515–48.

plane of rational reflection. Virtually every form of human activity could be intellectualized, even art.

During the early days of the war, these ruminations provided LeConte with an outlet for his frustrations and much-needed relief from personal problems. From that point on until 1866, however, he had neither the time nor the incentive to dwell on anything but more mundane subjects. In the midst of gloom, he pondered the future and the fate of his family. A "little ray of sunshine" broke through the dark clouds early in November, 1863, when Bessie gave birth to a fourth daughter. Christened Caroline Eaton, she soon became known by family and friends as "Carrie." On the day after her birth, Joseph wrote to Emma, who was away for a visit with Aunt Jane, now living in Savannah, that the blue-eyed baby was "smart & pretty" and looked like him. Frail like Josie, the new baby was a constant source of anxiety for her parents. Just after her first birthday, Carrie was inflicted with a serious illness, most likely poliomyelitis, which left her lame in one leg for the rest of her life. It was thus no wonder that Carrie, deprived of the normal "comfort & joy of childhood" and crippled by a mysterious malady, should become a hypersensitive and inward-looking woman who would eventually suffer abnormal psychological distress.[19]

Meanwhile, the Confederate military position continued to deteriorate. On September 2, 1864, after a long siege, Atlanta fell to Union troops, and General Sherman began his devastating march toward the sea. Virtually unopposed by Confederate forces, the "American Attila" moved across a sixty-mile front, destroying any property or supplies that might aid the rebels. As Sherman's mighty force rolled on, Joseph became increasingly worried about his daughter Sallie and his sister Jane and her two daughters, who were now on the Halifax plantation in Liberty County. Earlier in that year, Jane had given up her rented house in Savannah as it became more and more difficult to make ends meet. Sometime late in the year, probably in November, Joseph and Bessie had consented to let Sallie visit with her aunt and cousins in Liberty County. That decision now appears foolish, but at the time

19 Joseph LeConte to Emma LeConte, November 12, 1863, and Bessie LeConte to Emma LeConte, November 14, 21, 28, and [December, 1863?], all in McMillan Collection. Joseph's comments on the childhood deprivations of Caroline are given in the Autobiographical MS.; they were omitted by the editor of the *Autobiography*.

many southerners believed Sherman's troops would not advance so far south. At any rate, by early December it became obvious that a flank of Federal forces would extend as far as Liberty County, and the gravely concerned Joseph decided he must cross enemy lines and rescue his daughter, nieces, and sister. What ensued during the next few weeks was a valiant rescue effort, the details of which LeConte carefully recorded in a journal of his experiences.[20]

Hoping to reach his destination before Sherman's troops had thrust into Liberty County, LeConte took leave of his official post on December 9 and set out for Charleston. A day later he embarked for Savannah, but the train had to turn back when it was discovered that the enemy held the bridge over the Savannah River. The return was greatly delayed, however, and LeConte did not arrive back in Columbia until the fourteenth. On December 16, he caught a train to Augusta, where he saw Colonel St. John and received orders to remove the laboratory equipment of the Nitre and Mining Bureau from Columbia as soon as he returned. By the eighteenth he was on his way to Mayfield, a small terminal halfway between Augusta and Milledgeville. During the journey he met some friends who related woeful tales of suffering at the hands of Sherman's ravaging troops, and one person, recently returned from Liberty County, told Joseph that he believed Sherman had already overrun the area. LeConte's anxiety almost overwhelmed him, but little did he know of a graver crisis that lay in store. Nor was he aware at this point of the depth of his own courage, cunning, and determination.[21]

Since the railways from Mayfield had been torn up, LeConte was compelled to hire a horse and buggy to continue his journey. He arrived in Milledgeville on the evening of December 19, and from his father-in-law he received confirmation that "the Yankees had been in

20 John LeConte to William Sharswood, January 25, 1866, in William Sharswood Papers, American Philosophical Society Library, Philadelphia; Emma LeConte Furman, "Recollections"; LeConte, *Autobiography*, 185. The recorded account was published as *'Ware Sherman: A Journal of Three Months' Personal Experience in the Last Days of the Confederacy* (Berkeley, 1937). The published work faithfully follows the original manuscript (in LeConte Family Papers, UCB), but addenda written by LeConte are excluded. An important introduction by Caroline Eaton LeConte is included in the published version. She offered additional comments in letters to Samuel T. Farquhar on August 20, 1936, and to Harold Small on September 23, 1936, both in Caroline E. LeConte letters, Files of the University of California Press.

21 LeConte, *'Ware Sherman*, 1–17.

Liberty and destroyed everything." The next day he hastened on to Macon, and on the morning of the twenty-first he was able to catch a train for the south Georgia city of Albany. Time was now of the essence, and LeConte immediately took stagecoach passage for Thomasville, some sixty miles away. The cold, rainy night only heightened his misery. When the stage finally arrived in Thomasville at two o'clock the next afternoon, LeConte began to search frantically among the numerous refugees for anyone who might have news of his daughter. But all he heard were accounts of "the complete and wanton destruction of crops and stock of all kinds in the County." On the twenty-third, he decided to set out for Doctortown, about thirty miles southwest of his sister's home and a full day's trip by train from Thomasville. Weary and distraught, he arrived in Doctortown on Christmas Eve.[22]

Christmas morning broke in a chilling mist of gloom, but the day brightened when LeConte finally received news from his slave carpenter Lancaster that Sallie and the others were safe and unharmed. For the next several days, he managed to get letters to his sister by trusted Negroes who were free to move about the countryside. Jane returned messages to Joseph, informing him that she had no way out except by foot and advising him not to attempt a rescue "as the Yankees were still swarming in the vicinity of her house." For a week LeConte hung around the dismal town, waiting for an appropriate moment to take action. Anxious and distraught, he began "for the first time to doubt ... the final success of *our Cause*." He confessed, "I have wrestled in agony with this demon of despair. . . . For four long years the whole heart of the nation has beat for this Cause alone. . . . O God! and must it fail at last!" As he pondered these heavy matters, he received word that no Yankees had been seen in Liberty in the past two days. Joseph's heart quickened with joy, and he decided to go on.[23]

On the bitterly cold day of January 1, he arrived in Walthourville. Although disheartened by the "blackened ruins" of the little town, he could not be deterred, and early the next morning he walked on to Halifax, there to be warmly greeted by his sister, her two daughters, and Sallie. After his frightened relatives related their stories of Yankee intrusions, Joseph walked over to his own plantation, and there he

---

22 *Ibid.*, 17–21; Joseph LeConte to his wife, December 26, 1864, in McMillan Collection.

23 LeConte, *'Ware Sherman*, 21–23.

learned from his overseer Calder that his losses were heavy and that some of his slaves had become insubordinate. Returning to Halifax, Joseph decided it was too risky to bring everyone out, and, at Jane's urging, he chose to take only Sallie. He located "an old broken-down Yankee horse," placed his daughter on the back of the emaciated animal, and, with his manservant Joshua in the lead, set out early on the next day toward Walthourville. They departed none too soon, for Yankees returned to Halifax only minutes after they left. The little party got safely to Walthourville, but they soon discovered that most of the Federal troops were now on the road to Doctortown. He and Sallie must turn back![24]

For three days, LeConte waited for the departure of the troops, hiding out in the woods during daylight hours and returning to Jane's house each night. At last, on January 7, Lieutenant Colonel Arthur Hood of the Georgia Cavalry arrived under a flag of truce with permission to carry out all of the women and children in the area. After a burdensome journey, including a treacherous crossing of the heavily swollen waters of the Altamaha back swamp, the bedraggled party arrived in Doctortown on January 9. From there, Joseph, Sallie, and his two nieces caught a freight train to Thomasville and thence a passenger train to Thibeauville. Leaving the children there, Joseph retraced his steps back to Liberty County to rescue Jane. As he got to the swamp again, however, he "found the water far more swollen" than he had ever seen it. Unable to find passage across, the intrepid but virtually exhausted LeConte waded into the swift, waist-deep, and chilly water to cross the half-mile swamp. He arrived at Halifax early on January 16.[25]

Before departing, LeConte walked over to his own plantation once more to instruct his overseer Calder to sell much of the remaining corn and rice. He also "called up all the negroes" and asked them if they wished to leave with him. "If they desired to go," he said, "I would make some kind of provision for them even if it took my last dollar. . . . If they preferred remaining I had provisions aplenty, although no meat." All of them chose to remain, whereupon LeConte then lectured them on the "absolute necessity of organized work" directed by "a *head*." The slaves objected to Calder, however, and Joseph had to

24 *Ibid.*, 23–38.
25 *Ibid.*, 38–59.

promise that they would be supervised by his uncle, William Jones, whose plantation was nearby. To this arrangement they agreed, as they were fond of Jones.[26]

Accompanied by his devoted servant Henry, Joseph struck off with Jane and her four heavy trunks in a small cart pulled by an old horse they had rounded up. The weary party struggled onward, fighting fatigue, warding off apparitions of Federal troops, and fording treacherous swamps, until the tired old horse was completely "used up." Joseph then left Jane with Henry and set off by foot to Doctortown to find help. Again he had to wade the great swamp, but this time in pitch darkness. In the little town he secured a small mule, retraced his steps the next day, and brought the cart safely on to the back swamp, where the water had risen still higher. There Joseph found a canoe, and in ten trips across the strong current during heavy rainfall, he brought Jane, Henry, and all of the baggage over. Soon afterward the party stood on the banks of the swiftly flowing Altamaha. Joseph located a boat and, with the help of Henry, poled and rowed it across the dangerously flooded river. Many hours later, the ragged refugees arrived in Doctortown, soaked to the skin and fatigued to the bone. A full week had passed since LeConte had left this same place to rescue his sister, and it was now the twenty-third of January.[27]

Receiving news that Sherman's troops were headed toward Columbia, Joseph, Sallie, Jane, and her daughter Ada left Jane's daughter Annie to wait for her soldier-husband and quickly set off for Columbia. When their train could go no further because the tracks had been torn up, the weary group went on by wagon to Mayfield. Along the way a "Mr. Davis" attached himself to the party as an aide and friend. The young man claimed to be part of a Kentucky cavalry brigade and a Confederate spy. Although they never learned the truth about the mysterious fellow, Joseph and the ladies found his support invaluable. They would see him again in Columbia a short while later. Finally, on February 9, the weary refugees reached the LeConte home.[28]

26 *Ibid.*, 55–57. Little is known of LeConte's overseer J. R. Calder. He apparently served Joseph (and perhaps John) for several years, for he is mentioned in LeConte correspondence as early as 1858. After the war, Calder tried to make the former slaves work, since they were staying on the old plantations, but he did not succeed very well. He "died about the first of April [1865]" (William Jones to John and Joseph LeConte, July 3, 1865, in LeConte Family Papers, UCB).

27 LeConte, *'Ware Sherman*, 59–70.

28 *Ibid.*, 70–81.

As the reunited family rejoiced, they also considered their fate. There was no time to delay. The next day, Joseph received orders to pack up the laboratory equipment and take it to Richmond. He got the equipment ready for shipment and, after considerable delay because of the crowds of fleeing Columbians, he managed to find a place for the freight on a train that departed on February 15. Now he had to return home to attend to the safety of his family. In spite of official assurances that the city would be defended, LeConte was worried. The mysterious Mr. Davis had paid him a visit in the meanwhile, and his "rather gloomy account of the prospects of Columbia" heightened Joseph's fears. Indeed, Davis urged him to flee, for the Yankees could be in Columbia by the next day. The young man vowed that he would protect the homes of Joseph and John, and he asked the LeContes not to betray him if they saw him ride in with Federal troops. With this disturbing news ringing in his ears, Joseph hastened over to see his brother, who was at that moment talking with Major Allen J. Green, the post commander. To his great dismay, Joseph was informed by Green that the "military authorities had at last confessed that they could not hold Columbia."[29]

The three men and John's fifteen-year-old son Johnnie (his other son, Julian, was on active duty) decided to depart immediately, leaving their families behind. Joseph rushed home and began packing the family silver and jewelry and his manuscripts. That accomplished, he bade farewell to Bessie, Emma, Sallie, and the baby and went at once to the nitre factory. Sherman's guns were already booming in the background. At daybreak, the party embarked in a northeasterly direction toward Abbeville with their personal and the nitre bureau property in "two wagons, two carts, and one buggy, all heavily laden." In addition to the four white men, twenty-two blacks voluntarily accompanied the "cavalcade." Progress on the muddy, badly torn-up road proceeded at a snail's pace. By late evening, after frequently losing their direction, the little caravan had traveled only fifteen miles. "The roar of artillery" was "very distant and incessant all day," recorded Joseph in his journal. On the third day of their flight, the fugitive party sadly discovered that a contingent of Yankees was moving directly toward them. Unable to go forward and fearful of turning back, they decided to "turn into the woods and remain hidden" until the troops had passed. So into

29 *Ibid.*, 81–84.

the woods they went, carefully covering the wagon tracks to avoid detection.[30]

As the sun rose on "the fine bracing morning" of February 19, the little party's spirits lifted as they felt assured they had escaped. Alas, what appeared as a triumph quickly turned into a disaster. While Green was away on picket and the refugee party was at breakfast, Joseph's man Sandy shouted an alarm from Green. Joseph ran toward the wagons, which were parked only a few yards away, but when he arrived, he discovered that Yankees were "already swarming upon and pillaging" one of them. He watched in horror as they rifled the vehicles for valuables, but he had to withdraw for fear of being seen. Later he crept back, hoping to retrieve at least his manuscripts and papers. His spirits sank as he watched a second group of marauders pile the remaining contents of the wagons into great heaps and set fire to them. But time for grief over the fruits of his professional labors could not be spared, for the troops scattered out to search the woods, and Joseph had to run. He found a former hiding place and there lay concealed all day, agonizing over his loss, uncertain of the fate of his brother and nephew, and heartsick with worry over his family.[31]

As darkness approached, he returned to the campsite, where most of the Negroes had remained. He learned then that John had surrendered with Johnnie, probably because the young boy was still weak from an illness and could not hope to escape. The Yankees had carried them away along with five of the Negroes. Joseph also learned for certain that all of his possessions were gone. There was nothing he could do now but hide until it was safe to return to Columbia. Soon Green appeared. The exhausted officer had barely escaped capture by lying behind a fallen tree all day. Meanwhile, some of the blacks who had been forced to go with John and Johnnie returned and related to Joseph that his brother and nephew were not being badly treated by their captors. That was at least a bit of relief to Joseph. He then ordered all of the blacks to return to Columbia and with Green set out at dark for home.[32]

For five wretched days the weary pair plodded through chilling weather and icy rain toward their destination. After many a harrowing

30 *Ibid.*, 84–94.

31 *Ibid.*, 94–99.

32 *Ibid.*, 100–106; John LeConte to Josephine LeConte, March 4, 1865, in LeConte Family Papers, UCB.

moment, Joseph at last arrived in Columbia late in the afternoon of February 24, just two days short of his forty-second birthday. As his daughter Emma recorded in her diary: "What a scene! Embraces, kisses, weeping—he was wet through and in rags. We hurried him to the fire and listened to the story of his escape, an escape that seemed little short of miraculous."[33]

As the bedraggled Joseph warmed himself by the fire, he learned what his family had endured in his absence. Sherman's forces had marched into Columbia virtually unopposed on February 17, and on the second night of the occupation, some of his drunken troops set fire to the city. Within hours, the raging inferno consumed scores of buildings, homes, and churches. The flames licked at the roofs of the campus homes while Emma and her mother and sister watched terror-stricken at the great sea of fire that seemed to be engulfing the entire city. Fortunately, their home was spared, as were the other college buildings, partly through the efforts of the mysterious Mr. Davis, but primarily because the college facilities were still being used to care for wounded and sick soldiers.[34] Despite their losses, the LeContes were luckier than many of their fellow citizens. It had been a dreadful ordeal, however, and they would always remember their terrifying brush with disaster.

---

33 LeConte, 'Ware Sherman, 106–40; Emma LeConte, When the World Ended: The Diary of Emma LeConte, ed. Earl Schenck Miers (New York, 1957), 67. The original diary is in the Talley Papers.

34 Emma LeConte, When the World Ended, 41–52; Marion Brunson Lucas, Sherman and the Burning of Columbia (College Station, Texas, 1976), passim. Joseph LeConte blamed Sherman entirely for the holocaust (unpublished "Appendix [to the journal of experiences from December, 1864–February, 1865] written in 1878," in LeConte Family Papers, UCB). Joseph also served on a committee appointed in April, 1867, to investigate the matter by interviewing "eyewitnesses." The committee placed the blame upon Sherman (James Parsons Carroll, Report of the Committee Appointed to Collect Testimony in Relation to the Destruction of Columbia, S.C., on the 17th of February, 1865 (Columbia, 1893). Josephine LeConte wrote to William Sharswood on June 25, 1866, that her husband had been appointed to a committee "of the most distinguished and reliable intellects in the State to collect up all the testimony bearing upon the march of that fiercely incarnate fiend Sherman." She concluded that "a catalogue of crime will be laid at thy Door 'oh Union Savers'" (in Sharswood Papers).

Picking Up the Pieces: The Reconstruction Years

As General Sherman moved out of the shattered city of Columbia and pushed his way easily into North Carolina, the LeContes knew that the total collapse of the Confederacy was only a matter of time. With their city in shambles and their morale crumpled, the stunned inhabitants of Columbia began to meet their immediate task—survival. Joseph LeConte had lost almost all of his clothes, and he had to appropriate "the castoff blue of Federal soldiers who had died in the hospital." His wife and daughters were slightly better off, since they had not sent all of their clothes away on the ill-fated wagons. But Joseph still had a dozen blacks in Columbia, and he was compelled to cut up the carpets from his floors to clothe them. Bessie had lost virtually all of her silver, and she had to borrow some from her sister-in-law Josephine who had sent away all valuables except the family's table silver. John LeConte had entrusted his manuscripts and papers to the care of the local Catholic priest, whose quarters were also burned to the ground.[1]

For a week after the departure of the troops, Joseph and his family were fed by his slaves, who, along with other blacks, had pilfered and looted foodstuffs from the downtown area of the city during the dreadful night of the torch. Then, for two weeks, Joseph "drew rations from the city" that had been gathered "from the surrounding country." He and his family were also befriended by Francis S. Holmes, who had a reasonably good larder on his plantation in Edgefield, where Joseph

---

1 LeConte, *Autobiography*, 230; Emma LeConte, *When the World Ended*, 73; Emma LeConte Furman, "Recollections"; John LeConte to Lewis R. Gibbes, April 10, 1865, in Lewis R. Gibbes Papers, Library of Congress.

went to pick up provisions. During the whole time, Joseph worried that the government would recall him to duty, but little did he or anyone else in the city realize that the Confederate engines of war were damaged beyond repair.[2]

All throughout March and April Joseph continued to scrape together whatever resources he could, traveling here and there about the countryside to secure food for his family and slaves. From Augusta he obtained "flour, corn, and bacon—a few hams, but chiefly the sides" and rough homespun from which the ladies made underclothes. Amidst all of this struggle, the last glowing embers of the Confederacy were fading into powdery gray ashes, and in mid-April, Emma LeConte wrote with all the fervor and passion of her youth: "The South lies prostrate—their foot is on us—there is no help.... They say *right* always triumphs, but what cause could have been more just than ours? ... Is all this blood spilled in vain—will it not cry from the ground on the day we yield to these Yankees!" Like everyone else, she knew it was only a matter of days before the Confederate States of America would pass into oblivion. The end came in May, and a garrison of Federal troops was sent to Columbia to begin the long-anticipated occupation.[3]

Still pressed to find food, Joseph soon approached Colonel Nathaniel Haughton, the commandant of the garrison, for permission to take a flatboat down the Congaree River to bring up corn from the lower plantations. Haughton readily consented, but he offered to do more for the penniless professor. A number of northern scientists had asked the commandant to lend Joseph money in their name, and Haughton made this known to him. But the proud southerner declined, choosing to work with his own hands instead. For several weeks Joseph, dressed in the faded Yankee trousers and a "battered hat," plied the river with the old Nitre and Mining Bureau boat and secured "several thousand bushels" of corn, of which he was allowed to keep a hundred for himself. This he halved with his brother. "Poor Father is looking very badly," said Emma. "He can get no employment and not one cent of money in the house." Worse still, at least in the eyes of patriotic southerners, was the requirement that an oath of loyalty to

2 Emma LeConte, *When the World Ended*, 67–74; LeConte, *Autobiography*, 229.
3 Emma LeConte, *When the World Ended*, 83–91; Emma LeConte Furman, "Reminiscences."

the United States must be taken before one was "allowed to engage in any occupation." Joseph felt "it would be a most painful necessity that would compel him to such a humiliation." "If we could *only* leave the country," cried Emma. "We dream of this and make plans to emigrate—but the means are lacking now. We will have to wait." Surely Joseph thought of that, too, and he lamented "the humiliating annoyances of petty despotism by military Satraps dressed in brief authority."[4]

Meanwhile, LeConte had lost all of his slaves, though a few of those in Columbia continued to work for him as servants. But over his loss he claimed not to have lost a "wink of sleep." Owning slaves had always oppressed him, he said. Unlike many other plantation owners, LeConte even argued that the loss of slaves "was not necessarily any loss of property at all," for slaves "were not property; chattels, in the sense in which other things are." Holding that the only right claimed by slave owners was to the "labor" of slaves, he maintained the situation would simply be transformed from a "slave-system to a wage-system." Under the new arrangement, "if the negroes were reliable," plantation owners would suffer no loss of income, argued LeConte. His hopes were not realized in his own case, however, for the freedmen on his Liberty County property did not accept the proposition that only the system had changed; they rejected the notion of working for "Massah" in the same old way.[5]

For the first four years after the war, LeConte received a paltry income from his vast estate, and thereafter it earned him not a penny. Joseph had, of course, asked his uncle William Jones to supervise Syfax for him. On July 3, 1865, Uncle William wrote to his nephews in Columbia that some of their former slaves had ripped the bagging from their cotton bales and that he was trying to make the freedmen work, but to little avail. Then, only a few months later, Jane Harden, who had returned to Halifax, spoke of "the state of demoralization the negroes are in, in this part of the country." She urged her brothers to come down soon to try to get matters straightened out. Jane also said, "The stealing goes on worse than ever. . . . All of your [John's] cotton is gone (I am told) and very little of Joe's is left. The negroes seem crazy

4 LeConte, *Autobiography*, 231; Emma LeConte, *When the World Ended*, 105–107; Caroline LeConte, "An Introductory Reminiscence," in LeConte, *'Ware Sherman*, xv; addendum to LeConte's journal of experiences during the war, "2nd P.S. Six Months Later," in LeConte Family Papers, UCB.
5 LeConte, *Autobiography*, 232–34.

about cotton." To all of this she added that it was difficult to get cotton ginned, and since the railroad to Savannah had not been rebuilt, there was no way to get it to the port city.[6]

A short while later, Jane again wrote about the theft of cotton, noting that the blacks "escape punishment by our not being able to prove anything." At the end of the year, she reported to John that "your ne-groes had left you very little of anything. When your corn was divided with them, they liked to have had an open rebellion—they insisted that you should not have half the crop, but only one portion (or as much as a field hand).... At Joe's the negroes have acted somewhat better." She advised her brothers to get a white man to live on their plantations.[7]

The situation only worsened as time went on. Early in 1866, Uncle William sent a check in the amount of $610 to John, informing him that only $150 of it belonged to him and the rest to Joseph. "Joe's [Negroes] done better," than John's, leaving him a little more cotton and rice. The difference may well be attributable to the fact that Jo-seph had always been much more diligent in his concern for his slaves than his brother. At any rate, Uncle William expressed his regret that his nephews had not come to Liberty to give him "directions how to proceed" and let him know whether they "wished to go into con-tracts with the infernal negroes for another year or not." Later in the year, John received a request from six of his former slaves asking per-mission to rent his plantation, but John replied that he had already rented it to his nephew John Harden. Harden did not find his venture profitable, however, and he soon gave up his effort at farming, where-upon John rented land to some of the freedmen. But they refused to pay the rent when it came due, and so John made nothing from his land. The same was true for Joseph. In the meantime, taxes were still payable, and Uncle William reported to the brothers that he had paid it out of his own pocket.[8]

By January, 1868, Jones informed them that he could no longer rent the property, but that a neighbor, Samuel Varnadoe, would farm the

6 William Jones to John and Joseph LeConte, July 3, 1865, and Jane LeConte Harden to John LeConte, October 26, 1865, both in LeConte Family Papers, UCB.

7 Jane LeConte Harden to John LeConte, December 13, 1865, in *ibid.*

8 William Jones to John LeConte, January 2, 1866, and John LeConte to "Boston, Somerset, Tom, Hercules, Scipio, and Hamilton, My former Slaves," December 16, 1866, both in *ibid.*

two places for half the crops he could make. As for Uncle William, he had become disgusted with "negro labor since *'freedom come'*," and he intended to move to Athens. "I . . . expect to leave someone else to transact your business here, for my move will be a permanent one." Joseph and John consented to allow Varnadoe to farm on the half, and they gladly accepted Samuel Jones (another of William's sons) as their business agent. But as the year wore on, it became increasingly apparent that the once-productive estates would never again know the glory that Louis LeConte had brought to them. Cousin Samuel wrote to John in November that he could expect no more than $100 from the crop; but, he added, "a little more will be made at Cousin Joe's place." A few months later, Samuel rendered a final report: John's crop had brought only $23, and Joseph's, $65. "I am fearful," said the cousin, "you will never realize anything more from your land in Liberty." Indeed, this was the last money the LeContes ever earned from their vast landholdings.[9]

Back in Columbia, the prospects brightened for the LeConte brothers as the trustees of South Carolina College began to talk of reopening the institution. As early as June 19, 1865, John LeConte had received a letter from Brigadier General A. S. Hartwell indicating that the military authorities hoped the college would soon resume its activities. Maximilian LaBorde, still acting as chairman of the faculty, assembled the remaining four faculty members to consider Hartwell's letter. The faculty instructed LaBorde to discuss the matter with the local commandant and to try to regain possession of as many buildings as possible. Of course, the faculty's hands were tied until the board of trustees could take action and, even more so, until a legislature could convene and authorize funds for operating the school. These problems appeared on the verge of ready resolution when President Andrew Johnson reinstated South Carolina's local government and Ben-

9 Jane LeConte Harden to Josephine LeConte, October 30, 1867, William Jones to John and Joseph LeConte, January 14, 1868, and to John LeConte, January 31, 1868, Samuel J. Jones to John LeConte, March 23, 1868, November 9, 1868, and January 22, 1869, all in *ibid*. There are no records to indicate that Joseph ever earned anything more from his plantation, and in 1901 he wrote that "it has never made me a cent since the war." He noted that he held $8,000 in bonds at the end of the war, but apparently he was never able to redeem them (LeConte, *Autobiography*, 231, 234). See also Joseph LeConte to E. A. Nisbet, July 30, 1866, in Nisbet Manuscripts.

jamin F. Perry was appointed as provisional governor of the state.[10]

Joseph LeConte called his "very dear friend" Perry "a man of noble presence, untarnished integrity, and sterling character," and he was delighted over his appointment. On July 20, Perry, a long-time trustee of South Carolina College, issued a proclamation that authorized all state officials to carry out their functions, and he set a date for electing delegates to a constitutional convention. When the convention met in September, it drafted a constitution, and the new legislature began at once to take up its duties. Perry appealed to the legislature to give strong support to education, and he urged the body to consider the need for converting the college into a university. Meanwhile, he had met with the college trustees and sanctioned their proposal to reopen the institution in January, 1866.[11]

When Perry's appointment ended in November, 1865, James L. Orr succeeded him as provisional governor. Orr favored Perry's plans for the college, and he urged the trustees to model the school after his own alma mater, the University of Virginia. A proposal was submitted to the state legislature and approved on December 19, less than a month before the school was scheduled to resume classes. Under the act, eight "schools" were authorized, but they essentially amounted to little more than the old chairs of the antebellum college with the addition of modern languages and literature and civil and military engineering. The legislature appropriated $1,000 each for the eight professors, which was only half the salary they had earned before the war. The lean salaries were to be supplemented by student fees. But the legislature made no appropriations for building repairs. Yet it was a beginning, and the trustees convened at once to elect the faculty. John LeConte was appointed to the chair of the "School of Natural and Mechanical Philosophy and Astronomy," and Joseph, to the "School of Chemistry, Pharmacy, Mineralogy, and Geology."[12]

10 South Carolina College Faculty Minute Books, June 23, 1865; Hollis, *University of South Carolina*, II, 7.

11 LeConte, *Autobiography*, 237; Hollis, *University of South Carolina*, II, 7–9.

12 Hollis, *University of South Carolina*, II, 15–21; Joseph LeConte to Rev. C. Bruce Walker, Secretary, South Carolina College Board of Trustees, December 20, 1865, in LeConte Papers, SCL/USC. LeConte gave periodic reports to R. W. Barnwell, Chairman of the Faculty, on the status of his courses; see, for example, those dated May 7, 1866, May 4 and November 25, 1867, and May 4 and November 19, 1868, in LeConte Papers, SCL/ USC.

Although he felt it was "impossible" to teach his courses in pharmacy and agriculture in a "fully" adequate way, Joseph did the best he
could. In the pharmacy course, he dealt with "the preparation and
properties of the substances used in medicine," while in agriculture
he offered "six or eight lectures on the most fundamental principles
underlying the science and the art." Despite the inadequacies of these
"very meager" lectures, especially in agriculture, LeConte did not view
his efforts as a total waste, and in his later years, he proudly noted
that many of his former students had found the agricultural lectures
"of decided benefit to them." He could take special pride in the accomplishments of his son-in-law Farish Furman, whose development of a
valuable fertilizer formula represented the influence of his former instructor's emphasis on inductive reasoning and controlled experimentation.[13]

As a devoted father, Joseph was keenly aware that the war had badly
disrupted the lives of his teenage daughters, and he did his best to
give them what they had missed, encouraging both to become active
in dancing and other social affairs. Indeed, he often escorted Emma a
mile to a local dancing club, returning home to read and write in the
meantime and then going back to pick her up late in the evening. He
likewise encouraged regular church attendance for his daughters, not
only for religious reasons but also for social intercourse. In addition,
he invited the girls' companions, of whom there were many, including
John LeConte's children, those of his brother Lewis, and of his sister
Ann, to gather in his own home for dramatic readings. He even formed
a Shakespeare club, in which he took an active part. Such activities
gave him great pleasure, and he later claimed that he had never
known "so much real social enjoyment in Columbia as in the years
1866 and 1867. . . . Society was really gay, the necessary result of the
rebound from the agony and repression of the war."[14]

Yet tragedy and heartache were always around the corner for the
amiable LeConte. Frequently worried over Carrie's repeated illnesses

---

13 LeConte, *Autobiography*, 235; Lester D. Stephens, "Farish Furman's Formula: Scientific Farming and the 'New South,'" *Agricultural History*, L (April, 1976), 377–90. Le-
Conte wrote to Joseph Henry on September 24, 1866, asking Henry if he could comment
on "attaching" an A and M college to the university, and he wanted Henry's judgment on
the "best organized agricultural colleges in the country" (Record Unit 26, Box 9, Folder
1, in Smithsonian Institution Archives, Washington, D.C.; hereinafter cited as SIA).

14 LeConte, *Autobiography*, 236; Emma LeConte Furman, "Recollections."

and constantly concerned because his little daughter's leg seemed to be permanently crippled, he was beset with anxiety. Then, in July, 1866, he received word that his sister Ann was gravely ill with tuberculosis. Joseph had known for some time that the dissipating disease would sooner or later consume his sister's life, and so he was not surprised when he received word of her rapidly failing health. Soon after he arrived at the bedside of the dying Ann, he realized the end was near and quickly summoned Jane to come and make amends before it was too late. Since Jane could not even pay her own train fare, Joseph purchased a ticket for her. Joseph's efforts resulted in a partial reconciliation between the disaffected sisters just before Ann's death, but the poor fellow was never able to forget their tragic discord. As he pondered this unhappy event, Joseph soon realized that another tragedy lay in store when it became obvious that John and Josephine's daughter was also dying from advanced tuberculosis. To the very end, John refused to believe that his daughter had tuberculosis, although Joseph urged him to accept Lula's illness as a serious case of consumption and to seek medical help for her. Eventually, John sent Lula to Baltimore for treatment, but the afflicted young woman grew steadily weaker, and she died on March 21, 1868, after several months of suffering.[15] Although greatly pained by these tragedies, Joseph did not allow them to paralyze his personal strength.

For almost five years, LeConte had been virtually cut off from current scientific literature, and he had much catching-up to do. On October 3, 1865, he wrote to his old friend Spencer Baird, the assistant secretary of the Smithsonian Institution, to thank him for sending some "books and pamphlets" published during the war years. "I cannot express to you my gratification," said LeConte. "It is the first scientific news I have seen in 5 years." Then, lamenting his fate, he told Baird, "It seems to me that I can never again recover what I had lost. Not only the progress of science during the years of the war must be re-

---

15 Emma LeConte Furman, "Recollections"; Joseph LeConte to Bessie LeConte, July 11, 14, 17, 1866, and to Emma LeConte, July 21, 1866, all in Talley Papers and in LeConte-Furman Papers, Southern Historical Collection, University of North Carolina, Chapel Hill (hereinafter cited as LeConte-Furman Papers, SHC/UNC); Joseph LeConte to Matilda "Tillie" Harden Stevens, November 8, 1866, and Jane LeConte Harden to John Lawrence LeConte, April 12, 1868, both in LeConte Family Papers, APS; John LeConte to Isaac M. St. John, May 10, 1868, in Brush Family Papers, Yale University Library, New Haven; John LeConte to Lewis R. Gibbes, July 13, 1868, in Gibbes Papers.

covered; but every note, manuscript, reference to journal reading which I had collected together for the last 15 years was destroyed by Sherman and must be recovered. I almost despair." A few days later he wrote to Joseph Henry and asked the venerable scientist if he could inform him of the important scientific publications during the past five years, especially in chemistry and geology. "I have felt completely dispirited and unable to apply myself to any work," he said. But Joseph added that he now felt his "energy and spirits returning," and he expressed his determination to "devote" himself "to Science and Science alone."[16]

Shortly afterward, LeConte began to work on some scientific papers, but he found the task exceedingly difficult, for he simply could not get his hands on all of the literature he needed. He did exchange some material with Lewis R. Gibbes, an accomplished mathematician and astronomer of the College of Charleston, and, of course, he received some publications from Baird and Henry. Nevertheless, the selective nature of what he received dictated the course of his writing. Among the works he produced were a brief paper on sex, six lectures on coal and petroleum, and a number of original and truly first-rate papers on physiological optics. In addition, he managed to rewrite all of the chemistry textbook, the original manuscript of which had been burned. John never completed his portion of the chemistry text, however, and thus it was never published. Joseph even began to write a textbook on geology, but its completion was delayed by his move to California in 1869.[17]

The first paper published by Joseph after the war was a very brief piece on the determination of the sexes in plants and animals. It appeared in the *Nashville Journal of Medicine and Surgery*, a rather ob-

16 Joseph LeConte to Spencer Baird, October 3, 1865, Joseph LeConte to Joseph Henry, October 8 [1865?], both in Record Unit 26, Box 9, Folder 1, SIA.

17 Joseph LeConte to Lewis R. Gibbes, August 22, October 26, and November 3, 1866, October 19, 1867, February 22, July 25, October 15, and December 18, 1868, and February 14, May 19, June 4, and July 7, 1869, all in Gibbes Papers. On July 30, 1868, Joseph submitted his portion of the chemistry textbook to "Messrs. C. B. Richardson & Co." He added a postscript: "The physical part to be contributed by my brother ... will not be more than ¼ of the whole" (letter inserted in volume one of the chemistry ms., in Joseph LeConte, Miscellaneous Writings, Bancroft Library Archives, UCB). See also Joseph LeConte to John L. LeConte, February 12, 1866, and Joseph LeConte to Matilda "Tillie" Harden Stevens, November 8, 1866, both in LeConte Family Papers, APS.

scure publication but one of the few outlets open to LeConte at the time. Unfortunately, the short article was marred by some serious typographical errors, and LeConte was forced to publish a corrigendum in a later issue of the journal. The simple paper made little ripple upon the scientific waters, however, since it was mainly "a brief abstract" of the findings of three European scientists. In summary, LeConte accepted the notion that sex was determined by the stage of development of the ovum. If impregnated early, the sex would be female; if late, then male. He cited the experiments of some European scientists with cattle and chickens, which seemed to validate the hypothesis. From both "a scientific and a practical point of view," this should have significant implications for "the physiologist and the farmer," he said. Indeed, it would have if it had been true, but it was a fallacy. To his credit, LeConte insisted upon more experimentation and "patient observations."[18] Although the paper represented only a small contribution to the knowledge of selective breeding, it was a beginning in the resumption of Joseph's scientific career.

In February, 1867, LeConte delivered six lectures on coal and petroleum at the Peabody Institute in Baltimore. These quasi-popular addresses were merely a restatement and slight extension of the previous lectures he had given at the Smithsonian Institution a decade before. Nevertheless, they provided LeConte with an opportunity to get his feet back into scientific circles, and he wrote home to Bessie that his lectures had succeeded beyond his expectations. It was a small but much-needed boost to his ego. Both he and John had terminated their membership in the AAAS, and their scientific contacts with northern scientists were still very limited.[19] Since there were few

18 LeConte, "'On the Law of the Sexes,' or the Production of the Sexes at Will," *Nashville Journal of Medicine and Surgery*, n.s., I (October, 1866), 296–99; corrigendum, *ibid.*, n.s., II (April, 1867), 332–33.

19 LeConte, *Autobiography*, 240; Ernest Hilgard, "Memoir of Joseph LeConte," *National Academy of Sciences Biographical Memoirs*, VI (1907), 172; Joseph LeConte to John Lawrence LeConte, November 8, 1866, in LeConte Family Papers, APS; Joseph LeConte to Bessie LeConte, February 11, 1867, in Talley Papers. Meetings of the AAAS were suspended during the war years. Joseph was still listed on the membership roll printed in the *AAAS Proceedings* through 1868; termination of membership apparently came about because he discontinued paying dues. On July 14, 1861, John LeConte wrote to Lewis R. Gibbes that he had finished a paper "which was prepared for the *late* 'American Association for the Advancement of Science.' As this Association is now *extinct*, so far as we are concerned ... I have been thinking of sending it to London" (in Gibbes Papers).

first-rate scientists in the South, the LeContes had few opportunities for that kind of interchange of ideas and associations so crucial to scientific development.

During the war, Congress had enacted legislation establishing the National Academy of Sciences. Fifty Americans were designated as incorporators and founding members of the organization, but none of these was from the South. Thus, neither Joseph nor John was elected, and it would be a number of years before they were chosen as members of that prestigious body. The academy was the brainchild of Alexander Dallas Bache and a group of his associates who wished to enhance the status of science in America by creating an organization of the "leading scientific savants." Known as the "Lazzaroni," this group objected to the open membership requirements of the AAAS, and they sought the support of the federal government for their scientific activities and projects. The result was the formation of an official, government-sanctioned organization that reflected the bias of the Lazzaroni and subjectively decreed who was fit and unfit to join the ranks of *soi-disant* elite. Predisposed against all non-Union sympathizers, the Lazzaroni-controlled academy had no room for the LeContes after the war, despite the fact that John had moved in the inner circle prior to 1861, and, indeed, his beautiful wife was regarded as the queen of the group before the secessionist movement sundered the bonds of professional and personal association.[20]

But Joseph LeConte would not be deterred by this slight; he proceeded on his own to pursue his scientific interests. Since he was unable to conduct field work in geology and because he had no laboratory equipment, LeConte had to look elsewhere to satisfy his urge to resume scientific activity. He turned to an old favorite subject, research in physiological optics. Fascinated by visual phenomena since his childhood, the versatile scientist possessed a remarkable capacity for observing the functions of the human eye. Inasmuch as he had access to some British, French, and Swiss publications that carried

20 See Nathan Reingold (ed.), *Science in Nineteenth-Century America: A Documentary History* (London, 1966), 200–203; Sally Gregory Kohlstedt, *The Formation of the American Scientific Community* (Urbana, Ill., 1976), *passim*, especially 154–89. As an example of the close relationship between John LeConte and Bache before the war, see LeConte to Bache, April 22, 1860, in Alexander D. Bache Papers, Library of Congress. On June 4, 1864, however, Josephine LeConte wrote to N. A. Pratt that "Bache is against us in every way" (in LeConte Family Papers, UCB).

many articles on physiological optics, and because he needed little equipment to do research in that field, Joseph began a series of experiments on the phenomena of sight, especially binocular vision. Although his research and writings on vision have received little attention in modern times, they actually represented an important contribution to science. LeConte published over a dozen excellent articles on the subject and later reworked them and produced his masterful little book titled *Sight*, which was published in 1881.

Had events continued on the same course after the first two years following the war, LeConte probably would have remained in South Carolina, for, in spite of the numerous obstacles, life in Columbia was at least tolerable if no longer as desirable as it had been before Sherman cut his savage swath across the land. The provisional governors were acceptable to South Carolinians, and certainly LeConte welcomed their support of higher education. But trouble was brewing in Congress, and the outcome proved to be a bitter disappointment both to him and his brother. While the LeContes had been mildly upset by President Andrew Johnson's relatively lenient reconstruction plan, they were genuinely troubled when the Republicans eventually triumphed in their intention to reconstruct the South, both politically and socially.[21]

By March 2, 1867, the radicals had pushed through Congress an act establishing military districts in the southern states. This was followed by control of registration and voting and by the disenfranchisement of the old ruling class. Subsequent legislation extended the power of the military commanders over the appointment and removal of state officials, and, still later, another reconstruction act provided that state constitutions could be ratified by a majority of those who actually cast ballots, irrespective of any record of their registration as voters. The results of the radical efforts were eventually felt in South Carolina, where many carpetbaggers and scalawags assumed important state positions. In addition, a majority of those elected to the lower house of the South Carolina legislature were blacks. The consequences were not all bad; as a matter of fact, the reconstruction regime brought about a number of useful changes in the state. But, in

21 Hollis, *University of South Carolina*, II, 7–45; LeConte, *Autobiography*, 236–37.

the eyes of such diehard southerners as the LeContes, these events were absolutely unacceptable.[22]

The LeContes largely ignored the efforts of the radicals until they realized that their beloved college might be affected. The University of South Carolina had a shaky beginning, and its future was still somewhat unsound after the first two years. In 1867–1868 the university enrolled a total of 113 students, but only 57 students enrolled in October, 1868. This was largely due to a carpetbagger representative's proposal to close the university for one year, beginning in January, 1869. The proposal had been defeated in the General Assembly, but it nevertheless "undermined confidence in the University's continued existence." At issue was the admission of blacks to the state institution—a matter stoutly opposed by most white South Carolinians. The legislature had not ignored the needs of the university, however, and they had rather liberally appropriated operating funds and faculty salary increases. Meanwhile, several Republicans had received ex officio appointments to the board of trustees, and, although they were in the minority, their vocal support for admission of blacks agitated the feelings of the university's faculty.[23]

These actions had an electric effect upon all of the faculty, the LeContes included. Joseph and John began to cast about in search of positions outside the South. Their opportunities were severely limited, however, since the likelihood of appointments in the North was almost nil. Much earlier, both had entertained the possibility of migrating to Mexico when Emperor Maximilian was in control of that country, but they were discouraged by a letter from M. F. Maury, the erstwhile imperial commissioner in charge of the colonization office. Maury frankly told them that conditions were very backward in the country and that the emperor had been evasive when asked about the possibility of creating a department at the University of Mexico in which all instruction would be in English. This discouraging response further depressed the LeContes, and John spilled out his feelings to his old friend William Sharswood: "The *struggle for existence* has been and is likely to *continue* to be, *fearful* with us. *You can form no concep-*

22 Kenneth M. Stampp, *The Era of Reconstruction* (New York, 1965), 155–85; Joel Williamson, *After Slavery: The Negro in South Carolina During Reconstruction, 1861–1877* (Chapel Hill, N.C., 1965), *passim*; Thomas Holt, *Black over White: Negro Political Leadership in South Carolina During Reconstruction* (Urbana, Ill., 1977), *passim*.
23 Hollis, *University of South Carolina*, II, 44–60.

tion of it." He also told Sharswood that it was "impossible for any one living at the North to comprehend the *depth of the gloom* which shrouds our future!" The university, concluded John, is likely to be "broken up, and the officers turned out of houses and homes."[24]

Joseph felt the same way, and on November 18, 1867, he wrote to his old friend Joseph Henry that "the prostration of the entire South and particularly of the State in prospect of negro supremacy . . . is so great, that almost every one is looking forward to probability of being compelled to leave." Then he informed Henry that "my brother & myself have . . . very reluctantly determined in that event to leave our native & beloved South." Having heard of the possibility of the formation of an agricultural college in San Francisco, Joseph asked Henry to write "to some influential acquaintance of yours in San Francisco" to see if he could help them attain positions there. Although he reiterated his devotion "to abstract science rather than applied science," Joseph expressed a willingness to teach in such a technical institution if necessary. Continuing his long plea to his old friend, Joseph said: "I cannot express to you how reluctantly I have looked forward to the probability of my being compelled to leave this portion of our Country. My whole nature is powerfully conservative and clings strongly to old associations &c. I cannot but hope yet that the fearful prospect now before us may not be realized. I suppose a few months will decide." In a pathetic conclusion to his morose missive, Joseph told Henry, "It seems to me that the utter ruin of the South is inevitable. I have tried to be hopeful. I have tried to shut my eyes to the plainest indications but it is madness to do so any longer."[25]

Both Joseph and John continued their efforts to leave the South. Joseph wrote essentially the same letter to J. D. Whitney that he had penned to Henry. Since Whitney had served for a long time as the state geologist of California, Joseph thought his influence might help him secure a position on the West Coast. Many months later, however, the LeContes' prospects seemed no brighter, even though Agassiz, Benja-

24 M. F. Maury to "Mrs [Josephine] LeConte," February 7, 1866, in LeConte Family Papers, UCB; John L. LeConte to "Tillie" Harden Stevens, March 7, 1866, in LeConte Family Papers, APS; Josephine LeConte to William Sharswood, June 25, 1866, and John LeConte to William Sharswood, January 25, 1866, and September 1, 1867, all in Sharswood Papers.

25 Joseph LeConte to Joseph Henry, November 18, 1867, in Record Unit 26, Box 9, Folder 1, SIA.

min Peirce, and other leading scientists were working hard in their behalf. As the fortunes of the university continued to slide downhill, John wrote to his old friend and former chief, Colonel St. John, saying that "in view of the *Negro Rule*, which is being riveted upon us by bayonets, you will not be surprised to learn that, for sometime past, we have been looking *elsewhere* for some position *which will furnish us with bread.*" There was little hope, he said, "that this University will escape" the "blighting influence" of the degraded politicians now in control. Surely, he continued, "it is certain that the white race must *ultimately* rule the state," but in the meanwhile, for many years, South Carolina would be dominated by "an inferior and debased race." John also informed St. John that "we are liable to be turned out of our houses and homes in a few months."[26] The likelihood that the Le-Contes would be dismissed from the university was actually not great, but they felt it to be a strong possibility—and that is, of course, what counted at the moment.

As the future remained clouded by the events surrounding the university on into the summer of 1868, Joseph penned a letter to Henry, explaining that he had not written sooner because "the feeling of depression is so great among us that we shrink from intruding them [*sic*] upon others." He reiterated his intention to leave the South. The "legislature now sitting in this place," he said, "is composed entirely of the most ignorant men in the State and at least ¾ of them are *blacks*. It makes me sick to think of it." Stating that he must "get away from this state," he asked Henry to help him secure a position at the University of California. "I would be greatly obliged to you," he told his friend, "if you would write freely and candidly to any of the Regents whom you may know." Once more, he repeated his devotion to his beloved land: "You cannot imagine my dear Sir how it pains me to break the strong ties which bind me to my native South, but I see no alternative."[27]

A few weeks later, Joseph received a reply from Henry, informing

26 Joseph LeConte to J. D. Whitney, November 28, 1867, in Whitney Family Papers, Yale University Library, New Haven; Caroline LeConte, "An Introductory Reminiscence," xviii; John LeConte to "Dear General" [Isaac M. St. John], May 10, 1868, in Brush Family Papers.

27 Joseph LeConte to Joseph Henry, July 29, 1868, in Record Unit 26, Vol. 77, p. 455, SIA; Joseph LeConte to Alfred M. Mayer, April 28, 1868, in Alfred M. Mayer Correspondence, Princeton University Library, Princeton, N.J.

him that he had recommended the LeConte brothers for chairs at the University of California and that he had spoken to a U.S. senator and a representative from California in behalf of their applications. "I sincerely hope that your brother and you may secure the appointments," he said. "The change tho' painful will I think be the best for yourselves and your children, and perhaps for the cause of humanity," advised the old man who had never displayed the slightest resentment toward his professional peers in the face of their southern sentiments. He went on to suggest that "the South can never again be what it was. New habits, new thoughts—new men will have sway." Thus, he observed, "You can no longer be in unison with the times were you to remain, and therefore your energies would be best employed in a new sphere and under new conditions." Two months later, as Joseph received encouraging word from the University of California, he wrote to Henry, expressing thanks for Henry's support and confirming again his own desire to leave. "All my reasons for leaving this State increase in strength daily," said Joseph, "and all the information & news from California strengthens me in my determination to go there."[28]

The agony of waiting for word from California became virtually unbearable as the months passed. Rampant rumors of changes in the membership of the University of South Carolina board of trustees circulated freely, and the utterly detestable likelihood of the appointment of at least one black member to a smaller board had a palpitating effect upon the hearts of the faculty. Then, at last, on November 17, 1868, John received word from the University of California; he had been hired as the first faculty member of the newly founded institution and was instructed to report to Oakland in the spring of 1869. To his old friend Lewis Gibbes, John wrote on December 18, 1868, "It is a great relief to escape from the *despotism of ignorance*—to feel, in a measure, independent of the *ignoramuses* who are *now* swelling in our campus." Indeed, the faculty was terribly incensed that the General Assembly was now meeting in the university chapel, and the Senate, in the library. "I am afraid," said John, "thousands of books will

28 Joseph Henry to Joseph LeConte, August 13, 1868, in Record Unit 33, Vol. 2, pp. 431–32, SIA; LeConte to Henry, August 15, 1868, and October 25, 1868, both in Record Unit 26, Vol. 77, p. 419, SIA; Joseph Henry to S. F. Butterworth, July 23, 1868, in Regents' Correspondence File, Bancroft Library Archives, UCB. (Letters of recommendation from Louis Agassiz, Benjamin Peirce, and George F. Holmes are contained in the same file.)

disappear."[29] The charge was grossly exaggerated, of course, but the disaffected southerner could hardly speak rationally on a matter that troubled him so deeply.

Joseph rejoiced with John over the good news, but he was severely disappointed that he had received no word yet on his own application. The saving hope was Joseph Henry's letter of October 22, 1868, stating that one of the California trustees had noted that both brothers would probably be "offered positions in the University." For two weeks after John's appointment Joseph anxiously waited. Would he be appointed, or would he have to stick it out longer in South Carolina? Would he be separated from his dear brother and close companion? When would the telegram come, and what message would it bear? Already Emma had cried, "This uncertainty is horrid!" Then Bessie took up the lament. "I must confess," she wrote to Emma, ". . . to a deep disappointment as I had hoped the elections [for both the chairs of physics and geology] would take place at the same time. At least it would relieve our minds of this dreadful uncertainty, and allow your Father to make other arrangements." She continued, "Oh how sad for the two brothers to be parted. . . . The future dear Emma is full of dark forebodings." But the suspense ended on December 2 when good news arrived; Joseph had been appointed as professor of geology at the new school![30]

Elated over his election, although slightly disappointed that he would not begin his duties until September, 1869, Joseph sat down to write a letter to Emma, who was away on a visit to Georgia. "My Dear Dear Daughter," he began,

29 Hollis, *University of South Carolina*, II, 46–47; LeConte, *Autobiography*, 239; Joseph LeConte to Robert Means Davis, November 17, 1868, and to E. P. Alexander, November 18, 1868, both in Davis Papers; Joseph LeConte to Josephine LeConte, July 23, 1868, and to John LeConte, March 16, 1869, and July 4, 1869, all in LeConte Family Papers, UCB; Joseph LeConte to his wife, July 13, 1869, in LeConte-Furman Papers, SHC/UNC; Emma LeConte to Farish Furman, October 15 [1868], in Talley Papers; Emma LeConte to Farish Furman, March 11 [1869], in LeConte-Furman Papers, Georgia Historical Society, Savannah, Ga. (hereinafter cited as LeConte-Furman Papers, GHS); Minutes of the Board of Regents, University of California, November 17, 1868, in Regents' Office, University of California, Berkeley; John LeConte to Lewis R. Gibbes, December 18, 1868, and January 30, 1869, both in Gibbes Papers.

30 Joseph Henry to Joseph LeConte, October 22, 1868, in Record Unit 33, Vol. 2, p. 690, SIA; Emma LeConte to Farish Furman, October 15 [1868], and Bessie LeConte to Emma LeConte [November, 1868], both in Talley Papers; Minutes of the Board of Regents, University of California, December 1, 1868.

I suppose you have heard ere this of my election to the Chair of Geology in the University of California. I cannot feel the unalloyed joy which John and Josephine felt in John's election. I feel much sadness mingled with gratification. I sincerely feel for the old S. C. university. I very much fear we have unavoidably inflicted a serious blow to her interests. Under other circumstances she would easily recover but just now it will be hard. I feel sad to leave the State in which I have spent the best years of my life and which I really love.[31]

The blow to the university was serious. South Carolina had lost the services of two of its most able and widely recognized professors, and both the prestige and the development of sound scientific programs were severely diminished by this loss.

LeConte's "mingled" emotions were further exacerbated by the reality of an impending separation from his beloved daughter Emma, who planned soon to marry Farish Carter Furman and move to Georgia. A bright and talented young graduate of the University of South Carolina, Furman had inherited the two-thousand-acre plantation of his grandfather, Farish Carter, at Scottsboro, just outside Milledgeville and near Bessie's old home. Joseph was pleased over Emma's engagement to Furman, but he was greatly distressed to think of losing his daughter, especially since visits with her would be difficult and far-between. "I feel sad," Joseph told Emma, "that I will probably be separated at least for a brief time from you my dear daughter. But it must not be long, it will not be long. So keep a good heart my dear and think of the new world open before us."[32]

As the weeks rolled on into February, 1869, John LeConte received word that the University of California wanted him to report earlier than originally planned. He would be needed to help organize the curriculum and get the school ready for the fall term. In fact, the trustees suggested that they might make him acting president. Meanwhile, Joseph wrote to his niece Tillie to inform her of developments: "I feel only too surely that in leaving the Atlantic Coast I am leaving the best years of my life behind—the best years in vigour, energy, youth, and also the best years in affectionate association with loving hearts." Joseph dearly loved Emma; she was, in his own words, "one of the strongest and yet the gentlest most refined and beautiful characters I ever knew. It may seem strange, but is nevertheless true that I not only

31 Joseph LeConte to Emma LeConte, December 3, 1868, in Nancy Smith Collection.
32 Emma LeConte Furman, "Recollections"; Joseph LeConte to Emma LeConte, December 3, 1868, in Nancy Smith Collection.

love but actually reverence my own child." The feeling was mutual, and Emma often spoke of how she worshipped her father. As she said on the occasion when separation appeared imminent: "I thought I had found out all that was good and noble in him, but every month shows me something new to love and reverence. He does love me so tenderly—and with all my respect and reverence for him there has been such a close companionship between us." She concluded her tribute to Joseph: "He is so unselfish—so pure—so—oh! I cannot find words to say what I think of him and how sweet he is to me without the tears coming into my eyes."[33]

The marriage of Emma and Farish took place in March, 1869, and the young couple soon settled in the old Farish Carter plantation home at Scottsboro. Furman continued to read law, eventually passed the bar examinations, set up a practice in partnership with Daniel B. Sanford in Milledgeville, and became active in local politics. Within a few weeks after the marriage, Joseph wrote to Emma, telling her she could not "imagine how empty and desolate the house seems without you. . . . A pang shoots through my heart sometimes particularly when I think of California but I quickly strangle it in earnest work. But your mother seems to brood much." The pain of leaving Emma, of departing the South, of beginning a new life—all threatened to unnerve Joseph, and for a brief time he doubted that he had many productive years left. In reference to Emma's marriage, he said, it "marks a great era and turning point in my life. Until now I have been forming and rearing a family. Now the process of disintegration has commenced. I feel that I have reached the summit of life. I feel I must now commence to give place to others."[34]

Joseph's melancholy was deepened further when his daughter Sallie announced that she too was engaged to be married. She had fallen in love with another of Joseph's students, Robert Means Davis, an extremely bright individual who had considerable literary talent. Hap-

33 Emma LeConte to [Farish Furman], February 20, 1869, in Nancy Smith Collection; Farish Furman to Emma LeConte, February 9 [1869], and Emma to Farish, February 21 [1869], both in LeConte-Furman Papers, GHS; Joseph LeConte to Matilda "Tillie" Harden Stevens, February 21, 1869, in LeConte Family Papers, APS; LeConte, Autobiographical MS.; "Recollections of the personal characteristics of Joseph LeConte by his daughter, Mrs. Emma LeConte Furman," [no date; after LeConte's death in 1901] (MS in McMillan Collection).

34 Joseph LeConte to Emma LeConte Furman, April 25, 1869, in LeConte-Furman Papers, SHC/UNC.

pily for Joseph and Bessie, the young couple decided to delay their marriage until after the LeContes got to California. Means would follow as soon as Joseph could help him find a suitable job. Months later, after Joseph and his family were settled in Oakland, Davis did indeed come to California and remained there until his mother became desperately ill, forcing him to return to South Carolina. His marriage to Sallie was delayed until 1877. The story of their romance, the breaking of their engagement, the resumption of their courtship, and their eventual marriage is reminiscent of the tale of Jacob and Rachel. But in the spring of 1869 Joseph saw Sallie's engagement as a sign of the further "disintegration" of his little family, and it only added to his gloomy outlook.[35]

Yet there were so many things to do in preparation for the departure to California that Joseph could not dwell in despair. He desperately needed the back pay from the university, for he was short of cash and the overland trip would be expensive. Then he had to go down to his old estate in Liberty County to see if anything could be arranged with the freedmen. From there he had to travel to Savannah to consult with his sister Jane, who had left Halifax for good and was eking out a bare existence by running a boardinghouse with her daughter Ada. In the meantime, problems at the university worsened, and still no word came from the California regents regarding the exact time Joseph was expected to report to his new position. Bessie wrote to her sister-in-law Josephine in California, "I wonder how you can even think of us here in this Slough of despond." Finally, late in June or early in July, word came from California that Joseph should be there by September 1. A few days later, John sent Joseph eight hundred dollars "in part payment of his debt" to Joseph. "Things begin to look well," Joseph wrote to Bessie, who was visiting in Georgia, "so cheer up my dear—don't worry yourself about the future, but enjoy the present."[36]

Now definite plans could be made. Joseph began to make the arrangements for the long journey that would take him, Bessie, and their two younger daughters to San Francisco. Since it had been de-

---

35 See various letters in Davis Papers.

36 LeConte, Autobiographical MS.; Joseph LeConte to John Lawrence LeConte, July 24, 1869, in LeConte Family Papers, APS; Joseph LeConte to his wife, July 11 and July 25, 1869, both in LeConte-Furman Papers, SHC/UNC.

cided that Jane and Ada would accompany them to California, Joseph's plans included the travel arrangements for these two additional members of the party. The end of the first phase of Joseph's successful but troubled career had come to a close.[37]

37 Bessie LeConte to Josephine LeConte, May 12, 1869, in LeConte Family Papers, UCB; Emma LeConte Furman to her husband, June 25, 1869, in LeConte-Furman Papers, GHS; Joseph LeConte to John Lawrence LeConte, July 24, 1869, in LeConte Family Papers, APS.

—◄•❧ VII ❧•►—

Commencing "a New Life" in California, 1869–1875

By early August, the LeContes had completed their preparations to move to California. Originally they had planned to meet sister Jane and niece Ada in Charleston and there embark by steamer for New York, but Jane had not yet finished up her business in Liberty County, so Joseph and his family decided to go directly to New York by train, waiting there a few days until Jane joined them. When they arrived in New York, they were met by Joseph's old friend Lewis Sayre, who put them up in his home for several days. Jane did not arrive as scheduled, and the date for leaving had to be set back for more than a week. Finally, she and her daughter arrived, and the party set out for California on August 23, 1869, by way of the newly opened transcontinental railroad.[1]

From the bustling metropolis of New York to the great falls of Niagara, across the open prairie of Nebraska amid rumors of an attack by Chief Spotted Tail, on to the heights of the Rocky Mountains, the train rolled toward the West Coast, bearing the LeContes to their promised land. After traversing the magnificent Sierras, where Joseph would return again and again over the years, the great locomotive carried them into the Central Valley and arrived finally at Sacramento. From there the family took a riverboat to San Francisco. After crossing the bay, they finally arrived in Oakland and settled with John and his family in the luxurious home of the wealthy Charles Webb Howard, who was visiting in Europe at the time. In a letter to her fiancé Means Davis, Sallie spoke ecstatically of their splendid quarters and raved

1 Caroline LeConte, "An Introductory Reminiscence," xxv; "Report of Committee appointed to draw up a Preamble and Resolutions," Joseph LeConte to Joseph Henry, August 4, 1869, and R. Means Davis to a brother, August 23, 1869, all in Davis Papers.

over the beauties of California. Joseph's family shared this house with
John for awhile, then found another temporary home, and within a
year moved into yet another house with John and his family—the last,
a fifteen-room mansion on the shores of Lake Merritt.[2]

Finally Joseph could count on a steady income. He received an annual salary of $3,600, as compared with only $2,000 in South Carolina.
Years later, Carrie recalled how her father reacted when he brought
home his first month's pay: "He beckoned the family into mother's
room. Then pulling out the money, piece by piece, he flung it down
on the bed. 'Gold!' he shouted, 'Silver!' he proclaimed. And then
throwing up one hand, fingers outspread, and with a thrilling cry:
'Money.'" The event made an indelible impression upon Carrie, as did
her first Christmas in California and the "huge box of toys" she received. Soon Joseph was able to pay back the money he had borrowed
to move to the West Coast, and he was able to worry less about receiving all of the back pay of $1,600 owed him by the University of South
Carolina.[3]

Despite his nostalgia for old South Carolina, Joseph quickly found
himself falling in love with California. Less than a month after he arrived, he wrote to Emma of the invigorating climate, and two months
later he wrote to James Hall that he was "thus far greatly pleased" with
"this distant but delightful country." By the spring of 1870, he was able
to tell Emma that the society of Oakland was "much better . . . than
either Macon or Milledgeville." Even more encouraging to him was the
prospect that "at Berkeley . . . I have no dout [sic] we will collect about
the University a very choice society." Within two years, the University
of Georgia would try to lure LeConte back to his alma mater. "I have
much affection both for Georgia and the old College," wrote Joseph to
his friend William LeRoy Broun, who had returned to the institution
from which both he and Joseph had been dismissed many years before, "but it seems to me it would not be wise at present for me to

2 Sallie LeConte to R. Means Davis, September 5 and 7 [1869], December 13, 1869,
January 11 [1870], and March 5 [1870], and Sallie LeConte Diary, all in Davis Papers;
Caroline LeConte, "An Introductory Reminiscence," xxvi–xxix; Joseph N. LeConte, "Recollections: a few notes descriptive of a happy life" [written in 1942, with additions
through 1949] (MS in LeConte Family Papers, UCB).

3 Caroline LeConte, "An Introductory Reminiscence," xxviii; Joseph LeConte to William LeRoy Broun, June 28, 1872, in Broun Correspondence; Joseph LeConte to Maximilian LaBorde, January 30, 1873, in LeConte Family Papers, UCB.

leave this place." He added, "I have still and always will have strong yearnings toward my native South and I may eventually return." But Joseph had already become acclimated to California, although it took Bessie many years to accept Berkeley as her permanent home. Nostalgia is a sweet intoxicant, however, and Joseph would never completely shake off his remembrance of the golden years in the South. Still, California had not proved to be the foreign culture he had anticipated when he left the South.[4]

There were other reasons why the LeContes found joy in their adopted land. Joseph's two daughters adjusted very quickly to their new surroundings. The health of little Carrie had improved greatly in recent months, and Joseph happily told Emma that the child was in good spirits and that "her leg is slowly improving." Shortly thereafter, he informed Emma that Carrie was "busy all day long," playing, drawing, and "walking." "I rub her leg every night," he said. "She has gained strength constantly in her left leg. This leg is about ½ inch shorter than the right and it will be a long time before she gets complete &. equal use of the left, if indeed she ever does." A few months later, he wrote to Emma, saying, "We are strangers here," but he also declared that there was a "really good society here, just as good as one can desire."[5]

Emma was delighted to receive such encouraging words from her father, and she was even more thrilled when she learned that her mother was pregnant again. Both Joseph and Bessie had long hoped for a son, and they were greatly elated when Bessie gave birth to a boy on February 7, 1870. Christened Joseph Nisbet, he would perpetuate the name of his father and his mother. To Emma, the proud father wrote that the baby "is a joy and light in our household—I have really yearned for a son and am really very grateful and very happy." The infant, destined in stature and interests to be much like his father, grew strong and prospered, but in mid-December, 1870, he became extremely ill. He recovered from the illness but was stricken again at age three. Joseph and Bessie were shaken terribly when the child con-

4 Joseph LeConte to James Hall, December 29, 1869, in James Hall Papers, New York State Library, Albany; Joseph LeConte to Emma LeConte Furman, January 1 and March 31, 1870, both in LeConte-Furman Papers, SHC/UNC; Joseph LeConte to William LeRoy Broun, June 28, 1872, in Broun Correspondence.

5 Sallie LeConte to R. Means Davis, January 2, 1870, in Davis Papers; Joseph LeConte to Emma LeConte Furman, October 3, 1869, and January 1, 1870, both in LeConte-Furman Papers, SHC/UNC.

tracted an unknown illness that brought him near the point of death
in 1873. Within a week, however, the little boy had been nursed back
to health, and Sallie recorded in her diary: "Our darling Dodo . . . is
restored to us from the grave."[6]

By mid-February, 1870, Means Davis had come to California to be
near Sallie until he could earn enough money for them to get married.
Joseph worked hard to find a suitable position for his future son-in-
law, but he did not succeed until many months later, when he finally
found a high school teaching job for the young man. Davis did not fall
in love with California, however, and he found his stay there very dis-
tasteful. Moreover, he did not enjoy teaching high school students,
especially since he was assigned the "scrubs" rather than the scholars.
In addition, his mother became gravely ill after he got to the West
Coast, and the homesick young man had to worry about that, too. He
blamed his mother's illness upon the "miserable Yankees" who had
ruined his father's fortunes, forcing his mother to work in the family
mill from daylight to dusk. In 1871, Means found a better teaching
position in Santa Rosa, but he was constantly separated from his beloved Sallie, and he could neither get his heart into teaching nor warm
up to California.[7]

In November, the distraught lover received word that his mother
was not expected to live. He gave up his job and returned to South
Carolina; unfortunately, his mother died several days before he arrived. Means and Sallie continued to correspond for several months,
but eventually they broke off their engagement. It was just as well, for
Means seemed to be getting no closer to his goal, and Sallie was still
unready for marriage. Her letters clearly indicate that she had not yet
matured fully. In addition, Josephine LeConte had sought to influence
Sallie against Means, thereby creating hard feelings. But Joseph always admired Davis, and years later he would gladly welcome him to
the family. Davis went on to make a successful career as a newspaper
editor and as a professor of history at the University of South Carolina.[8]

6 Joseph LeConte to Emma LeConte Furman, February 20, 1870, in LeConte-Furman
Papers, SHC/UNC; Sallie LeConte to R. Means Davis, December 17 [1870], and Sallie
LeConte Diary, February 14 [1873], both in Davis Papers.
   7 See numerous letters between R. Means Davis and Sallie LeConte from September
5, 1869, to June 6, 1871, all in Davis Papers.
   8 Correspondence between Sallie LeConte and R. Means Davis, September 27, 1870,

Meanwhile, Joseph suffered from occasional moods of homesickness as he read letters from his dear Emma. Life in an old plantation home far removed from the center of a bustling social life ill-suited the young bride, and she often wrote to her father of her isolation and loneliness. Her psychological stress was deepened by the still-birth of twins in January, 1870. Moreover, her husband no longer talked about moving to California, and as Joseph continually urged them to come west, Emma became terribly depressed. Her nostalgia and longing for a return to the place of her youth led her to take a trip to Columbia in the summer of 1870. She sent ivy leaves from the old home to her parents, and they wept. Joseph wrote back to her: "I can hardly think of any place as home to you & to myself except Columbia and the old ivy-covered house in the campus."[9] It would be six years before Emma saw her parents again, and the homesick young woman continued in the meantime to pine for reunion with her revered father.

The demands upon Joseph's time were enormous. In addition to his regular duties at the university, he kept busy with his writing, continued his studies of physiological optics, and began to concentrate upon geology once more. As he put it later in life, "Those early years in California were very active ones for me. . . . Coming to a new country, I had to make myself known to the people, so accepted invitations to lecture on many occasions." In addition, like all members of the university faculty, LeConte was required by the regents to lecture at the San Francisco Mechanics Institute, which had been formed to fulfill the university's land-grant requirement and to provide additional preparatory work for students planning to enter the university. The faculty, including Acting President John LeConte, had not been pleased over the arrangement, principally because it demanded so much extra effort by an already heavily taxed faculty. But the regents adamantly insisted on "popularizing" the university, and their wishes prevailed.[10]

The regents exercised much direct control over the university dur-

---

to June 1, 1873, and Sallie LeConte Diary, entries from July 10 [1871], to April 30 [1873], all in Davis Papers.

9 Joseph LeConte to Emma LeConte Furman, March 31 and June 7, 1870, and March 27, 1871, all in LeConte-Furman Papers, SHC/UNC.

10 LeConte, *Autobiography*, 243–44; Minutes of the Board of Regents, University of California, June 21, 1870; Verne A. Stadtman, *The University of California, 1868–1968* (New York, 1970), 54–57.

ing its early years. In 1868, long before the school was scheduled to open, they had elected the well-known Union officer General George B. McClellan as president of the fledgling institution, but McClellan had declined to accept. Thereupon the regents decided to elect three of their own members as "executive head" of the university until another candidate for the presidency could be agreed upon. Eventually, they asked John LeConte to take the office as acting president at a handsome salary of $6,000 a year. John accepted the position, although he declined the additional salary of $200 per month, settling instead for his regular professorial salary of $300 per month. The triumvirate of regents continued to exercise control over all matters except "faculty and student affairs," so John's authority was severely limited. This situation continued on into mid-1870, when the regents drew up a new list of presidential candidates, among whom was Joseph LeConte. They finally chose Daniel Coit Gilman, a brilliant and able member of the Sheffield Scientific School of Yale University, but for personal reasons Gilman declined. The regents then elected Henry Durant, an experienced but aging man who had previously directed the Contra Costa Academy and the College of California. Durant served for two years and then retired from academic life.[11]

As the fall term of 1870 commenced under its new president, Joseph and John LeConte expressed displeasure over two actions of the board of regents, the first of which permitted women to enroll in the university. Means Davis wrote to Sallie regarding the matter: "I agree with you that it is a nuisance; and I am fast learning to think the University a humbug, not in regard to the professors, but the management." Sallie reported that her father vehemently criticized the "confounded" regents for their foolish decision. She also indicated that a number of women had applied, but she was unsure of "how many of these 'neuters,' as Uncle John contemptuously designates them, have already entered." Joseph further complained that women would "swarm" the university, but within a year he had changed his tune. "Surely before long," he wrote to Emma, "female talent will be so far recognized that the Professors chairs may be occupied by them."[12]

11 Stadtman, *University of California*, 18–50; San Francisco *Daily Examiner*, June 15, 1869; San Francisco *Evening Bulletin*, March 2 and June 22, 1870.
12 R. Means Davis to Sallie LeConte, October 26 and September 10, 1870, both in Davis Papers; Joseph LeConte to Emma LeConte Furman, June 19, 1871, in LeConte-Furman Papers, SHC/UNC.

More disturbing to LeConte, however, was the regents' decision to dismiss Robert Fisher, the professor of chemistry and dean of the faculty. In October, 1870, without so much as a warning or even the courtesy of prior notice, Fisher learned through a newspaper article that he had been dismissed. The report indicated that he was fired because of "economic retrenchment," but, as Sallie told Means, the real reason was that the luckless professor had been secretly accused of "some dishonorable conduct in some business transactions with Mr. Finley, one of their body." Whatever the case, not a single shred of evidence ever came forth, and the plain fact of the matter is that the regents acted in high-handed fashion. Fisher was never officially accused of any wrongdoing; no charges were brought against him, and he was given no hearing or recourse to due process. In fact, President Durant did not even notify him officially of his termination until twenty days after it had been announced in the newspaper.[13]

"We are all indignant," said Sallie, "and Mother more set against California than ever." Sallie also reported that Joseph "warmly espoused his [Fisher's] cause" and that he worked hard to get the professor reinstated. The event came about just at the moment Joseph received a letter from Charles Venable, his old friend of Georgia and South Carolina academic association, inquiring if Joseph would like to apply for a post at the University of Virginia, where Venable had been since the end of the Civil War. Joseph replied that he was indeed interested, and, as Sallie phrased it, "he says that things have come to such a pass that a man had better begin looking around him for something else to do." As the LeContes gathered round to discuss the sordid situation, they expressed a wish that all of the faculty would resign in protest, and they felt as if they were "walking over a mine which may burst out at any moment."[14]

The crisis subsided fairly quickly, however, and nothing more came of it. In spite of his objection to the Fisher affair and his stated desire to return to the South, Joseph was finding a number of advantages in the University of California. Durant's presidency, although not outstanding, set the university on an even keel. Then, in 1872, upon Durant's retirement, the regents turned once again to Gilman. This time

---

13 Stadtman, *University of California*, 58–59; Sallie LeConte to R. Means Davis, September 10 and November 3, 1870, both in Davis Papers.

14 Sallie LeConte to R. Means Davis, September 10 [1870], in Davis Papers.

Gilman accepted, much to the good fortune of the university. Gilman
was an able administrator, a forceful leader, and a genuinely gifted
innovator. Joseph strongly admired him and got on very well with the
new president. Under Gilman's leadership, Joseph threw himself into
his work. He also found great stimulus from the Berkeley Club, of
which he and a handful of others were founders along with Gilman,
the instigator of the idea. In Joseph's view, the Berkeley Club was an
ideal organization combining both social and intellectual purposes.
Intended as a club of "diverse spirits," the organization brought together men from numerous pursuits, both from the academic and the
civic communities. As Joseph saw it, the club served to "mitigate" the
"evil" of "overspecialization." The topics of discussion covered politics,
social concerns, scientific subjects, and any other matter of intellectual interest. In nearly three decades of membership in the organization, Joseph presented about two dozen papers of his own, and he
actively participated in the discussion of others.[15]

In 1873 two buildings were completed on the Berkeley site, and
classes commenced on the new campus. There was still neither a
town nor any faculty or student residences around the campus, so for
many months both faculty and students had to travel from Oakland
by a horse-car line. By 1874 some living accommodations had been
constructed, including two residences owned by the university, one of
which was rented to Joseph. In Joseph's opinion, "the site of the University is certainly one of the most beautiful in the world."[16] It was a
scene that eventually came to mean as much to LeConte as the old
South Carolina campus. He was becoming a Californian at heart.

Perhaps as much as anything else in his adopted land, it was the
grandeur of the Sierras that captured LeConte's spirit. From his first
excursion into the mountains in 1870 to his death in Yosemite in 1901,
LeConte reveled in the delights of mountain-climbing, geological investigations, and the worship of God as manifested in the natural
wonder of the West Coast cordillera. He recorded scores of observations and reflections in his little pocket notebooks, and he published

15 Stadtman, *University of California*, 61–66; Fabian Franklin, *et al.*, *The Life of Daniel Coit Gilman* (New York, 1910); Hugh Hawkins, *Pioneer: A History of the Johns Hopkins University, 1874–1889* (Ithaca, N.Y., 1960); LeConte, *Autobiography*, 257, 261–64.

16 LeConte, *Autobiography*, 251–52; Joseph N. LeConte, "Recollections"; Joseph Le-Conte to Sallie LeConte, December 10, 1874, in Davis Papers.

both popular and professional pieces growing out of his sojourns. Certainly one of the most revealing of these observations is found in the little journal he wrote in the summer of 1870. So impressed by the felicitous record were the students who accompanied him on the trip that they arranged for the journal to be privately printed a few years later. It has since been reprinted several times by the Sierra Club, of which Joseph was a founding member.[17]

In later reference to his first excursion, LeConte said, "I never enjoyed anything so much in my life—perfect health, the merry party of young men, the glorious scenery, and, above all, the magnificent opportunity for studying mountain origin and structure." He related also that "everything was so new to me and so different from anything I had previously experienced." His observations on this first trip subsequently served as "the basis for ten or eleven papers" on mountain formation and related topics of landforms, diastrophism (or the deformation of the earth's crust), and historical geology. The excursion was instigated and arranged by Frank Soulé, Jr., a young professor of mathematics and engineering at the university.[18]

Beginning on July 21, the six-weeks' trip carried LeConte, Soulé, and eight students through Yosemite and across the High Sierra into the regions of Lakes Mono and Tahoe. As Joseph departed for the trip, Sallie declared that her father was "as excited about it as a boy." She also related that she wanted to go, "but petticoats are excluded entirely from this party." On July 29 the travelers arrived in the "Big Trees," where Joseph marveled at the wonder of the giant sequoias. "The whole forest is filled with magnificent trees," he recorded in his journal. Especially impressed by the famous "Grizzly Giant," Joseph described the monumental specimen and compared it to "the type of a great life, decaying, but still strong and self-reliant." Then, he jestingly added, "perhaps my own bald head and grizzled locks—my own top, with its decaying foliage—made me sympathize with this grizzled giant."[19]

17 The notebooks are in the LeConte Family Papers, UCB; Joseph LeConte, *A Journal of Ramblings Through the High Sierra of California By the "University Excursion Party"* (San Francisco, 1875). The Sierra Club edition of 1960 is used throughout the following account.

18 LeConte, *Autobiography*, 247.

19 Sallie LeConte to "Mrs. Davis" [mother of R. Means Davis], July 19, 1870, and to R. Means Davis, July 22, 1870, both in Davis Papers; Joseph LeConte to Bessie LeConte, July

For over a week, Joseph and his friends camped in Yosemite, making excursions to points of special interest each day. Swimming in the frigid waters of the Merced River and in the pools at the foot of the great falls, Joseph was in a state of ecstasy. At Bridalveil Fall he watched "the wavy, billowy, gauzy veil" as it swayed in the wisps of wind in its six-hundred-foot descent, and at Vernal Fall he climbed to the top to catch the sight of the crescendo of water. "Oh, the glory of the view," he exclaimed, "the emerald green and snowy white of the falling water; the dizzying leap into the yawning chasm; the roar and foam and spray of the deadly struggle with the rocks below." But it was the Nevada Fall that he denoted as "the grandest" he had ever seen.[20]

In camp at night, the happy group drew together around their campfire and idled away the evening hours by singing and talking on such topics as art and literature. Even more exciting were the visits of the great naturalist and "son of the wilderness," John Muir. Joseph described Muir as "a gentleman of rare intelligence," and he considered him the greatest authority on the geology and botany of the Sierras. The two men were destined to maintain close contact for the rest of Joseph's life. Muir agreed to accompany the party to Mono Lake, and for several days he and Joseph discussed the effects of glaciation upon the mountains and valleys.[21]

On August 8 the group struck out for Tenaya Lake. After a brief sojourn there, they ambled on to Soda Springs, through Tuolomne Meadow, and up Mount Dana. LeConte's vigor seemed to increase as the hard journey continued, and he easily beat most of the young men up the mountain. By August 13 the party had come over Mono Pass, down through Bloody Canyon, and finally to Mono Lake, the great basin of which "has been scooped out by a glacier." LeConte's interest in glaciation was greatly stimulated in this area. He and Muir agreed in their main hypotheses on the eroding effects of glaciers, although they differed in their views on some of the volcanic effects evident in the area.[22]

---

24 and 28, 1870, both in LeConte Family Papers, UCB; LeConte, *Journal of Ramblings*, 3–36.

20 LeConte, *Journal of Ramblings*, 36–56; Joseph LeConte to Bessie LeConte, August 1, 1870, in LeConte Family Papers, UCB.

21 LeConte, *Journal of Ramblings*, 56–66; Linnie Marsh Wolfe, *Son of the Wilderness: The Life of John Muir* (New York, 1945), 133–36.

22 LeConte, *Journal of Ramblings*, 66–111. This section of the journal contains Le-

Muir left the group on August 14 to go on to visit some of the higher volcanic cones. Joseph wanted to accompany him, but a hand he had accidentally burned at a campfire grew too inflamed for him to attempt any rugged climbing. He was sure this would not be the last time he would converse with Muir, however, and, as he bade farewell, the two promised to write to each other. Joseph returned to the shores of the lake to observe the "thousands of birds" flocking about the water's edge. He was particularly struck by the sandpipers that alighted on the water and never on the shore. "They swam, rose in flocks, settled on the water exactly like true ducks," he recorded in his journal. The question he then asked is most revealing: "Will not these in time undergo a Darwinian change into web-footers?" He later learned that the birds were really partially web-footed phalaropes, but his question nevertheless revealed a marked change that was taking place in his mind.[23] He was swinging over to the theory of organic evolution. This was indeed a crucial point in his intellectual transformation.

After several days in the Mono region, the party moved on to Lake Tahoe, where they ended their excursion. As Joseph reflected upon the wonders he had viewed, he could not refrain from revealing his religious sympathies. Said he, "Natural beauty is but the type of spiritual beauty," and earlier he had declared in the midst of a clear Yosemite night, "I lifted my heart in humble worship to the great God of *Nature*."[24] His already liberal religious views had been expanded further, and soon he worked out a systematic theory for himself that would reconcile the tenets of orthodoxy with the facts of nature as he saw them.

The excursion increased Joseph's desire for further geological field work, of which he was in great need if he was to make any contributions to his favorite field of study. But, although ideas on mountain formation were already germinating in his fertile mind, he first had to expand his field studies to other regions. The opportunity arose in the summers of 1871 and 1873 when LeConte traveled to the states of the

---

Conte's extemporaneous lectures on glaciation and carbonate springs deposits. On the agreement between LeConte and Muir regarding glaciation in the Sierras, see Wolfe, *Son of the Wilderness*, 135, 160.

23 LeConte, *Journal of Ramblings*, 111–16.

24 *Ibid.*, 117–48; Joseph LeConte to Bessie LeConte, August 15 and 18, 1870, both in LeConte Family Papers, UCB.

Pacific Northwest. The first trip took him to the mouth of the Columbia River and on to Olympia and the surrounding area, where he made numerous geological observations. He also visited Seattle and Mount Rainier and then moved on to the Fraser River, whose beauty caused him to weep with joy. "O! God! help me to be worthy of this glorious nation," he cried. "Oh! this is finer than the Columbia River. It is the finest I have ever seen." He continued his investigations on into British Columbia, returning to Berkeley in early September.[25]

The latter trip proved especially worthwhile to LeConte because it gave him much-needed field knowledge and stimulated a hypothesis on Miocene lava floods. But the excursion was too brief to allow him to gather all the information needed for a definitive paper on the subject, and thus he had to wait until he could return to the region. He did so two years later, visiting on this occasion the John Day River valley and the Cascade Range in northern Oregon. In the summers of 1872 and 1874, he again explored the Sierras. These excursions provided valuable information for LeConte, and he began to publish papers on geological topics.[26]

The incentive for LeConte's rejuvenated interest in scientific research certainly sprang both from the vitality of his character and from the fresh opportunities that challenged his inquisitive mind, but it was heightened by a visit from his old teacher Louis Agassiz. Almost completely broken down from his excessive labors, Agassiz came to California in 1872 for recuperation. During his brief vacation on the West Coast, the great naturalist spent much time with LeConte. Eager to meet the famous man who used to play with her when she was a small child, Sallie found Agassiz "grey and broken, his mind too much enfeebled with softening of the brain." But, despite his poor health, Agassiz was still intellectually active, and Sallie later amended her first impression. After a visit with him, her father, and Edward Tompkins, a wealthy regent of the University of California, Sallie exclaimed, "*A perfect day!*" She had been sitting and listening for hours to "all these three great spirits." As she recorded in her diary, "It was very pleasant

25 Sallie LeConte Diary; Joseph LeConte, notebook labeled "Oregon & Brt. Columbia," in LeConte Family Papers, UCB.

26 Joseph LeConte, notebook labeled "Oregon 1873 & Tahoe 74," in LeConte Family Papers, UCB.

to see Father and him [Agassiz] talking their glacial theories over to-gether—it seemed again the master and the pupil."[27]

Agassiz returned to Cambridge in October, 1872, and plunged again into his work, once more devoting much of his energy to fund-raising activities in behalf of scientific projects. The stout old opponent of the Darwinian conception of evolution had begun to moderate his objections to the theory of the transmutation of species. But Agassiz would never abandon his belief that the theory was synonymous with materialism, and he remained unbending to his final day of life, which came in December, 1873. Among the many memorial tributes to the famous man was a paean of praise by LeConte, delivered before the California Academy of Sciences a few days after Agassiz's death.[28]

By now LeConte was virtually convinced of the validity of the evolutionary theory. He therefore found it necessary to rationalize his mentor's opposition to the idea of organic evolution. To do so, LeConte credited Agassiz with laying "the whole foundation for the modern doctrine of evolution." He argued that, since Agassiz had advanced the idea of "development from lower to higher, from simple to more complex, from general to special by a process of successive differentiation" among the earth's species, the esteemed scientist was truly "the great apostle of evolution." Although he admitted that Agassiz had rejected the transmutational theory in favor of "*an intelligent plan*" of "*substitution* of one species for another," LeConte maintained that the deceased naturalist had nevertheless demonstrated the crucial concept of graded development of successive forms of species.[29]

On several later occasions after he had become a full-fledged proponent of evolutionary theory, LeConte returned to the subject of Agassiz's contribution. Most of those comments were repetitive, but they all iterated LeConte's belief that it was Agassiz who "*established the laws of succession of living forms*" and thereby made the breakthrough necessary for the development of the true theory of organic

27 Lurie, *Louis Agassiz*, 368–77; Sallie LeConte Diary, September 17 and 21 [1872].

28 Lurie, *Louis Agassiz*, 381–87; "Agassiz Memorial Address," 13–19. The address was delivered on December 22, 1873.

29 LeConte's belief in Agassiz's direct contribution to the development of the theory of evolution was repeated often in his publications on the subject. For an effective analysis of LeConte's dubious argument, see Arthur O. Lovejoy, "The Argument for Organic Evolution Before the Origin of Species, 1830–1858," in *Forerunners of Darwin: 1745–1859*, ed. Bentley Glass, Owsei Temkin, and William L. Strauss, Jr. (Baltimore, 1959), 356–414, esp. 357–58.

evolution. "Without Agassiz (or his equivalent)," argued LeConte in his later defenses of the man, "there would have been no Darwin." It was he, said LeConte, who "advanced biology to the *formal* stage," while it was Darwin who "carried it forward, to some extent at least, to the *physical* stage." According to LeConte's scheme of the history of science, first came observations, then "*laws* of phenomena," and finally "the *causes* or explanation of these laws." Agassiz, insisted LeConte, clearly belonged to that second stage.[30]

LeConte's defense of Agassiz was a Procrustean endeavor, but it clearly harmonized with his steadfast belief in the necessity for reconciliation of opposing theories. The conflict between the Agassizian and the Darwinian theories had been boiling about in his mind for some time after he had read Darwin's *Origin of Species*, and by the late 1860s LeConte had worked out a reconciliation that satisfied him. It was to serve as his reference point in all of his extensive commentaries on evolution in later years, no matter to what subject it was applied—botany, zoology, geology, religion, education, sociology, and philosophy. LeConte had in fact first presented his theory to his classes at the University of South Carolina and, just before he left the South, as a popular lecture. He delivered the lecture again in San Francisco on May 8, 1870, and in 1871 he published his theory as the "Natural Law of Circulation." This paper represented LeConte's first systematic effort to deal with the theory of evolution. Although it was written before he fully accepted the transmutational theory, which was not evident until 1873, it was a milestone in the process of his conversion into a "reluctant evolutionist," as he later called himself.[31]

Nearly two decades earlier, LeConte had argued that "each species was introduced by the direct miraculous interference of a personal intelligence," and he had rejected the Lamarckian view, holding that "physical conditions cannot change one species into another." But his later encounter with Darwin's theory and the writings of Herbert Spencer ultimately began to shake his faith in the old dogma. Yet the

30 Joseph LeConte, "Review of *Louis Agassiz: His Life and Correspondence*," *Overland Monthly*, 2nd ser., VII (January, 1886), 103–105; Joseph LeConte, "Agassiz and Evolution." *Popular Science Monthly*, XXXII (November, 1887), 17–26.

31 LeConte, "Natural Law of Circulation," *Proceedings of the California State Teachers' Institute, September 13–16* (Sacramento, 1871), 54–67. The paper had been delivered as an address under the title of "The Law of Circulation in Nature" in 1868; cited in Hollis, *University of South Carolina*, II, 41.

die-hard conservative clung tenaciously to the views of Agassiz until he could no longer reject the transmutational theory. From this point he began to grapple with efforts to reconcile the Agassizian, the Darwinian, and the Lamarckian views, finally arriving at his "law of circulation, or cyclical movement."[32]

To illustrate this law, LeConte cited the "circulation of air and water," the formation and melting of glaciers and icebergs, the chemical processes of organic life, and the motions of planets and stellar bodies. In his explanation, LeConte maintained that as each phenomenon completed its cycle, it did not "close perfectly" but instead left "a small residuum, which accumulating from cycle to cycle, enters into the composition of another cycle of longer period." Thus, the circulation of air and water, for example, always resulted in a slight modification of geological structure, while circulation in the organic kingdom left residual characters that led to the development of species "into higher and higher life," eventually producing successive species "more diverse in form and complex in organization." As LeConte saw it, even civilizations experience the same cyclical movement: as a cycle closes, a "residuum of ideas and principles . . . enters into the progress of the race." In other words, there is an evolution or successive development of both natural and social phenomena in a sort of slowly spiraling ascent, and over long periods of time new forms evolve from the "infinitesimal" residua, although these forms always contain elements of their predecessors.[33] Certainly, LeConte did not refer to the transmutation of species in his paper, but clearly his scheme of the evolutionary progression of species came close to the idea.

Shortly after Agassiz's visit in the fall of 1872, John Muir came to call upon his friend LeConte. Ever since their meeting in August, 1870, the two had continued to correspond. Muir spoke of how he "enjoyed exceedingly" the company of his new friend, and he sent Joseph much useful information on glaciation in the Sierras, which helped LeConte to formulate an article on the topic. Actually, it was Muir who first argued in favor of living glaciers in the High Sierras, but LeConte first put the idea into print, properly crediting Muir with the discovery. In a letter to LeConte, dated December 11, 1871, Muir provided a lengthy and lucid account of the deformation of the Sierras by gla-

32 LeConte, "Lectures on Coal," 168.
33 LeConte, "Natural Law of Circulation," 54–67.

ciers, and he drew sketches to illustrate his points. His theory conflicted with that of J. D. Whitney, who argued in favor of water erosion.
Whitney continued to oppose the idea of glacial erosion, but LeConte
had observed the evidence for himself, and he had been aided directly
by Muir in working out the details. The versatile Muir likewise provided LeConte with detailed information on the flora of the region.[34]

After Joseph published his first paper on orogeny, or mountain-formation, Muir commended him for his original ideas. In fact, LeConte
had pointed out a number of things that he felt Muir should observe
more closely to confirm their investigations. Their common interests
thus established, it was only natural that LeConte should seek out
Muir at every opportunity when he visited the Sierras, and, of course,
it only followed that Muir should visit his friend on those rare occasions when he left his beloved mountains for a sojourn into the city.
Sallie recorded her impressions of Muir during his visit to the LeConte
home: "His face is ruddy, hair curly brown falling in rather an unkempt
style over his forehead. . . . His eye is the most wonderful eye—it looks
straight at you like a child's eye, even as there is something of a child's
innocent nature is [sic] all his talk and manners." She continued: "He
worships Nature as a child would a surpassingly beautiful Mother—
he speaks of her and his eye glows and his words become poetical
and inspiring." Joseph was equally impressed, and he spoke with
pride when he told his cousin John Lawrence LeConte that Muir "is a
very good friend of mine."[35]

The visits of Agassiz and Muir, the field excursions, and the stimulating associations with Gilman and the Berkeley Club all increased
LeConte's desire to publish more and more of his ideas. It was as
though he was determined to make up for the lean years of 1861–1865.

34 John Muir to Mrs. Ezra S. Carr, August 20 [1870], and December 11 [1871], both in
William Frederic Badè, *The Life and Letters of John Muir* (2 vols.; Boston, 1924), I, 348,
382; Muir to Mrs. Carr, October 8, 1872, and March 30, 1873, both in Robert Engberg and
Donald Wesling (eds.), *John Muir: To Yosemite and Beyond* (Madison, Wisc., 1980), 131–
33, 143; Muir to LeConte, December 17, 1871, and April 27, 1872, and LeConte's notebook
labeled "Yosemite 1872," all in LeConte Family Papers, UCB; Joseph LeConte, "On Some
of the Ancient Glaciers of the Sierras," *American Journal of Science and Arts*, 3rd ser., V
(May, 1873), 325–42.

35 LeConte, notebook, "Yosemite 1872," and John Muir to Joseph LeConte, April 20,
1872, both in LeConte Family Papers, UCB; Wolfe, *Son of the Wilderness*, 164, 172; Sallie
LeConte Diary, November 25 [1872]; Joseph LeConte to John Lawrence LeConte, June 9,
1873, in LeConte Family Papers, APS.

He wrote to his nephew William Louis LeConte in February, 1872, telling him that he had begun "to feel the shortness of life," and when he reflected upon that, he realized "how little really effective work" he had done. He added that he "must lose less time . . . and do more & better work during the time yet left." Indeed, he published more papers during his first five years in California than he had written in his entire previous career. Among them were eight papers on physiological optics, eleven on geology, and five on miscellaneous topics. Moreover, he also published two books during the same period—*A Journal of Ramblings in the High Sierra of California* and *Religion and Science*.[36]

In addition to these publications, LeConte continued to work on his geology textbook, having completed roughly two-thirds of it by 1873. He also found time to revise the old manuscript of the chemistry text. All of these writings the industrious LeConte completed while engaged in regular class duties, numerous extracurricular lectures, field trips, and a very busy personal life. Emma wrote to her father in 1873 expressing fear that he was overtaxing himself. To this Joseph replied that he felt restless and uncomfortable unless he was constantly occupied by work. Later Sallie stated the same concern, but Joseph assured her that she "need not fear I shall hurt myself by work." Emma had also urged him to give up teaching for full-time research, but LeConte had no desire to follow her suggestion. He not only enjoyed his classroom work, for which he was already held in high esteem by his students, but also found the preparation and delivery of lectures an essential breeding ground for the production of formal papers. "We never know any subject perfectly," he said, "until we teach it." He also declared that "investigation ought not to be separated from teaching, as many suppose."[37]

With respect to the chemistry textbook, Joseph encountered more obstacles. He had rewritten the entire manuscript after the destruc-

36 Joseph LeConte to William Louis LeConte, February 28 and May 31, 1872, in possession of Mrs. Joseph Nisbet LeConte, Greenville, N.C.; Emma LeConte Furman to Farish Furman, September 25, 1875, in McMillan Collection.

37 Joseph LeConte to Emma LeConte Furman, February 3, 1873, in LeConte Family Papers, SHC/UNC; Joseph LeConte to Sallie LeConte, January 20, 1875, in Davis Papers; Joseph LeConte to E. P. Alexander, April 24, 1874, in E. P. Alexander Papers, Southern Historical Collection, University of North Carolina, Chapel Hill; LeConte, *Autobiography*, 257.

tion of the original, but by 1873 his brother had yet to do anything on his share of the project. Hoping that a promise of publication might stimulate John, Joseph sent his portion to a publisher, but he was discouraged by the response. He subsequently told Emma, "I don't believe the chemistry will ever be published," noting that the publisher considered it "too advanced" and pointing out that John would not write his part on physical chemistry in the absence of a guarantee of publication. In all likelihood, the LeContes could have found a publisher if John had completed his share, but his drive simply could not match that of his brother. Joseph had also received an unsolicited inquiry from Barnes and Company about publishing his "miscellaneous articles," but the publisher subsequently advised Joseph that they were too scholarly. To be sure, some of the specialized papers on optics and geology were too technical for the ordinary reader, but those on education and the fine arts would likely have appealed to an educated readership. The time was not far away, however, when LeConte's popular writings would attract a large audience, especially those interested in the reconciliation of religion and science.[38]

LeConte had first presented his lectures on religion and science to a Sunday school class of the Columbia Presbyterian Church in the late 1860s. He delivered them again after moving to California. Some of the members of his small Sunday school class in Berkeley liked them so much that they asked LeConte to "repeat them to a larger audience." He agreed to do so if they would arrange for a literal stenographic report. That was done, and LeConte then made some "verbal alterations and corrected some infelicities and redundancies of style." In 1873 he submitted the manuscript to D. Appleton and Company, who subsequently published it as *Religion and Science*.[39]

Originally LeConte drafted the lectures for the purpose of combating "a constantly-growing feeling among intelligent people, that there is an irreconcilable antagonism between science and revelation." Since he was personally struggling more with that very problem in the early 1870s than ever before, his *Religion and Science* reflected as much a personal testament as it did a public treatise. He had not yet become an avowed evolutionist, and, therefore, the small volume represented

38 Joseph LeConte to Emma LeConte Furman, February 3, 1873, in LeConte Family Papers, SHC/UNC; Joseph LeConte to William J. Rivers, May 17, 1872, in Rivers Papers.
39 Joseph LeConte, *Religion and Science* (New York, 1873), 3–4.

little of the more serious reconciliation he strove for in his later work, *Evolution and Its Relation to Religious Thought*. Nevertheless, *Religion and Science* contained some hints toward the acceptance of evolutionism, and it certainly indicated that LeConte had taken a step further away from traditional religion. Yet the whole book clearly demonstrated that LeConte was a man of such deep religious leanings that he was unlikely ever to abandon the broad framework of Christian doctrine.[40]

LeConte's attainments in the early 1870s did not escape professional notice, and in 1873 he was elected to membership in the American Philosophical Society. Less than a year later, he received a letter from Joseph Henry requesting that he send "lists of your various contributions to science to be presented to the National Academy of Sciences at its next meeting." Soon thereafter, he received word that he had been elected a member of the NAS. At last LeConte had gained national recognition. John LeConte was not admitted to membership in the NAS until 1878. Joseph was also elected a member of the American Academy of Arts and Sciences in 1875 and of the New York Academy of Sciences in 1876. Although John was also elected to the latter two, it had become clear that Joseph's reputation had finally exceeded that of his brother.[41]

---

40 *Ibid.*, 10; Timothy Odum Brown, "Joseph LeConte, Prophet of Nature and Child of Religion" (M.A. thesis, University of North Carolina, Chapel Hill, 1977), esp. 108–18. *Religion and Science* was not a very controversial book, and although it apparently sold a goodly number of copies, it received neither special acclaim nor criticism by reviewers. As one reviewer wrote, the author "carefully avoided" or "cautiously approached" scientific evidence that seemed to run counter to traditional religious views (*New York Times*, February 14, 1874, p. 9). Other reviews of the work appeared in the *Nation*, XVIII (January 15, 1874), 46; and in the *Methodist Quarterly Review*, 4th ser., XXVI (July, 1874), 520. The conservative theologian John T. Duffield called it a valuable contribution to the discussion of the relationship between science and religion ("Evolutionism Respecting Man and the Bible," *Princeton Review*, 4th ser., I [January, 1878], 155–56). See also William I. Gill, *Evolution and Progress* (New York, 1875), in which an entire chapter (pp. 150–62) is devoted to an analysis of *Religion and Science*. Gill found the work admirable, but he opposed LeConte's "didactic" argument and rejected the too-free tolerance of the nontheistic evolutionists.

41 J. Peter Lesley to Joseph LeConte, April 18, 1873, in LeConte-Furman Papers, SHC/UNC; Joseph Henry to Joseph LeConte, August 5 [?], 1871, and Joseph LeConte to Joseph Henry, September 8, 1874, respectively in Record Unit 33, Vol. 40, p. 311, and Record Unit 26, Vol. 145, p. 416, SIA; *Proceedings of the American Academy of Arts and Sciences*, XI (1875–76), 317, 322–23, 378; *Transactions of the New York Academy of Sciences*, III (1883), 132.

His reputation was further enhanced by the wide acclaim given to his articles on mountain-formation, and in April, 1874, Joseph was offered "the place of Geologist" in the state of Georgia. Although one newspaper erroneously reported that he had accepted, LeConte declined because the salary was only one-half that at the University of California. In addition, he noted: "I don't know if the work [would] be as pleasant as what I now have." Although he admitted to his old friend E. P. Alexander that he did "indeed long after the old country," he believed it unwise "to change just now." Over a year later, the University of Georgia offered faculty posts to both Joseph and John, at an annual salary of $2,000 each. "If Georgia appreciated such men," Emma Furman proclaimed indignantly, "they would offer them $10,000." Although her protest was somewhat overstated, it did hit the point, especially because the offer had come through a letter addressed to John. This, said Emma correctly, "shows how little people in Georgia know of the comparative reputations of the two brothers. Uncle John has done absolutely nothing in science since he came [to California], whereas Father's reputation is not only American but European."[42]

Meanwhile, Daniel Gilman had left the presidency of the University of California to accept the same post at the newly established Johns Hopkins University in Baltimore. A number of southerners hoped to secure positions there when the institution opened in the near future. As Emma observed, "Southern professors all seem to be looking with longing eyes to the Hopkins." Both she and her mother hoped that Joseph might get a chair at the new school. "The only trouble about that," Emma wrote to Farish, "is that it may be several years before the University gets into operation." Joseph himself seems to have entertained some hope of going to Johns Hopkins, but no letters exist to reveal his true thoughts on the matter. However, when Sallie told Means that her father would take a trip to the East in 1876, she said this would "decide many important things ... especially we are hoping it will decide whether Father's future home will be in the East or the West." Later John Lawrence LeConte said he had heard that Joseph "had been offered a place in the Johns Hopkins University," to

42 Joseph LeConte to E. P. Alexander, April 24, 1874, in Alexander Papers; R. Means Davis to Sallie LeConte, March 28, 1874, in Davis Papers; Emma LeConte Furman to Farish Furman, July 23, August 22, October 5 and 7, and November 21, 1875, all in McMillan Collection.

which Joseph replied: "This is *not* exactly true. Gilman spoke to me on the subject ... as a possibility [and] I believe that the Trustees have thought of it." He went on to say, however, that he planned "to inquire of the prospects of the University while I am in the East," but nothing more came of the matter.[43] At any rate, it is obvious that Joseph had now established a national reputation, but it is likewise apparent that he was not overly eager to leave California.

By 1875, the fifty-two-year-old LeConte had received the recognition that probably would have come much sooner had not the disruptive war set him back. As the year progressed, other personal pleasures buoyed his spirits. Early in the year, Sallie informed him that she and Means had resumed their engagement, and Emma announced that she would come to California for an extended visit. Joseph and Bessie were "set on fire" at the thought of seeing their oldest child after six long years. Now the mother of two daughters, Emma was, in Joseph's words, "in very feeble health," and she had "so many misgivings about Furman." In fact, Emma told Farish that she felt she was becoming a semi-invalid, but from her correspondence it is obvious that their marriage was under severe strain. On one occasion after she came to California, she wrote to Farish that she had "dreamt twice of going home but they were not pleasant [dreams] by any means as in neither of them were you at all glad to see me."[44]

Emma and her daughters came to Berkeley in July, 1875, while Joseph was away in the High Sierra. Since Emma had not known the exact time she would arrive, Joseph had gone on the excursion, wait-

43 Emma LeConte Furman to Farish Furman, October 7 and November 12, 1875, both in McMillan Collection; Joseph LeConte to Daniel C. Gilman, June 11, 1875, in Daniel Gilman Papers, Johns Hopkins University Library, Baltimore; Sallie LeConte to R. Means Davis, December 26, 1875, in Davis Papers; Joseph LeConte to John Lawrence LeConte, March 21 and May 4, 1876, both in LeConte Family Papers, APS.
44 Sallie LeConte Diary, entries between 1872–76, and letters between Sallie LeConte and R. Means Davis, [1869?], November 10, 1870, October 6, 1872, August 26, 1874, April 9, May 10 and December 26, 1875, January 9 and 31, 1876, Joseph LeConte to Sallie LeConte, December 10, 1874, and January 20, 1875, and Joseph LeConte to R. Means Davis, March 12, 1876, all in Davis Papers; Joseph LeConte to Emma LeConte Furman, February 3, 1873, and January 11, 1875, in LeConte-Furman Papers, SHC/UNC; John LeConte to E. P. Alexander, December 8, 1874, in Alexander Papers; John LeConte to John Lawrence LeConte, January 1, 1875, in LeConte Family Papers, APS; LeConte, Autobiographical MS.; LeConte, notebook, "Yosemite 1875," in LeConte Family Papers, UCB; Emma LeConte Furman to Farish Furman, September 8 and 20, November 30, and December 22, 1875, all in McMillan Collection.

ing word to return when his daughter arrived in Berkeley. As it turned out, he had to shorten his trip anyway, for he suffered a hard fall when his horse threw him. In addition to some very bad bruises, he also severely dislocated his right thumb. When he received word that Emma had arrived, he headed straight home, and, as he recorded it in a notebook, his reaction upon entering the door was: "Oh dear children—Sallie, wife. 'Where is Emma? how is Emma?' Ah! Poor child in bed alas & looking so badly—oh dearest I will nurse you & I know you must be better." His beloved daughter stayed with her parents for seven months.[45]

It had been nearly six years since LeConte left South Carolina for the strange and distant land of California. During that time he had gone through a transitional stage in his life. He had overcome large obstacles, and he had started a new life when many men his own age were already thinking of retirement. In the face of numerous difficulties, LeConte had maintained an optimistic attitude and a persistent desire to explore new intellectual frontiers. From his "second birth" in Cambridge in 1850–1851, he had come to a "third birth" in California.

45 LeConte, notebook, "Yosemite 1875," in LeConte Family Papers, UCB; Joseph Le-Conte to Henry Edwards, August 6, 1875, in Henry Edwards Papers, Museum of Comparative Zoology, Harvard University.

Joseph LeConte, *ca.* 1861. *Courtesy Carolyn Shaw McMillan*

Joseph LeConte, *ca.* 1867.
Courtesy Carolyn Shaw McMillan

Joseph LeConte as a professor at the new
University of California, 1869. *Courtesy University
of California Archives, Bancroft Library*

Joseph LeConte, *ca.* 1875. *Courtesy University of California Archives, Bancroft Library*

Joseph LeConte in his study, *ca. 1895. Courtesy Carolyn Shaw McMillan*

Joseph LeConte in a classroom decorated by his students in honor of his seventy-sixth birthday, 1899. *Courtesy Carolyn Shaw McMillan*

Joseph LeConte, *ca.* 1900.
Courtesy Carolyn Shaw McMillan

Joseph LeConte in his study, *ca.* 1895. *Courtesy University of California
Archives, Bancroft Library*

Excursion to Yosemite, 1897. Left to right: Joseph Nisbet LeConte, Helen Gompertz,
Joseph LeConte, Caroline LeConte, Anita Gompertz. *Courtesy, Bancroft Library*

Benjamin Ide Wheeler, President of the University of California, and
Joseph LeConte at Commencement Exercises, 1900. *Courtesy Bancroft
Library*

John LeConte, *ca.* 1867.
Courtesy Carolyn Shaw McMillan

Officers of the American Association for
the Advancement of Science, 1892. Left to
right: Joseph LeConte, president; F. W.
Putnam, permanent secretary; H. L.
Fairchild, local secretary. *From* Scientific
Monthly, *XVII (August, 1923), 188.*

Caroline Elizabeth ("Bessie") LeConte and Joseph LeConte at their
home in Berkeley in the 1890s. *Courtesy Carolyn Shaw McMillan*

Caroline Elizabeth ("Bessie") LeConte,
wife of Joseph LeConte, *ca.* 1867.
Courtesy Carolyn Shaw McMillan

"Bessie" LeConte, *ca.* 1875.
Courtesy Carolyn Shaw McMillan

Emma Florence LeConte, oldest child of
Joseph LeConte, *ca.* 1867. *Courtesy
Carolyn Shaw McMillan*

Sarah Elizabeth ("Sallie") LeConte, second child of Joseph LeConte, *ca. 1867. Courtesy Carolyn Shaw McMillan*

Caroline Eaton ("Carrie") LeConte, fourth child of Joseph LeConte, *ca. 1883. Courtesy Carolyn Shaw McMillan*

Joseph Nisbet LeConte, fifth child of Joseph LeConte, *ca. 1885. Courtesy Bancroft Library*

—◆◆❧ VIII ❧◆◆—

In the Study and in the Field:
Physiologist and Geologist

During its early years, the University of California encountered a number of problems stemming primarily from conflicting views on the
purposes of the school, political machinations, and personal ambi-
tions. From the outset the regents took a strong hand in the operation
of the fledgling institution, and until the late 1880s they continued to
exercise such direct control of the institution that most of the presidents were virtually powerless. The administration of Daniel Coit Gilman was the sole exception, and that was of brief duration.[1]

Much more pragmatic in orientation than their eastern counterparts, many Californians minimized the role of the liberal arts in the
training of college youth. The purpose of a state university, in their
opinion, was to provide practical training, especially in the agricultural and mechanical arts. But Gilman believed that a true university
should be founded upon liberal studies in the arts and sciences, with
professional programs added as part of its broader mission. Although
he was not averse to scientific training in agriculture, Gilman maintained that most practical subjects could be acquired in direct apprenticeship. His philosophy thus clashed with the views of a large
number of Californians who favored a curriculum devoted to useful
training. This conflict ultimately led to controversy, and Gilman resigned at the end of 1874.[2]

The vacuum created by Gilman's departure was hard to fill. Keenly
aware that their chances of attracting another leader of national
prominence were greatly diminished and conscious of the criticisms

1 Stadtman, *University of California,* 47–68.
2 *Ibid.,* 68–80.

135

leveled at them for turning over the reins of authority to the president, the regents decided to take more direct control of the administration of the university, reverting to the practice followed during the first few years of its operation. Where better could they turn than to a man who had already proven his willingness to serve as president under such circumstances? Thus it was that on March 21, 1875, the regents once again appointed John LeConte as acting president. The elder LeConte labored under the tentative appointment for fourteen months, until he was appointed as permanent president on June 1, 1876.[3]

Joseph must have been happy over the appointment of his brother, but, since he was aware of the strong role of the regents, he probably also had great misgivings. Nevertheless, the younger LeConte had faith in his brother's ability to administer the institution in an even-handed manner.[4] John LeConte was an old-fashioned administrator schooled in the philosophy that the president was a statesman, not an innovator. He was thus unable to generate any strong movement toward the advancement of the university as Gilman had done. Although a number of important gifts, such as the Lick Observatory, were bequeathed to the University of California during LeConte's presidency from 1875 to 1881, they represented no real initiative by John to secure them. In 1879 the university faced its first financial crisis. After enduring round after round of withering criticism from students, parents, and regents, the president decided to throw in the towel in June, 1881, and resume his old duties as professor of physics.[5]

The full impact of this matter upon Joseph is not clear. He had developed a strong affinity for the University of California, and he doubtless did not relish the thought of picking up and leaving at this stage of his life. He was extremely busy with his writing, and he enjoyed teaching in his adopted university. Nevertheless, he wondered if he ought to look for a position elsewhere. Bessie was still eager to move closer to her old home, and she seems to have encouraged Joseph to explore the prospects of a post in the eastern part of the coun-

3 *Ibid.,* 88–89.

4 For comments on Josephine LeConte's reaction to the appointment of her husband John to the presidency, see Jane LeConte Harden to Matilda "Tillie" Harden Stevens, June 25, 1876, in LeConte Family Papers, APS; Bessie LeConte to Emma LeConte Furman, April 11, 1877, in LeConte-Furman Papers, SHC/UNC.

5 Stadtman, *University of California,* 88–92; San Francisco *Evening Bulletin,* May 30, 1879, p. 2.

try in general and in the South in particular. Her hopes had already been raised by the strong feelers Joseph had received from Johns Hopkins and the Universities of Georgia and Virginia. After the initial query, LeConte heard no more from Johns Hopkins, and the Georgia trustees gave up after Joseph and John declined their offer, settling instead for a final gesture of recognition to the native sons they had driven out in the 1850s. The LeContes' old alma mater bestowed "the Honorary Degree of Doctor of Laws" upon the two brothers in August, 1879.[6]

In the same year, the University of Virginia again expressed an interest in hiring Joseph, and, as Joseph told Emma, he was "greatly tempted to go" to the reputable school. He declined, however, for he "hardly" thought "it would have been wise under the circumstances."[7] A year later, he received an offer from the University of South Carolina. As he put it, "The offer . . . has disturbed my equanimity greatly, in fact has filled me with anxiety & uncertainity." He confessed to Emma that he felt "strongly drawn eastward by my heart but my prudence holds me back." The matter of "prudence" was not merely one of moving; it also involved salary and duties. South Carolina wanted to bring him back as president of the school for a salary of $2,500 to $3,000 per year, although he had gone to California over a decade before at a salary of $3,600 per year. But it was the administrative post that disturbed him most. "Now I might as well say," he declared, "I don't want to have anything to do with a Presidency." John's misfortunes at that very moment were enough to turn Joseph against the office. In addition, Joseph correctly perceived that "it would be an end of my . . . scientific work—in other words a laying me up on the shelf."[8]

Some of LeConte's South Carolina supporters would not give up easily, however, and they kept on trying to induce Joseph to accept the presidency. To his son-in-law Means Davis, Joseph wrote in early 1882 that "the inducements are not sufficient for so great a change. I cannot express to you how much it pains me to be obliged to put aside such repeated kind offers." He added that "it really looks as if all

6 P. H. Mell, Chancellor, University of Georgia, to Joseph LeConte, August 18, 1879, in LeConte Family Papers, UCB; Joseph LeConte to William L. Mitchell, Secretary, Board of Trustees, October 23, 1879, in E. Merton Coulter Collection, University of Georgia Library.
7 Joseph LeConte to Emma LeConte Furman, March 15, 1879, in LeConte-Furman Papers, SHC/UNC.
8 Joseph LeConte to Emma LeConte Furman, March 10, 1880, in *ibid.*

chances of my return are slipping away. But I do not think it would be wise for me to break up here unless inducements were much stronger." One of the trustees of the University of South Carolina told Davis that "a desperate effort would be made to get back the Le-Contes," and he applied pressure on Davis to use his influence. In fact, Davis was himself under consideration for a faculty position at the school, and a couple of the trustees informed him that his chances of being hired would certainly be helped if he could persuade his father-in-law to return.[9]

As late as July, 1882, Davis was still trying to convince LeConte that he should come back to South Carolina. Joseph replied that "the worry and responsibility are too great for a man of my students [*sic*] habits. And I am quite sure it would be the end of my literary career." He reminded Means that his "whole ambition is as a writer and thinker in Science & philosophy—and to be successful one must be comparatively free from anxiety & responsibility." That ended the matter with South Carolina, but Joseph had one more possible offer to consider. In December, 1883, he received notice that the University of Texas was interested in him. Joseph wrote to Emma that he had been "written to" regarding the chair of chemistry that was vacant in Texas, but he said the position would "never suit" him "because all my reputation is now in another department & I cannot afford to throw that away." He noted also that Texas was talking of "soon establishing [a] Chair of Geology," and he said, "If they do I suppose I can have it if I desire.... I don't know what to think about it."[10] But LeConte really did know, for he was too much rooted in California to entertain serious thoughts of moving.

His record of activities during the years from 1876 to 1885 offers sufficient testimony that LeConte was generally satisfied with life in California. His scholarly production, his numerous field trips, and his correspondence with Sallie and Emma, as well as with fellow scientists all reveal that the displaced southerner had found the California climate extremely conducive to his intellectual development. In fact, he

9 Joseph LeConte to R. Means Davis, January 2 [1882?], Joseph W. Barnwell to R. Means Davis, January 10, 1882, John Bratton to R. Means Davis, January 30, 1882, W. J. Duffie to R. Means Davis, May 11, 1882, all in Davis Papers.

10 Joseph LeConte to R. Means Davis, July 15, 1882, in Davis Papers; LeConte to Emma LeConte Furman, December 15, 1882, in McMillan Collection.

published over four dozen articles, several book reviews, and three books during that single decade, or slightly more than in all of the previous thirty years of his career. Among his major works were *Elements of Geology* and *Sight*, the latter of which grew out of a series of articles begun before he left South Carolina.[11]

Although he had always maintained an interest in the physiology of vision, LeConte was too preoccupied with other matters before 1866 to give the subject any sustained attention. Then, after the Civil War, he read a paper by the famous German physiologist Hermann von Helmholtz on the topic of eye movement. Shortly thereafter, he read Édouard Claparède's article on binocular vision. These two papers revived his latent interest. As he noted many years later, he immediately perceived "that both these papers were all wrong in their interpretation of the phenomena described," and he soon "wrote three articles" in response. After he moved to California, LeConte read a series of articles on the physiology of vision that he considered "fundamentally wrong" but "suggestive." As a consequence, he "wrote an elaborately illustrated paper," because he "saw plainly that a new mode of diagrammatic representation of binocular phenomena was necessary." LeConte claimed that he might have ended his studies of physiological optics then, but soon thereafter he read a paper by the Swiss physicist Raoul Pictet that precipitated another series of articles on binocular perspective. He continued to delve into the subject, publishing additional articles and eventually his book on human vision.[12]

11 Joseph LeConte to James Hall, September 20, 1878, January 4, 1879, and December 1, 1881, all in Hall Papers; LeConte to Alexander Agassiz, September 28, 1878, in Alexander Agassiz Papers, Museum of Comparative Zoology, Harvard University; LeConte to Spencer Baird, December 21, 1878, in Record Unit 27, Box 12, Folder 19, SIA; same to same, January 21, 1880, in Record Unit 28, same to same, January 20, 1880, Record Unit 33, vol. 91, p. 329, SIA; LeConte to Spencer Baird, September 27, 1883, in George P. Merrill Collection, Record Unit 7177, Smithsonian Institution Archives, Washington, D.C.

12 Hermann von Helmholtz, "On the Normal Motions of the Human Eye in Relation to Binocular Vision," *Proceedings of the Royal Society of London, 1863–1864*, Vol. B (April, 1864), 186–99; Édouard Claparède, "Quelques mots sur la vision binoculaire et stéréoscopique et sur la question de l'horoptre," *Archives des Sciences physiques et naturelles*, n.s., III (1858), 138–68; Joseph Towne, "The Stereoscope and Stereoscopic Results," *Guy's Hospital Reports*, 3rd ser., X (1864), 125–41, and XI (1865), 144–80; Joseph Towne, "Contributions to the Physiology of Binocular Vision," *Guy's Hospital Reports*, 3rd ser., XII (1866), 285–301, XIV (1869), 54–84, and XV (1870), 180–212; LeConte, *Autobiography*, 240–41, 244–46, 270; William James to Joseph LeConte, March 12, 1881, in LeConte Family Papers, UCB; LeConte to James, March 22, 1881, in William James Letters, Houghton Library Manuscripts, Harvard University, Cambridge, Mass. See also Lester D. Ste-

Once he had begun, LeConte could not stop, and he thoroughly immersed himself in his work on sight. As he told Sallie on one occasion, "I have been very busy recently writing, writing [on vision]—and this is my only excuse . . . for not writing sooner to you." Bessie confirmed that her husband was devoted to his study, telling Emma that her father was "still hard at work on his crosseyed paper" and that she "could not get a word out of the Tycoon of the Inner Temple in his binocular abstraction." A few weeks later, Bessie jested to Emma that "Father has not yet reached the grand climax of getting his eye rotating around his head. No doubt additional light will be thrown on the Binocular when he arrives at that point."[13]

Within a few weeks after LeConte's *Sight* appeared in print in 1881, it was reviewed by William James. Lamenting the fact that English physiologists had failed to keep pace with their German and Dutch counterparts in the study of optics after Helmholtz's volumes were published, James said: "It is pleasant, under these circumstances to find an American book which can rank with the very best of foreign works on the subject." James criticized LeConte for his failure to deal with the psychological aspects of optics, and he noted that LeConte was obviously "quite unversed in the German part" of the literature on the subject. Nevertheless, he commended the versatile scientist for his ability to deal with visual phenomena and concluded that LeConte had written a substantial work. LeConte later admitted to James that his "knowledge of German language is very imperfect," and that he regretted the fact that he had to depend upon "French & English translations & French & Swiss reviews." The book was also applauded by the *American Journal of Science* as "a popular and readable, and at the same time scientific, exposition of the subject of vision." The *Journal* reviewer was especially impressed with the lengthy section on binocular vision, "a subject to which the author has contributed much by his own researches."[14]

---

phens, "Joseph LeConte and the Development of the Physiology and Psychology of Vision in the United States," *Annals of Science*, XXXVII (1980), 303–21.

13 Joseph LeConte to Sallie LeConte, December 10, 1874, in Davis Papers; Bessie LeConte to Emma LeConte, November 12 [?], 1868, in Talley Papers; Bessie to Emma, December 2 [1868], in Nancy Smith Collection.

14 *Nation*, March 17, 1881, pp. 190–91. The review is unsigned, but Ralph Barton Perry attributes it to James in his *Annotated Bibliography of the Writings of William James* (New York, 1920), 12. Joseph LeConte to William James, March 22, 1881, in James

The small volume was likewise praised by a reviewer for *Popular Science Monthly*, who declared that LeConte's "explanations are so clear, and his facts and principles so interesting, that they will be sure to engage the attention of ordinary readers." He conjectured, moreover, that LeConte's treatment of important topics would "appeal to instructed critics as new contributions to the subject." In addition, he noted one of the most unique features of the work, namely that it was "largely a book of experiments." *Sight* was also favorably reviewed in *Mind*, although that journal was justifiably critical of LeConte's neglect of the psychological aspects of vision. It was not that LeConte refused "to take account of psychology," said the journal, but that he tended to emphasize "simple sensations" and ignored the problems of psychological phenomena. On the other hand, *Mind* called LeConte's book a "noteworthy" contribution to knowledge and a source of "original" modes of illustrating visual phenomena.[15]

Although LeConte attempted to cover the entire subject of sight, he devoted the larger portion of his work to binocular vision. Two of the most important functions of binocular vision are axial and focal adjustments of the eye (or, in current scientific terminology, "convergence" and "accommodation"). Many contemporary physiologists believed that the two types of adjustment could not be "dissociated," and for several years LeConte accepted the validity of Helmholtz's theory of eye adjustments. As he stated it, LeConte first set out to confirm the theory, but he discovered in the process that there are "three adjustments of the eye, viz., the axial adjustment, the focal adjustment, and the contraction of the pupil." In his opinion, these three adjustments had been "so associated through successive generations" that individuals accomplished them in "a single act of volition" and are thus unaware that at least the focal and the axial could in fact be entirely separated. After observing the discrete functions of focal and axial adjustments, LeConte then set about to find if they were "intimately associated" with pupillary contraction. From the results of his observa-

---

Letters; Joseph LeConte to John Lawrence LeConte, April 23, 1881, in LeConte Family Papers, APS; *American Journal of Science*, 3rd ser., XXI (May, 1881), 405.

15 *Popular Science Monthly*, XIX (June, 1881), 272; *Mind*, VI (1881), 439. *Sight* was translated into German as *Die Lehre vom Sehen* (Leipzig, 1883). This edition was reviewed in *Literarisches Zentralblatt*, March 24, 1883, p. 443. See also J. Hirschberg, *Geschichte der Augenheilkunde* (9 vols.; Leipzig, 1915), III, 192. "I find it very noteworthy," said Hirschberg, "that his book was so little mentioned in the American literature."

tions, LeConte correctly concluded that the contraction of the pupil is more closely connected with focal than with axial adjustment (that is, more with accommodation than convergence). It is now known, however, that LeConte was only partially correct in his findings, for although the processes of accommodation, convergence, and pupillary contraction "are closely associated functionally" and "normally occur together," their interconnections are not necessarily unified. Thus, LeConte's findings did not completely refute the Helmholtzian hypothesis, but they nevertheless raised the topic to a new level of investigation.[16]

Helmholtz had also exposited certain laws of ocular motion. "I took issue with Helmholtz ...," said LeConte, "and showed that his views are not only wrong but self-contradictory." Helmholtz argued that "when the eye turns from its primary to any secondary position, it turns 'on a fixed axis which is normal both to the primary and to the secondary visual line.'" By numerous experiments and by observations of his own eye and of those of five young subjects, LeConte arrived at a contrary view. Helmholtz acknowledged that the eyes do rotate on the optic axis to a limited degree, but he contended that they do so in accordance with a law established by the Dutch oculist, F. C. Donders—that is, "that the eye returns always into the same position when the visual line is brought into the same direction." But LeConte found that "Donders' law" of eye movement did not apply in every instance of convergence. Instead, he discovered that the eye sometimes rotates outward in tests of visual convergence. LeConte's experiments thus established that Donders' law, and hence Helmholtz's explanation, were limited because "normal" eyes vary in their response to certain visual images.[17]

At the same time, LeConte dealt with Helmholtz's treatment of "Listing's law," which held that when the eye moves from the primary (or normal) position to any other position, it rotates on an axis at right angles to the first visual line and the new point of observation. Through numerous experiments, LeConte discovered Listing's law to be true only when one eye moves in parallel direction with the other

16 Joseph LeConte, "On Some Phenomena of Binocular Vision: Adjustments of the Eye," *American Journal of Science*, 2nd ser., XLVII (January, 1869), 68–77.

17 LeConte, *Autobiography*, 270; Joseph LeConte, "On Some Phenomena of Binocular Vision: Rotation of the Eye on the Optic Axis and the Horopter," *American Journal of Science*, 2nd ser., XLVII (March, 1869), 153–78.

but invalid when the eyes "move in *opposite* directions, as in strong convergence." Thus, LeConte concluded that Helmholtz was only partially correct in his verification of eye movement, and he showed that both Donders' law and Listing's law applied according to degrees of torsion on the eye.[18]

In binocular vision, the function of eye-rotation is associated with the phenomenon of unification of two corresponding images. Each eye receives its own separate impression, at least within "a limited part of the binocular field." This "singleness" of binocular vision is achieved through a plane of vision commonly called the horopter. LeConte noted that all previous descriptions of the horopter had been derived by mathematical calculations. In 1855, however, Georg Meissner, a German physiologist, began to investigate the matter through experimentation. Meissner concluded from his experiments that the horopter was "a plane at right angles to the visual lines" when a person looks straight ahead at an infinite distance, but that in all other instances (*i.e.,* measurable distances) it is "a straight line passing through the point of sight and increasing in inclination to the visual plane as the convergence of the optic axes increases." Moreover, Meissner found that when the visual plane is turned downward, the horopter is inclined decreasingly in relation to the plane until eventually a perpendicular position is achieved at forty-five degrees, whereupon the horopter begins to expand "into a plane at right angles to the median line of sight." When the visual plane is elevated to its maximum point, however, the horopter inclines increasingly toward the visual plane.[19]

After using different means to test Meissner's findings, LeConte decided they were generally correct, and he concluded that Helmholtz was wrong in rejecting them. "It has been with much hesitation," said LeConte, "that I have ventured to criticize the conclusions of so distinguished a physicist," but he felt that Helmholtz had ignored the matter of the rotation of the eyes on their optic axes. LeConte went further, however, by showing that Meissner was not entirely correct either. First, he demonstrated that the horopteric line inclines increasingly in relation to the degree of convergence. Then he showed that when

18 Joseph LeConte, "On Some Phenomena of Binocular Vision: Laws of Ocular Motion," *American Journal of Science*, 3rd ser., XX (August, 1880), 83–93.
19 LeConte, "Rotation of the Eye on the Optic Axis and the Horopter," 153–78.

the eye is turned upward, its rotation increases under all circumstances in relation to the increased inclination of the horopteric line.
Finally, he declared that he believed Meissner was wrong in stating
that the horopter becomes a plane when the eyes are turned downward beyond forty-five degrees; instead of a plane, the horopter becomes a "*surface.*" In final analysis, LeConte surely had not exhausted
the subject, and he had also failed to realize that Helmholtz was not
completely in error either. Nevertheless, he had certainly shown that
studies of the horopter were not yet complete, and he had helped to
clarify the nature of that aspect of binocular vision.[20] Later, he dealt
with the subject in a different manner when he took up the topic of
double images and stereoscopic relief.

The phenomenon of stereoscopic vision had long attracted the attention of both professional physiologists and interested amateurs. Sir
Charles Wheatstone, an English physicist and inventor of the stereoscope in 1838, insisted on a "*complete union* of all parts of a stereoscopic picture of an object," but others disagreed with him, including
Ernst Wilhelm von Brücke, a German physicist and physiologist, Alexandre-Pierre Prévost, a Swiss physiologist, and LeConte himself. It
was during a visit with his Uncle Jack in Philadelphia back in 1854
that LeConte had his first chance to see a stereoscope. Having already
read much about Wheatstone's theory of mental fusion of two corresponding images, he anxiously examined the much-talked-about instrument. Because of his superior skills of observation, LeConte noted
immediately that mental fusion did not in fact explain stereoscopic
images. He thus announced to some scientists gathered in Major
LeConte's home that Wheatstone was wrong. "Exclamations of surprise and dissent were heard on every side," he said, and "I was unanimously set down on every side as a very conceited and disputatious
young man." But, even though silenced by those who were perturbed
by his statement, LeConte knew he was right, and one year later
Brücke published a theory that coincided with LeConte's ideas, although, of course, LeConte did not publish his own views until many
years later.[21]

20 *Ibid.*
21 Alexandre-Pierre Prévost, "Note sur la vision binoculaire," *Archives des Sciences
physiques et naturelles*, n.s., IV (1859), 105–11; Ernst Brücke, "Über die stereoskopischen

Brücke and later Prévost disagreed with Wheatstone regarding the fusion of corresponding images, insisting instead that stereoscopic relief occurred by means of optic convergence. On the other hand, Heinrich Dove, a German physicist, cited experiments in which stereoscopic relief was achieved through the flash of an electric spark. LeConte argued that such relief could be achieved "while looking steadily at one point," and he and his brother John conducted several experiments to test that hypothesis. The results of their investigation led LeConte to conclude that "stereoscopic relief can be perceived instantly and without change of optic convergence," thus confirming Dove and refuting part of Brücke's and Prévost's contrary view that binocular relief was perceived by a process of trials in which the eyes move repeatedly and unconsciously "between foreground and background" of a stereoscopic scene until the two distances are combined by alternating optical convergences.[22]

To explain his point satisfactorily, LeConte turned to a discussion of the two major ideas concerning sense impressions, namely "the nativistic and the empiristic theories." The former held that corresponding points of impression upon the horopter are received automatically through inheritance—that is, by the congenital nature of the human eye. The empiristic theory, on the other hand, posited that sense impressions occurred through repeated trial and error—that is, through experience. LeConte believed that neither theory was entirely correct and that the only true explanation lay in a reconciliation of the opposing views. Thus, regarding the "law of corresponding points," he stated that the ability to perceive relief is "acquired by the experience of successive generations transmitted by the law of inheritance, and made more perfect by individual experience." In lower animals, "inherited experience is greater," whereas in man, "individual experience is greater." Therefore, he concluded, "binocular vision is ... to a large

Erscheinunger und Wheatstones Angriff auf die Lehre von den identischen Stellen der Netzhaute," *Archiv für Anatomie und Physiologie* (1841), 459–76; LeConte, *Autobiography*, 159–60; Joseph LeConte, "On Binocular Vision," *Philosophical Magazine*, 5th ser., V (January, 1878), 27–29. LeConte's discussion of the theories of binocular perspective in his *Sight* are found on pp. 145–55 (1st ed.) and pp. 168–75 (2nd ed.).

22 Joseph LeConte, "On Some Phenomena of Binocular Vision: So-called 'Images of Illusion,' and the Theory of Binocular Relief," *American Journal of Science*, 3rd ser., II (November, 1871), 315–23; Heinrich Dove, "Über Stereoskopie," *Annalen der Physik*, 2nd ser., X (1860), 494–98.

extent, instinctive even in man, and much more so in lower animals."[23] LeConte's argument had come far afield of purely physiological considerations, but the question of the nature of sense impressions was an important topic in the late nineteenth century, and, later, when it became a highly controversial subject among psychologists, LeConte was compelled to deal with it at greater length.

In order to offer an explanation that would reconcile the counter-vailing theories of Brücke and Dove regarding "the instantaneous perception of relief," LeConte considered how stereoscopic images were actually perceived in binocular vision. He maintained that by instinct an observer knows "without trial" whether double images "will be united by greater or less convergence," and therefore "the observer never makes a mistake nor attempts to unite by a wrong movement of the optic axis." He concluded that "the eye perceives *instantly*, by means of double images," but he argued that "perception is made much clearer by *changes* of optic convergence by ranging the eyes back and forth foreground to background and *vice versa*, and the successive combination of different parts of the object or pictures." LeConte thus not only demonstrated the basic validity of Brücke's hypothesis but also showed how Dove's modification of it could be reconciled into a suitable explanation regarding the absolute necessity for optic convergence in the perception of relief. In addition, he dealt a blow to the theory of fusion, although the idea persisted on into the twentieth century before it was finally abandoned in favor of better explanations. In "the true language of binocular *vision*," LeConte averred, "I have expressed what we *see* rather than what we know."[24]

LeConte conducted many other investigations on the nature of vision, including "illusive images," "star rays," optical illusions, the structure of the crystalline lens, the position of the eyes in sleepiness, and comparative binocular sight in animals. Some of these were written as letters to journal editors, and others were published as articles.[25] In

23 Joseph LeConte, "On Some Phenomena of Binocular Vision," *American Journal of Science*, 3rd ser., II (December, 1871), 417–26; Joseph LeConte, "On Some Phenomena of Binocular Vision: The Mode of Representing the Position of Double Images," *ibid.*, 3rd ser., I (January, 1871), 33–44; Joseph LeConte, "On Some Phenomena of Binocular Vision: Stereoscopic Phenomena," *ibid.*, 3rd ser., II (July, 1871), 1–10.

24 LeConte, "The Mode of Representing the Position of Double Images," 33–44.

25 Joseph LeConte, "On an Optical Illusion," *Philosophical Magazine*, XLI (April, 1871), 266–69; Joseph LeConte, "On Some Phenomena of Binocular Vision: Position of the Eyes in Sleepiness," *American Journal of Science*, 3rd ser., IX (March, 1875), 159–64; Joseph

the 1890s, a hot controversy developed over the question of "upright vision" that pitted LeConte against a number of able psychologists and physiologists. The range of LeConte's interests and his corresponding ability to deal with those interests on a level equal with the specialists in each field of investigation were marks of the man's versatility.

While LeConte kept busy with these investigations, he continued to teach his classes, read professional papers, give popular lectures, take field trips, and attend to personal affairs. In the summer of 1876, he set out for New York to consult with D. Appleton and Company on the publication of his geology textbook, after which he went to Philadelphia to visit his cousin John Lawrence LeConte and to see the Centennial Exposition. He then went on to Emma's home in Scottsboro to join the rest of the family who were gathered there. The desire to see his old plantation was too great to suppress, so in July Joseph made his way down to the country. There he found his former slaves eking out a bare existence. "The whole country once so active, thriving, moral," reported LeConte, "is dead, dead, dead." He empathized with the poverty-stricken blacks and wondered if they were better off than in the days of bondage.[26]

Joseph, Bessie, and the two younger children returned to California in early September, but Sallie remained at Scottsboro, awaiting the day of her marriage to Means in January, 1877. Upon his arrival in Berkeley, Joseph found his sister Jane very ill. Although she was only sixty-two years old, Jane had been in feeble health for some time. The unfortu-

---

LeConte, "In Binocular Vision the Law of Corresponding Points May Be Opposed to the Law of Direction," *ibid.*, 3rd Ser., IX (March, 1875), 164–68; Joseph LeConte, "Comparative Physiology of Binocular Vision," *ibid.*, 3rd Ser., IX (March, 1875), 168–71; Joseph LeConte, "On Some Phenomena of Binocular Vision: The Structure of the Crystalline Lens and Its Relation to Periscopism," *ibid.*, 3rd ser., XIV (September, 1877), 191–95; Joseph LeConte, "Some Peculiarities of Phantom Images Formed by Binocular Combination of Regular Figures," *ibid.*, 3rd ser., XIV (August, 1877), 97–107; Joseph LeConte, "Note on the Binocular Phenomena Observed by Professor Nipher," *ibid.*, 3rd ser., XV (April, 1877), 252–53; Joseph LeConte, "A Singular Optical Phenomenon," *Science*, III (April 4, 1884), 404; Joseph LeConte, "Double Vision," *ibid.*, VII (June 4, 1886), 506; Joseph LeConte, "Star Rays," *ibid.*, IX (January 7, 1887), 14; Joseph LeConte, "On a Curious Visual Phenomenon," *American Journal of Psychology*, III (September, 1890), 364–66; Joseph LeConte, "Shadow Images on the Retina," *Psychological Review*, VII (January, 1900), 18–28.

26 *Berkeleyan*, August, 1876, pp. 1–3; LeConte, *Autobiography*, 254–55; Sallie LeConte to R. Means Davis, July 22, 1876, in Davis Papers; Joseph LeConte, appendix to ms. of Civil War journal, written in 1878, in LeConte Family Papers, UCB.

nate woman had never really regained her psychological composure since the traumatic days of the Civil War and the terribly humiliating experiences of the early Reconstruction period when she had to take up the ignominious occupation of running a boardinghouse. Once in California, Jane never had an opportunity for even a brief visit to the home of her birth. Moreover, her dispute with her sister Ann and her daughter Tillie had left her with bitter feelings and, almost certainly, great remorse over her imperious actions toward her close relatives. In late October, 1876, the lonely woman died.[27]

Joseph had the sad duty of reporting Jane's death to Tillie, for Jane's daughter Ada had been so embittered by her mother's ill feeling toward her sister that she was not on speaking terms with her. Later, Joseph assured Tillie that Jane's "mind had greatly relented toward Sumner," and he said he had "often heard" his sister "speak kindly" of Tillie's husband and "with great affection" for her and her daughter Annie. But Joseph knew these weak and carefully worded assurances were not very consoling, and he told his niece that she must accept the disagreement between her and her mother as a simple difference of philosophical views of life. He concluded with the exhortation that "cleaning our minds of all prejudice is the very highest ... of all virtues." For him, Jane's death was another serious personal loss, and he reminded Tillie that now there were only two of the Louis LeConte family left.[28]

On the last day of December, 1876, Joseph wrote a long, affectionate letter to Sallie. He wished her the best in her forthcoming marriage and praised Means as an excellent choice for a husband. Only a few days before he wrote to her, Joseph had climbed Grizzly Peak, just outside of Berkeley, with Joe and Carrie, and he related the event to Sallie. Young Joe, only six years old, was getting an early taste of mountain climbing, an activity in which he would eventually become an expert. "It was a severe trial for [Carrie's] weak leg," reported Joseph. He was becoming concerned over more than Carrie's crippled leg, however, for his young daughter was showing signs of excessive introversion. But the young girl was very bright, and Joseph called her "a little genius." He was especially impressed with her study of botany:

27 Joseph LeConte to Matilda "Tillie" Harden Stevens, October 30, 1876, in LeConte Family Papers, APS.
28 Joseph LeConte to Matilda "Tillie" Harden Stevens, December 17, 1876, in *ibid.*

"she is deeply interested and shows really remarkable acuteness of observation," said the adoring father. Shortly thereafter, he purchased a microscope for his studious teenager.[29]

In 1877 LeConte had to forego his usual lengthy trip to the mountains, except for a brief excursion with Lieutenant G. M. Wheeler of the U.S. Army Engineers, which he considered an excellent opportunity for improving his practical knowledge of geology. Because of the detailed and specialized engravings to be included in his textbook on geology, LeConte had to spend most of the summer in New York overseeing the production of the numerous illustrations. It was a burdensome task, and at one point he wrote to Othniel C. Marsh that "the strain upon me has been such that I have hardly been myself." But the special attention LeConte gave to his textbook paid off, and his *Elements of Geology* quickly received favorable notice. Bessie wrote to Emma that "everyone seems delighted with your Father's book, it has been splendidly reviewed, and so far not an instance of adverse criticism. I think it will be obliged to become a leading authority on Geology." She also predicted it would make its way as "a paying book as soon [as] it can be generally introduced in colleges." Her estimation ultimately proved to be correct, and LeConte's book became the primary introduction to the field for a generation of geologists and general students alike. Joseph worried, however, because he believed the work might not be so well received in his native region. In October, 1878, he asked Emma whether she thought the book was "too radical in its views on evolution. I have sometimes suspected it. The South is singularly conservative in every respect."[30]

Containing well over six hundred pages and upwards of nine hundred figures and woodcut illustrations, LeConte's *Elements of Geology* provided a comprehensive introduction to the discipline. As he

29 Joseph LeConte to Sallie LeConte, December 31, 1876, in Davis Papers. Joseph and Bessie were unable to attend Sallie's wedding in South Carolina (Joseph LeConte to Emma LeConte Furman, December 29, 1876, in LeConte-Furman Papers, SHC/UNC). At the precise time of the wedding, Joseph took his watch from his pocket and announced to his family at his home in Berkeley that the wedding ceremony was under way (Bessie LeConte to R. Means Davis, January 28 [1877], in Davis Papers).

30 LeConte, *Autobiography*, 256; LeConte, Autobiographical MS.; Joseph LeConte to Emma LeConte Furman, April 28, 1877, and October 27, 1878, both in LeConte-Furman Papers, SHC/UNC; Joseph LeConte to O. C. Marsh, July 29 and August 5, 1877, both in O. C. Marsh Papers, Yale University Library, New Haven; Bessie LeConte to Emma LeConte Furman, April 7, 1878, in LeConte-Furman Papers, GHS.

noted in the preface, the author did not intend to offer "an exhaustive *manual* to be thumbed by the special student" but instead to present "a suitable text-book for the higher classes of our colleges" and a volume useful "to the intelligent general reader." Dana's *Manual of Geology*, he readily conceded, ably served the special student, but the current textbooks on geology, he argued, were "either too elementary ... or else adapted as manuals for the specialists." Sir Charles Lyell's three-volume work on the subject he admired but found deficient in its "treatment of *American* geology" and wanting in its organization, merging as it did dynamic with structural and historical geology. LeConte thus drafted a work that he believed would supply "a distinct want."[31]

LeConte recognized the difficulty of organizing the content in a general geology textbook, but his decision to structure the work into three major parts ultimately proved to be both sound and useful, at least according to the contemporary state of knowledge in the discipline. In Part One he covered "Dynamical Geology," including chapters on atmospheric, aqueous, igneous, and organic agencies. The first of these was very brief, but all three of the latter contained lengthy sections that served amply to explain and illustrate erosive forces, igneous activities, crustal deformations, and chemical reactions at work in the earth's surface. In Part Two he treated the topic of "Structural Geology," including sections on sedimentary, igneous, and metamorphic rocks. In addition, he discussed at length the nature of mineral veins and mountain origins, which were, of course, two of his more special interests. On the whole, however, the second part of his *Elements* received far less space than the first and third parts, although he did expand it in the fourth edition as his knowledge of the subject increased. The great bulk of illustrations in this part he drew from the American geological scene, thus satisfying his intention to emphasize the study of the physical structure of his own nation. This was on the one hand a useful instructional device, but on the other, it tended to distort the value of the work as a general, elementary textbook. The choice thus reflected in part a parochial point of view, and it represented some of the limitations of LeConte's knowledge of the universal study of geology.[32]

---

31 LeConte, *Elements of Geology* (New York, 1877), iv–vi.
32 *Ibid.*

Part Three of *Elements* received by far the greatest amount of attention, covering over 50 percent of the space in the book. In this part on "Historical Geology," LeConte devoted considerable attention to geological and organic evolution, and he included over 650 figures and illustrations, the great majority of which depicted fossilized flora and fauna. Such a disproportionate emphasis might well have been expected in light of his evolutionary views. In fact, in his preface, LeConte freely admitted that he made "*evolution* the central idea" of the part, viewing it as "a thread running through the whole history, often very slender—sometimes, indeed, invisible—but reappearing from time to time to give consistency and meaning to the history." Nevertheless, in the first two editions of *Elements*, LeConte avoided the use of the word "evolution" in the introduction to Part Three, bringing it in at a much later point. Instead, the prolegomenon to this all-important part was couched in terms of "insensible graduations" of eras and life forms into modified structures. By the time of the third edition, however, the author had cast aside his reticence and begun the initial sentence of the third part with the concept of geology as "the history of the evolution of earth-forms and earth-inhabitants." His reputation was so sufficiently secure and his commitment to evolutionary theory so firmly fixed by then that he could eschew timid caution. After all, it mattered little at that point if southern colleges elected not to adopt his *Elements* for their geology classes.[33]

As Ernest Hilgard claimed, LeConte's *Elements* was "probably the most widely used text-book of geology in the English language." That judgment was confirmed by other geologists, among them Charles Schuchert, John C. Merriam, and Herman L. Fairchild. Fairchild was certainly in a position to know, for he revised the work after LeConte's death. Although it was used primarily in the United States, *Elements* also gained recognition abroad, and the Scottish geologist Sir Archibald Geikie cited it as one of the "more important" textbooks still available in 1903. Moreover, Geikie's colleague Joseph Prestwich frequently referred to the book in the second volume of his own *Geology*. Thus, from its original publication in 1877 through its fifth revision by Fairchild after LeConte's death, *Elements* received wide acclaim, and it continued to be used until the mid-1920s.[34]

33 *Ibid.*
34 Hilgard, "Memoir of Joseph LeConte," 177; Charles Schuchert, "Lower Devonic As-

Shortly after the initial publication of *Elements*, journal reviewers began to praise the work. A writer for the *Popular Science Monthly*, for example, declared that it was the first "good college text-book of geology" for American colleges. He added that LeConte had produced "a volume of great value as an exposition of the subject," and he commended the book "without qualification to all who desire an intelligent acquaintance with the science, as fresh, lucid, full, authentic, the result of enthusiastic study and of long experience in the art of teaching mature classes." The *Nation* also liked LeConte's book, noting that "nearly all that can be said in favor of a text-book may be said of this" and claiming "it is up to the level of Dana's great manual." Although the reviewer for the *Nation* criticized the work for giving "a little too much affirmation" to debatable topics and for omitting "a good many important opinions," he asserted that these "slight defects" were compensated overall by the masterful coverage of topics not included in other geology texts.[35]

When the second edition of *Elements* appeared in 1882, a reviewer for the *American Naturalist* praised the revised work for its "well-proportioned" presentation and generally up-to-date findings. The book also received the highest commendation from Major John W. Powell, who cited it along with Dana's *Manual* and Asa Gray's famous work on botany as "unequalled" textbooks. After the third edition of *Elements* was published in 1891, the *American Geologist* wrote that the revised edition "of a widely used text-book will be welcomed by multitudes of teachers and students," and it hailed LeConte for including many "important additions" and for rewriting sections to incorporate the latest findings in geology.[36]

pect of the Lower Helderberg and Oriskany Formations," *Bulletin of the Geological Society of America*, XI (May 10, 1900), 242; John C. Merriam, *Published Papers of John Campbell Merriam* (4 vols.; Washington, D.C., 1938), IV, 2005–2006; Herman L. Fairchild, "Memoir of Joseph LeConte," *Bulletin of the Geological Society of America*, XXVI (1915), 47–57; Herbert E. Gregory, "Steps of Progress in the Interpretation of Land Forms," in *A Century of Science in America*, ed. Edward S. Dana (New Haven, Conn., 1918), 143; Archibald Geikie, *Textbook of Geology* (4th ed.; 2 vols.; London, 1903), I, 7; Joseph Prestwich, *Geology: Chemical, Physical and Stratigraphical* (2 vols.; Oxford, 1886–88), II, 141, 310, 487, 539.

35 *Popular Science Monthly*, XII (February, 1878), 501–503; *Nation*, June 20, 1878, pp. 408–409.

36 *American Naturalist*, XVII (April, 1883), 395–96; *ibid.*, XVII (October, 1883), 1095; *Dial*, V (February, 1885), 273; *Popular Science Monthly*, XXVI (January, 1885), 413–14; *American Geologist*, VII (April, 1891), 260–61; *Polybiblion* (August, 1888), 112.

A similar commendation came from the *American Geologist* reviewer in 1896 when the fourth edition of *Elements* appeared. Said the reviewer, N. H. Winchell, a leading American geologist: "The well-known author of this popular work has rendered geology, and especially American geology, a service by revising it. . . . It is to-day probably the best known of American geological text-books, and the improvements impressed on this edition will tend to keep it in the lecture rooms and libraries of most American geologists." Winchell offered some criticisms, however, one of which was what he considered Le-Conte's overuse of "graphic illustrations." More telling perhaps was his argument that LeConte should have written an entirely new book because the framework of the original was no longer entirely adequate to treat the newer dimensions of geology. He correctly noted that LeConte continued to view his subject in much the same light that he had twenty years before, and, as Winchell observed, he failed to comprehend "some of the important fields of recent research in American geology." But, despite his pointed and valid objections, Winchell concluded that the "author writes interestingly and forcibly on many general topics. That is one of the excellencies of the volume."[37]

Grove Karl Gilbert made similar comments upon the fourth edition of *Elements*. Reviewing it in *Science*, Gilbert said, "For nearly twenty years LeConte's *Elements of Geology* has stood side by side with Dana's Manual in the working libraries of American geologists and teachers." Then he added that because of the "enviable position" it held, *Elements* needed "neither introduction, encomium nor criticism." Gilbert applauded LeConte for his "wise choice of material" and his ability to explain the principles of geology without obscuring them "by clouds of detail." In addition, Gilbert praised the author for his use of "alternative theories" to explore certain difficult problems in geology and for his emphasis on methodological inquiry as opposed to mere memorization of material. Yet he objected to LeConte's treatment of some debatable topics from a "personal standpoint," calling these the "least convincing and satisfactory" aspects of *Elements*. Although Gilbert cited some "flaws" in LeConte's treatise—most of which were simply attributable to the old textbook writer's inability to keep abreast of all current findings in the rapidly expanding field of geology—he nevertheless concluded that "these and other blemishes

37 *American Geologist*, XVIII (December, 1896), 384–86.

may be freely forgiven to a book that sets forth the broad generalizations and fundamental principles of its particular science in orderly
and attractive form, and at the same time illustrates and embodies the
true and essential spirit of all science."[38] Over the years, LeConte's geology textbook won the general esteem of scientists, and it played a
long and influential role in the education of literally hundreds of college students.

With the favorable reviews of the original edition of *Elements* spurring
him on to further geological investigation, LeConte soon planned another excursion into the Sierras. Typically, he combined his field work
with personal recreation. In the summer of 1878, he struck off for the
mountains in the company of Bessie, Carrie, and several acquaintances. The group camped in Yosemite and the Calaveras Grove in the
"Big Trees" for almost five weeks in June and early July. A year later, he
took Bessie and Joe along with him for an extended excursion into
Oregon, Washington, and British Columbia. This was, of course, the
country that Joseph had admired so much on his previous trip a half-
dozen years before. He also utilized his time for further geological
investigations, recording information and sketching various phenomena for future study. These recreation-research trips added much
to LeConte's knowledge of geology, but they were rarely long enough
to qualify him as a genuine field student. His work in the field never
compared in intensity with that of many of his peers in geology, and
he always remained much more of a theoretical than a field geologist.[39]

In the summer of 1880, LeConte had to go to New York again, this
time to work with the publisher of his *Sight*. When his task was finished, LeConte entrained for South Carolina to visit Sallie and there to

38 *Science*, n.s., IV (October 23, 1896), 620–21; *National Geographic*, XIV (November,
1903), 425–26; *Engineering and Mining Journal*, LXXII (July 13, 1901), 34; *American Journal
of Science*, 4th ser., XII (October, 1901), 248.

39 LeConte, *Autobiography*, 267–69; Joseph N. LeConte, "Recollections"; Caroline
"Carrie" LeConte's notebook "Yosemite 1878," in LeConte Family Papers, UCB; Carrie
LeConte to Emma LeConte Furman, June 23 and October [n.d.], 1878, both in LeConte-
Furman Papers, SHC/UNC. Earlier in the summer, Joseph also traveled to Nevada City
(June 12–20); then in October he took a trip to Sulphur Banks in central California to
study "sulphur and cinnabar works" and old lava flows. Along with the Oregon and
Washington trips, these are recorded in the notebooks of Joseph LeConte, in LeConte
Family Papers, UCB.

meet Bessie and the children, who had arrived much earlier. Several weeks later, the LeContes went on to Georgia for a month's visit with Emma. In Joseph's words, it was "a delightful month" in which he not only enjoyed swimming but also taught the skill to his young children. During the following summer, however, LeConte stayed at home to direct the construction of a house. It would be the first he had owned since he had built one in Macon back in 1849. Located on Bancroft Way, the ten-room, two-story house cost Joseph $4,800. By now he had accumulated some savings, and the old penurious days of Reconstruction were all but forgotten. He was admittedly a Californian, and, except for toying a bit with the idea of moving to Virginia or to Texas, he was obviously committed to ending his career in Berkeley.[40]

As the summer of 1882 approached, LeConte planned a special trip to Yosemite for the benefit of Emma and Cousin John Lawrence and his son John, or "Jack" as he was commonly called. Jack was suffering from what Joseph diagnosed as a collapsed lung, but the symptoms indicate he had a bad case of tuberculosis. His father had sent him to California much earlier to recover his health, but the trip had done little good. Thus, upon Joseph's recommendation, John L. decided to take the young man on the excursion into Yosemite. Emma was likewise again in bad health, and after it became apparent that Joseph would not return to the East Coast, she once more urged her husband to move to California. Although Farish entertained the idea, he could not be budged. Georgia was his home, and he would stay there until he died. Emma's only recourse therefore was another visit with her parents, and she came out early in the summer.[41]

While he was in Yosemite, LeConte received word of the discovery of "wonderful footprints of man and animals . . . in the prison-yard at Carson Nevada." He desired to see these for himself, and as soon as he got Emma and Carrie back to Berkeley, he set off in late July to make

40 LeConte, *Autobiography*, 270–71; LeConte, Autobiographical MS.; Joseph LeConte to John Lawrence LeConte, June 19, 1880, in LeConte Family Papers, APS; Joseph LeConte to Emma LeConte Furman, September 17, 1880, in LeConte-Furman Papers, SHC/UNC.

41 LeConte, Autobiographical MS.; Joseph LeConte to Farish Furman, December 6 and 31, 1880, and Joseph to Emma LeConte Furman, October 27, 1878, January 18, March 4, and September 29, 1881, all in LeConte-Furman Papers, SHC/UNC; Emma LeConte Furman to Farish Furman, November 13, 1881, and February 16, 1882, in LeConte-Furman Papers, GHS; Emma LeConte Furman's journal of a camping trip in Yosemite, 1882, in McMillan Collection.

a firsthand observation. In the company of his colleague Professor
W. B. Rising, LeConte went first to Steamboat Springs, because he was
interested in the formation of mineral veins, evidence of which was
prominent in the geyser-riddled area. This was a topic in which Le-
Conte had shown increasing interest in recent years. In fact, he had
visited a number of mines over the past decade, and within a year he
would present one paper on metalliferous veins to the National Academy of Sciences and publish another paper on the topic in the *American Journal of Science*.[42]

On July 22 LeConte arrived at the Carson prison, and for two days
he examined the numerous tracks that had been uncovered in digging
operations by the prisoners. After returning to Berkeley, LeConte read
a paper on the Carson footprints before the California Academy of
Sciences in August. Unfortunately, the academy did not publish the
paper for several months, and in the meantime the paleontologist
O. C. Marsh visited the Carson site and published his findings. Prior
claim should have gone to LeConte: he certainly visited the site before
Marsh, and he delivered his paper to the CAS before Marsh published
his own. But LeConte held no animosity toward the esteemed paleontologist, whose work he greatly respected and many of whose illustrations he used in the historical geology section of his *Elements of Geology*. Nevertheless, LeConte was somewhat perturbed over the delay
of the CAS in getting his paper published in their proceedings. In fact,
before it came out, he decided to submit it to *Nature*, which published
the article in May, 1883. The tracks were eventually determined to be
those of a ground sloth, one of the views that LeConte had favored. He
did not accept it over the "large-bear theory," however, until Marsh
provided more convincing evidence.[43] Nonetheless, LeConte had once
more demonstrated his wide-ranging versatility.

42 LeConte, *Autobiography*, 272–75. The papers he wrote were "The Phenomena of
Metalliferous Vein-formation Now in Progress at Sulphur Bank, California," *American
Journal of Science*, 3rd ser., XXIV (July, 1882), 23–33 (coauthored with W. B. Rising); "On
Mineral Vein Formation now in Progress at Steamboat Springs Compared with the Same
at Sulphur Bank," *ibid.*, 3rd ser., XXV (June, 1883), 424–28; "On the Genesis of Metalliferous
Veins," *ibid.*, 3rd ser., XXVI (July, 1883), 1–19. The last was read before the NAS on April 17,
1883.

43 Joseph LeConte, notebook labeled "Carson Footprints," in LeConte Family Papers,
UCB; Joseph LeConte, "On Certain Remarkable Tracks, Found in the Rocks of Carson
Quarry," *Proceedings of the California Academy of Sciences*, August 27, 1882, pp. 1–10 &
plates; Joseph LeConte, "Carson Footprints," *Nature*, XXVIII (May 31, 1883), 101–102, and

LeConte continued his research and writing at a pace exceeded by only a handful of scientists, and certainly by few of his own age. In 1883 he was sixty years old, but LeConte seemed to ignore the encroachments of time upon his life. He kept up his activities without a hitch, served as chairman of the local Darwin memorial fund, and continued to correspond with fellow scientists, despite the great tragedy of losing all in the same year his young second cousin Jack LeConte, his cousin John L., and his son-in-law Farish. LeConte had always admired his former student in spite of the problems between Emma and her husband, and he was greatly distressed over the young man's death, the moreso perhaps because it threw Emma entirely on her own. He need not have worried, however, for Emma proved to be a resourceful person. Indeed, she continued to run the Scottsboro farm successfully until the end of the century.[44]

But Joseph LeConte was no stranger to tragedy, and he had long since learned to shake off its oppressive effects and look ahead to better things. Both his *Sight* and his *Elements of Geology* were doing well. He had already revised the latter (in 1882), and the publisher was calling for yet another edition. Moreover, he was now condensing and modifying the work as a high school textbook, which was due off the press in 1884. Also there were still many papers he planned to write in geology and on other topics that had greatly interested him for a long time. The busy scientist was also contemplating a comprehensive treatment of evolutionary theory and religion—a subject that was to bring him even greater recognition and to have a powerful impact upon both the popular and the academic mind.

Independent, XXXIV (September 28, 1882), 7–9, and XXXV (December 13, 1883), 6–7; LeConte to O. C. Marsh, October 27, 1882, in Marsh Papers; O. C. Marsh, "On the Supposed Human Footprints Recently Found in Nevada," *American Journal of Science*, 3rd ser., XXVI (August, 1883), 139–40.

44 Joseph LeConte to Alexander Agassiz, January 4, 1883, in Agassiz Papers. LeConte also worked hard to secure a good mineral collection for the University of California; see his letters to James Hall, May 9, 1882, and February 20 and March 5, 1883, all in Hall Papers. Joseph had written to his cousin John Lawrence LeConte on October 7, 1882, stating that, in spite of Jack's "solidified" right lung, the young man's general health appeared good (in John Lawrence LeConte Manuscripts, American Entomological Society Papers, MS E76, Academy of Natural Sciences of Philadelphia). Furman died from what was described as an attack of malarial fever. On December 15, 1883, Joseph sent Hugh Colquitt a strong appraisal of Furman's ability for a memorial booklet (in McMillan Collection). See also Mrs. Furman's ledger books, in LeConte-Furman Papers, GHS.

—◆◆≤ IX ≥◆◆—
The Roads to Religion and Evolutionism Converge

Soon after the appearance of Charles Darwin's *Origin of Species* in 1859, Asa Gray published a favorable review of the work in the *American Journal of Science*. In all likelihood, LeConte read Gray's review, and it is certain that he soon purchased a copy of the book that would ultimately have a profound impact upon science—and indeed the whole world of thought. Yet, at least initially, the path-breaking work made no striking impression upon LeConte. But the same was true for many of his contemporaries as well. The American response to the theory of evolution came slowly at first, but eventually it accelerated with the force of a freshly loosened avalanche.[1]

Louis Agassiz stood before the snow-tide of the theory at the outset and tried, vainly it proved, to hold it back with the feeble strength of his argument for the creation of new species by special acts of divine intelligence. He warned Gray that Darwin's theory must be stopped before it engulfed science with a materialistic conception of life. Gray was opposed to Agassiz's "philosophical idealism," however, and he found no disturbing implications for theism in the theory of natural selection. On the other hand, he was bothered by the absence of a theistic interpretation in Darwin's carefully constructed argument, and subsequently he expressed alarm over "the implications of natural selection for religious philosophy."[2]

1 Paul F. Boller, Jr., *American Thought in Transition: The Impact of Evolutionary Naturalism, 1865–1900* (Chicago, 1969), 4–7; A. Hunter Dupree, *Asa Gray, 1810–1888* (Cambridge, Mass., 1959), 244–47; Bessie LeConte to Emma LeConte Furman, August 5, 1866, in Talley Papers.

2 Boller, *American Thought in Transition*, 6–69; Dupree, *Asa Gray*, 264–383; Lurie, *Louis Agassiz*, 259–302; Edward J. Pfeifer, "United States," in *The Comparative Reception of Darwinism*, ed. Thomas F. Glick (Austin, 1974), 168–206.

The concept of the theory of evolution was not new, of course; it had been introduced by earlier thinkers. Most scientists were already familiar with the ideas of Jean Baptiste Lamarck, and undoubtedly many of them, including LeConte, had read Chambers' *Vestiges of the Natural History of Creation* and the works of the English philosopher Herbert Spencer, who popularized Lamarck's ideas. As James R. Moore has so aptly summarized Lamarck's views, the French naturalist developed a transmutational theory based upon "two causal factors," the first of which was "an innate complicating power" bestowed by God on species, and the second, "an inner adapting disposition." The "innate power . . . tends to produce a series of plants and a series of animals, each of which shows an orderly progression in complexity and perfection among its major taxonomic groups," while the "inner disposition" of all species "assures the performance of actions sufficient to meet new needs . . . created by a changing environment." In time, such "actions" develop into habits, and eventually these habits evolve into instincts and thus lead to "organic change." Such changes are passed on to the progeny of the species.[3]

But Darwin added something new to Lamarck's theory of transmutation by posing that evolution occurred by natural and sexual selection and accidental variations among species. He had originally accepted the Lamarckian theory that species evolved through inherited characteristics acquired in an adaptation to their environment and according to the law of use and disuse of organs. In general, however, Darwin "was sharply critical of Lamarckian evolution," although, as Moore has indicated, "he seems never to have understood it well." By the time Darwin had published the sixth edition of *Origin of Species*, he had, in the face of an onslaught of criticisms against his theory of natural selection, given greater credence to the Lamarckian explanation, yet he continued "to believe that the heritable effects of the environment and of use and disuse either supported or were subsidiary to natural selection." In effect, Darwin "assimilated the objections of his critics"—a response that resulted in a number of "striking inconsistencies" in the structure of his theory.[4]

Meanwhile, some opponents of Darwinism insisted that the theory

3 Daniels, *American Science in the Age of Jackson*, 57–62; Richard Hofstadter, *Social Darwinism in American Thought* (Rev. ed.; New York, 1955), 13–50; Pfeifer, "United States," 168–206; James R. Moore, *The Post-Darwinian Controversies* (Cambridge, 1979), 142–43.
4 Moore, *Post-Darwinian Controversies*, 136–52.

represented a serious materialistic threat to traditional religion. Among them was Agassiz, who acted as the leading spokesman for these critics. Asa Gray, on the other hand, maintained his defense of the Englishman's provocative theory and published several popular articles in support of it. Nevertheless, this esteemed botanist embraced a theistic philosophy, arguing that natural selection functioned according to divine purpose. In short, Gray took a teleological stance, holding that there was "design in nature." His position thus avoided the materialistic interpretation of life and made it possible for one to remain theistic while supporting evolutionary theory. Briefly stated, Gray argued that God, or a Supreme Intelligence, had worked out a plan for the universe through natural processes. At least one critic has termed the idea "supernatural selection," but what Gray advocated has been more appropriately designated as "supernatural variation."[5] Gray therefore smoothed the way for the acceptance of the transmutational theory by many American scientists and other intellectuals. Yet the theistic interpretation of the theory left a welter of unsolved problems about the role of faith and prayer for divine intervention in human affairs, and it spun a web of questions regarding evil, the biblical story of creation, miracles, and immortality.

LeConte's initial interest in evolution had germinated in the Lamarckian theory, and it had grown under the Agassizian view that each species was individually created—that is, that the differences among species occurred through cycles of new creations. Then, some time after learning of the theory of natural selection, LeConte started to work out a reconciliation of the opposing views, beginning with his law-of-circulation explanation in the late 1860s. Yet he was still basically a Lamarckian in his idea of the accumulative effects of acquired characteristics and the transmission of infinite residua from one generation of species to the next. But, like all Lamarckians, LeConte accepted the transmutational theory, firmly believing that a species could be changed completely from the structural and organic forms of its earliest predecessors after the steady accumulation of a sufficient amount of tiny residual characteristics resulted in a total transformation.[6]

5 Boller, *American Thought in Transition*, 9–11; Dupree, *Asa Gray*, 277–78 and *passim*. I am indebted to an anonymous reviewer for the suggestion of "supernatural variation."
6 LeConte, "Natural Law of Circulation," 54–67.

Thus, by 1870 LeConte had moved far away from the theory of Agassiz, but his complete conversion to evolutionism did not come until a few years later. Once convinced of the validity of the theory of evolution, he never looked back. In fact, he subsequently declared that the theory was "the grandest of modern ideas," and he applied it consistently to his interpretation of virtually every facet of thought. The process of converting to evolutionary theory was long and arduous for Joseph LeConte, however, for, as a deeply religious man, he had to work out an acceptable reconciliation between what appeared to him to be serious conflicts. The position taken by Gray helped to facilitate his conversion, but it was made all the more easy by his liberal religious views. He had, of course, already endeavored through his *Religion and Science* to show the compatibility of revelation and science, but he found the next step more difficult, for the theory of the transmutation of species raised serious ontological questions.[7]

Over the years LeConte had grown more critical of organized religion and correspondingly inclined toward "natural" religion. He never rejected the former, but he increasingly questioned its tenets as he encountered conflicts between the geological and biblical interpretations of the age of the earth and between reason and faith. By the time he published his *Religion and Science*, he was already saying that the language of the Bible—for instance, the "Mosaic cosmogony" or the Genesis account of creation—had to be taken as allegorical. Science did not supplant but rather supplemented scriptural truth, and reason did not overthrow the authority of the Divine Word but merely extended its validity through more rational forms. The laws of nature therefore complemented the statements of revelation, and whenever the serious inquirer examined the two forms of truth he would eventually discover their compatibility. Organized religion, said LeConte, tended to resist change, to cling tenaciously to traditional beliefs, thereby petrifying into an institution that could no longer grow or receive *new* truths.[8]

7 Joseph LeConte, "Relation of Biology to Sociology," *Berkeleyan*, XXIII (1887), 123. Earlier he had said, "Evolution is certainly the grandest idea of modern science, embracing alike every department of nature" (Joseph LeConte, "Man's Place in Nature," *Princeton Review*, 4th ser., II [November, 1878], 776).

8 LeConte, *Religion and Science*, 256–65. In his later books on geology and evolution, LeConte virtually ignored the "Mosaic" record and its disparity with geological accounts. Although that subject had in the 1830s and 1840s served as a source of great contention,

LeConte had, of course, already indicated his acceptance of the theory of evolution in his memorial address on Agassiz in late 1873, and in the same year he republished his old paper on the correlation of vital with chemical and physical forces, with a significant revision of the concluding paragraph. Initially, he had ended the paper with a statement that "the question of the origination of species ... is utterly beyond the limits of science." In its place he substituted a weak warning against taking anything he had said as an acceptance of "a pure materialism." By 1874, however, he was certainly calling himself an evolutionist in his classroom. In his lecture on instinct to his class in comparative physiology, for example, he clearly revealed that he belonged to the group called "evolutionists."[9]

In the published version of that lecture, titled "Instinct and Intelligence," LeConte maintained that scholars had created a fallacious dichotomy between physiology and psychology, some arguing that learning was purely a matter of brain chemistry, and others, that it was solely a function of experience. For LeConte, such a division was wrong, and he sought to reconcile the two theories by maintaining that physiological functions of the brain operate fullest in lower forms of life, while psychological functions ascend in the higher forms. The latter, or "mental forces," are derived from the lower "cosmic energy," or "vital forces," but they go beyond them. "In a word," he said, "intelligent conduct is *self-determined* and becomes wise by individual experience. Instinctive conduct is *predetermined in wisdom by brain-structure*." Thus the higher animals are free to choose their actions,

it gradually subsided, and LeConte apparently saw no reason to deal further with a matter which, among many scientists at least, had to be discarded in the face of geological discoveries. Thus, his statement in *Religion and Science* continued to serve as his explanation throughout the remaining years of his life. On the nature of the controversy over the conflict between the Mosaic record and the geological account of the early stages of the earth's history, see Herbert Hovenkamp, *Science and Religion in America, 1800–1860* (Philadelphia, 1978), especially pp. 119–45.

   9 Joseph LeConte to Sallie LeConte, January 20, 1875, in Davis Papers; Joseph LeConte to Emma LeConte Furman, December 29, 1876, in LeConte-Furman Papers, SHC/UNC; LeConte, notebook, "Oregon 1873 & L. Tahoe/74," in LeConte Family Papers, UCB; Joseph LeConte to Matilda "Tillie" Harden Stevens, December 11, 1878, and January 18 and March 24, 1879, all in LeConte Family Papers, APS; Edward J. Pfeifer, "The Genesis of American Neo-Lamarckism," *Isis*, LVI (Summer, 1965), 156–67; Robert Scoon, "The Rise and Impact of Evolutionary Ideas," in *Evolutionary Thought in America*, ed. Stow Persons (New Haven, Conn., 1950), 4–42; George W. Stocking, Jr., *Race, Culture, and Evolution* (New York, 1968), 234–69; Joseph LeConte, "Correlation of Vital with Chemical and Physical Forces," in *The Conservation of Energy*, ed. Balfour Stewart (New York, 1874), 201.

although "liable to mistakes and stumblings and hurtful falls," whereas the lower forms of life function largely by "automatic" responses or "like the motion of an engine laid upon a track which bears it swiftly and surely to its destined goal."[10]

Critical of "the *old* theology" for disposing of the question of the origin of instincts "in the most summary way," LeConte rejected the notion that instincts "are miraculously given in perfection to the first individuals of the species, to each species its several kind." From that point he proceeded to show how instincts developed in evolutionary fashion among species. This he did on the basis of his interpretation of the Lamarckian theory of evolution. As each form of life engaged in a trial-and-error pursuit of some activity, such as food-gathering, locomotion, and nesting, it gradually accumulated fixed habits that proved successful. The residua of these habits were passed on to generation after generation until the functions became automatic. In other words, these characteristics had become "permanent" or "*petrified in brain-structure.*" For higher animals the implications were clear: the repetition of a "volitional act" would eventually become a fixed characteristic.[11]

This interpretation bordered upon materialism, of course, because it ostensibly effaced the role of divine intelligence in the creation of new forms of life, leaving the process of evolution instead to the voluntary action of organic forms. Yet LeConte as a theist adamantly insisted that life emanated from a Cosmic Being. Once life had been created, however, it evolved on its own. Certainly, this raised troublesome questions about salvation, sin, and divine intervention, but LeConte was not yet ready to deal with all of the ramifications of his interpretation of the transmutation of species. Nevertheless, he was clearly now an evolutionist, and the theological implications would have to be worked out in his own mind over a period of time.

The question of biological evolution now partially resolved, LeConte turned to the subject of geological transformation. Naturally, the kind of evolutionary process he advocated would take so many eons to develop that the age of the earth itself must be pushed back by millions of years, making it necessary for LeConte to explain away the

10 Joseph LeConte, "Instinct and Intelligence," *Popular Science Monthly*, VII (October, 1875), 653–64.

11 *Ibid.*

creation of the world in six days. This he had to do by arguing that
the days of the Genesis account were not to be reckoned as man measures time. Even then, the slow process of accumulating residua
would make it necessary to accept an almost unfathomable conception of time. The same applied to the epochs of the earth's history. To
get around the problem, LeConte adopted the theory of "critical periods in the history of the earth," which became the title of an article
published first in 1877 and revised and republished again in 1895.

Long before the nineteenth century, some geologists had recognized that the earth had gone through a number of epochal changes.
But these early investigators argued that such changes occurred cata-
strophically. Later thinkers, like the Scotsman James Hutton, argued
for a uniformitarian view—that is, that changes in the form of the
earth occurred slowly, gradually, and uniformly over the ages.[12] Typically, LeConte avoided a hard-and-fast acceptance of either of these
opposing theories, striving instead to work out a sensible reconciliation. In the process, he selected from each theory to suit his purposes,
and although the practice of choosing what one will from extreme
theories is fraught with dangers, it proved in this case to move Le-
Conte to a more plausible explanation of epochal changes.

In general, LeConte argued that the earth's transformation had
come about "at a *uniform rate* of evolution," but he believed there were
obvious cases of "unconformity" in which the earth's surface strata
bore no record of fossils, while there were other cases in which a
multitude of new forms evolved. This could be explained only by accepting that both organic forms and the earth's surface sometimes
experienced "periods of *rapid movement*" and at other times "periods
of *comparative repose*." LeConte referred to the former as "paroxysmal" development.[13] All of this came before the discovery of muta-
tional changes, of course, and LeConte had to find an explanation to
satisfy the flaw in the Lamarckian theory. He had done so in his own
mind by showing how evolution was effected through the ebb and

12 Edward Battersby Bailey, *James Hutton: The Founder of Modern Geology* (London,
1967); Karl von Zittel, *History of Geology and Palaeontology* (London, 1901; reprint ed.,
New York, 1962), 298–304.
13 Joseph LeConte, "On Critical Periods in the History of the Earth and Their Relation
to Evolution; and on the Quaternary as Such a Period," *American Journal of Science*, 3rd
ser., XIV (August, 1877), 99–114.

flow of a changing environment in which the pressures were some-times enormous and at other times gradual and uniform.

Although LeConte had become an evolutionist by the mid-1870s, he had yet to address himself directly to the question of the relationship between religion and the theory of evolution, only hinting at it in his first forays into the field of organic and geological change. His first serious attempt to face the question came in an address before the Chit-Chat Club in 1877. To the San Francisco equivalent of the Berke-ley Club, LeConte gave a candid statement of his position:

First of all I wish frankly to acknowledge that I am myself an evolutionist. I may not agree with most [evolutionists] that evolution advances always *cum aequo pede*. On the contrary, I believe that there have been periods of slow and periods of rapid, almost paroxysmal, evolution. I may not agree with most [that] we already have in *Darwinism*, the final form, and in *survival of the fit-test*, the prime factor of evolution. On the contrary, I believe that the most important factors of evolution are unknown.... Nevertheless, evolution is a grand fact ... and the origin of species by derivation must be regarded as an established truth of science.

At last, LeConte was a full-fledged evolutionist. Certainly, he was not a Darwinian; for, like many of his fellow American scientists who were "Neo-Lamarckians," he rejected the idea of natural selection as the major force of organic change. But he was now forthrightly an avowed evolutionist, and evolution was now for him a *fact*, not merely a theory.[14]

LeConte's simple address was designed only to show that the idea of evolution did not conflict with the idea of God-As-Creator, not to attempt an extended discussion of evolution or of theology. Indeed, the new convert merely assured his audience that "evolution is *one* thing and materialism *another* and quite a different thing," pointing out that the former is "a sure *result of science*," whereas the latter is "a doubtful *inference of philosophy*." To this he added that he could find no "paradox" between the idea of a Divine Creator and the evo-lution of species by transmutation. Evolution, he said, is "only an *old*

---

14 *Proceedings at the Annual Dinner of the Chit-Chat Club*, November, 1877, copy in Joseph LeConte, Miscellaneous Writings, Bancroft Library Archives, UCB. The title "Evo-lution in Relation to Religion" is written above the address and initialed "JCR." In his "Index Rerum," LeConte listed it as "Evol^n in Rel^n to Materialism." The material quoted is from pages 3–4.

truth in a *new form.*" The old truth was evident in the *"evolution of the individual* by a slow process from a microscopic germ," which nobody could honestly deny. This concept, he concluded, does not interfere "with a belief in an intelligent Maker of each of us," and neither should the theory of organic evolution.[15]

At this point, LeConte turned to his stock-in-trade idea of reconciliation. "In all vexed questions ... there are three views," he said. Two of those are "opposing, partial, one-sided views, and a third, more rational and comprehensive, which combines and reconciles them." He rejected the "old adage that 'truth lies in the middle,'" however, because it contained the "pernicious error" that truth was "a kind of gray or neutral tint." A proper reconciliation of obstensibly opposing truths was not to be found in "a mere *mixture*" but in "a more comprehensive view which combines or reconciles opposing partial views." True reconciliation was "a stereoscopic combination of *two* partial surface-views into *one* objective reality." In the instance of the origin of species, therefore, the true stereoscopic combination was neither *"miraculous* creation" with evolution nor *"evolution-process"* without "Divine agency" but rather *"Divine agency by evolution-process."* New truth, then, does not destroy old truth; the former simply provides a more rational explanation for the latter. In brief, the *"new* must include the *old*—the old must incorporate and assimilate the new, and each must modify and be modified by the other."[16]

As a reconciliationist, LeConte belonged to that school of evolutionists that James R. Moore has appropriately labeled "Christian Darwinisticism." Thus, unlike the "Christian Anti-Darwinist" who rallied round the banner of orthodox theology and "sought certainty through inductive inferences and ... believed in the fixity of biological species," LeConte embraced the doctrine of the Christian Darwinisticists who hastened, primarily through Lamarckian evolutionary theory, to "reinterpret" Darwinism within the framework of Christian teachings.[17] In reality, the reinterpretation worked the other way around, and LeConte and other evolutionists of this persuasion actually began from the premise of the validity of evolutionism, not from the verbal accuracy of the Scriptures. What he and his kind did therefore was to transform their interpretation of Christian doctrine, although in the

15 *Ibid.,* 4–6.
16 *Ibid.,* 6–12.
17 Moore, *Post-Darwinian Controversies,* 193–251.

process they were necessarily forced to select and emphasize those elements of evolutionism that allowed a reconciliation of the conflicting ideas. Since the Darwinian theory stressed natural selection, which clearly made it difficult, at least in the traditional sense, to assign an intercessory role to the Divine Power, LeConte and the Christian Darwinisticists seized upon the Lamarckian theory of environmental adaptations to effect their reconciliation.

Among the serious questions facing the reconciliationists was that of the meaning of evolution for man as an animal. Soon after his declaration of faith in the theory of evolution, LeConte addressed this topic in a paper titled "Man's Place in Nature," which was published in 1878 in the *Princeton Review*. Although LeConte later modified and republished the article, his original statement on man-as-animal remained basically the same throughout the rest of his life. A comprehensive, if not always consistent, blend of LeConte's disparate thoughts, the article represented a systematic philosophical statement in which the concept of theism, the idea of correlation of chemical, physical and vital forces, the theory of organic evolution, and views on the spiritual destiny of man were merged into a teleological whole.[18] LeConte could not forget the old; he was theistic by nature, a physiologist, zoologist, and geologist by training, and an Agassizian by sympathy. Yet he could not ignore the new; he was an evolutionist by genuine conversion. Above all, LeConte was a philosopher at heart, and large questions so challenged his mind that he felt impelled to amble about over the mountains of science, philosophy, and religion until he had discovered sufficient answers to draw a general map of the terrain.

In LeConte's view, man differed from the lower animals in spirit, not in general physiological structure. In that regard, he said that "the gap between man and animals is simply infinite; for the difference is not only one of degree but also of *kind*, and therefore incommensurable in terms of degree." The major problem facing LeConte at this time, therefore, was the relationship of man's spirit "to the *anima* of sentient nature, the *vital force* of living nature, and the *chemical and physical forces* of dead nature." Distinguishing between "matter" and "force," he hypothesized five graded planes of the former, to wit (in ascending

18 LeConte, "Man's Place in Nature," 776–803. See also his "The Psychical Relation of Man to Animals," *Princeton Review*, 4th ser., XIII (May, 1884), 236–61; and "From Animal to Man," *Monist*, VI (April, 1896), 356–81.

order): "chemical elements," "chemical compounds" (the "mineral kingdom"), "plant life" ("the vegetable kingdom"), "sentient life" (the "animal kingdom"), and "moral life" (the "human kingdom").[19]

As he had stated in his earlier paper on the correlation of forces, LeConte again argued that life forms ascended through each of these steps by means of vital force. Now, as an avowed evolutionist, LeConte maintained that this process was achieved by means of "the law of the transmutation and successive elevation of matter and force," but the law operated *"not by infinite gradation* . . . not by sliding scale, but always by *paroxysms."* In other words, the process inevitably resulted in a "series of platforms raised one above another with blank spaces between them." To give this explanation meaning in terms of the origin of higher forms of life, LeConte had to posit a "peculiar condition" necessary to the transformation of matter and force into animal form. This he called *"living matter,"* or a "super-physical condition." It was a pre-existing vital force (or "life-force") that emanated from God and was an "insoluble mystery."[20]

Although already deep into a discussion of the nature of evolutionary processes and their significance for spiritual matters, LeConte had yet to offer a working definition of evolution. He promised to do so in his article, but instead he merely resorted to an explanation of the characteristics of the process. "Every system of related parts," he stated, "may be studied from two distinct points of view, . . . [the] *static* and [the] *dynamic."* The first deals with the mutual functions and relationships *"within* the system," whereas the latter covers the growth and development of forms into more complex structures, and it includes a study of the means of "progressive change in the whole system." However, LeConte was not concerned here with the proof of evolution; he merely contented himself with a few positive assertions to show that the process constituted an established fact and that it could be correlated with spiritual life. What he wanted to prove was that the theory was a grand discovery that opened new avenues to still higher forms of existence.[21]

LeConte regarded the "forces of nature" as "an effluence from the Divine Person—an ever-present and all-pervading divine energy." But

19 LeConte, "Man's Place in Nature," 776–80.
20 *Ibid.,* 780–83.
21 *Ibid.,* 784–86.

this was not to be taken as a deistic philosophy in which God created life and then left it to work out its own course by the energy or forces released. Such a mechanistic philosophy was anathema to LeConte's scheme. "God is ever present and ever working in nature," he declared, and the "laws of nature are naught else but the *regular* modes of operation" of the divine energy. Natural phenomena were thus "the acts" of an omnipresent, omnipotent, and invariable Deity. In the beginning, said LeConte, the "Divine Spirit" had "energized dead inert nature" with "the force of evolution of the cosmos." In due time, the divine energy became more and more "individuated," first through physical and chemical forces, then through the "vital forces of plants or non-sentient living beings." The process increased until individual forms of animal life were created and, further, until "still higher new properties and powers" appeared. Finally, this "progressive individuation of force" resulted in the development of man, and it continued in human beings to the highest forms of consciousness, intelligence, moral sense, and a "true" individual personality. This highest representation of divine energy possessed a "completed kinetic individuality," or what is commonly known as "the spirit of man."[22]

Once the process of individuation was completed in the development of true man, however, a new process must begin. As complete man, the individual is a true image of God. But with all of his free will, intelligence, and freedom of action, man still struggles with his "inherited animality," and, as an individuated personality, he is also separated from God. A new, and final, plane must therefore be achieved before the process of evolution is complete. Man must strive for the ideal—that is, the goal of the "*divine man*." This LeConte viewed as the culmination of the law of the circulation of energy. Every person leaves upon his death all of the force and matter of his body, which is recirculated in the great cycle of the natural world. But every man also carries away with him at death "just so much of the sum of natural energy." Eventually, of course, this lack of "refunding" will exhaust "the bank of natural forces" and bring an end to all material forces. In keeping with the law of conservation of energy, however, the exhaustion of one form is the creation of another; thus, material forms are converted into spiritual.[23]

22 *Ibid.*, 795–96.
23 *Ibid.*, 796–802.

By now LeConte was absorbed with the ramifications of evolutionism for other facets of human activity, and over the next two decades he explored its meaning in terms of racial development, psychology, education, sociology, and other areas of thought. Above all, he devoted considerable attention to its implications for religion, beginning with his "Evolution in Relation to Materialism," published in 1881. In that paper, LeConte greatly expanded upon his theoretical explanation of the direct association of the ideas of evolution and theism. In fact, he now identified himself as a "theistic evolutionist"—an apt term that succinctly captured the character of his reconciliationist philosophy.[24]

Since the theory of evolution had become a focal point of controversy among many intellectuals, LeConte believed it was necessary to clear up some "misconceptions" about the idea. He identified three particular misconceptions, the first of which dealt with the "scope" of the theory. Maintaining that "the popular mind" viewed evolutionism as applicable only to organic forms and, at that, only in the Darwinian conception, LeConte denied that the phenomenon was restricted to "one special department of nature." He thus sought to disabuse his readers of the belief that Darwinism and evolution were strictly "interchangeable terms." A second error he identified as the belief that scientists had acted too hastily in adopting the theory as "an unverified hypothesis." Not so, countered LeConte, for the theory was an excellent "example of inductive caution." The final misconception was, in LeConte's estimation, the old bugbear of equating the theory with materialism. Such misunderstanding was to be expected, he argued, for materialism was part of the current trend that "impregnates the thought and permeates the literary atmosphere of the age." Indeed, he confessed, the movement had nearly swept him off his own feet, but careful reflection had saved him from the fad.[25]

Once more LeConte returned to his belief that extreme theories had to be reconciled and dogmatism avoided. A solution to the "sphinx-enigma" of theism versus evolution could never result from a one-sided view; neither the philosophical idealism of the religionists nor the materialistic conception of the extreme evolutionists or agnostics

---

24 LeConte, notebook, "*1878 McGillivray, 1879 Hughes, 1879 Oregon & Brit. Col.,*" in LeConte Family Papers, UCB; LeConte, untitled notebook in possession of Helen Malcolm LeConte; Joseph LeConte, "Evolution in Relation to Materialism," *Princeton Review,* 4th ser., VII (March, 1881), 149–74.

25 LeConte, "Evolution in Relation to Materialism," 150–54.

constituted a full statement of the truth. Those Christians who believed man was created "directly, miraculously, or without natural process" were taking the view of a small child, whereas the materialists who argued against the creation of life were taking the view of an older but "untrained child." Only the theistic evolutionists embraced a mature, rational view by maintaining that species "were made by a process of evolution." The intermediate position thus correctly combined and reconciled the one-sided explanations. With that said, he launched into a discussion of four areas in which materialism seemed at greatest odds with theism, seeking to show that a reconciled position was not only tenable but in fact correct.[26]

The first concept was that of primary causation. As a conscious being, man observes that there are forces that make things happen. From this state of "fetichism," man advances to "polytheism" and finally to "monotheism." Thus, in the most advanced societies, man conceives a "First Cause," or a Divine Power. This is the position held by religionists and philosophical idealists. Materialists, on the other hand, ascribe force solely to the inner nature of a thing and reject the notion of external causation. But, said LeConte, these opposing views are not really contradictory, for there is a First Cause or Creator who imbues objects with energy or force, and in turn those objects change or modify themselves.[27]

This conception was, in LeConte's mind, tied to a second—that is, to teleology, or to the "argument from Design." If causes result from will, they are carried out according to an intelligent purpose. For the philosophical idealist, God is not only First Cause but also the Master Designer. For the materialist, however, design is inherent in nature itself, and its end cannot be known because there is no purposeful transformation; things simply evolve as they do. LeConte objected to both positions because he felt they were too narrow. He did support the theory of design in nature, but, like several thinkers who had a decade or so earlier begun to accommodate the idea of development in the design argument, he rejected the traditional theistic view. The "evidence of design," he explained, "is not in the *materials*, but in the *use* of them; not in the *parts*, but in the adjustment of parts for a purpose." In brief, LeConte came down more on the side of the reli-

26 *Ibid.*, 157–63.
27 *Ibid.*

gionists in the question of design in nature, much as Asa Gray had done, but he construed the concept to fit the theory of evolution. The theory, he said, is merely a revelation of "the ceaseless activity of Deity" and His *"eternal unfolding* of the original conception." God is "anxious to teach us by revealing His processes to us in proportion to our efforts to learn," he concluded.[28]

If the first two problems of the relation of evolution to theism were aimed primarily at intellectuals, the other two were calculated to reach the popular mind. Although ordinary folk might not comprehend the complex conceptions of causation and teleology, they certainly could understand the more personal notions of immortality and miraculous intervention. Thus, borrowing from the ideas of Immanuel Kant, LeConte maintained that man's sense of external objects is correlated directly with conscious perception of "essential activity or of spirit," and he argued that, since moral activity depends upon a belief in immortality, "the belief must have a corresponding reality." In terms of evolutionary theory, the meaning was clear: man has a higher nature than animals, and just as organic evolution occurs so does moral evolution, leading eventually to the loss of "conscious personality" through progress to a "higher plane" of immortality. The concept of immortality, he insisted, is "not inconsistent with evolution, but is the crowning act, the only rational completion of evolution."[29]

From there, LeConte moved to the related topic of miracles and revelation. The problem for LeConte, of course, was that he had accepted the idea of God as First Cause and Great Designer who worked out His plan through evolutionary processes that functioned, if not independently, at least according to known laws. God was, of course, immanent in nature, but He was impersonal in that form. How, then, could one pray for a child's crippled leg to be healed or for blindness to be cured? LeConte really had no answer, and therefore he simply resorted to the argument that some phenomena "lie beyond" the "domain" of science. The goal of science, he admitted, is to "reduce as much as possible to material laws," but he confessed that the "region of activity of pure spirit" was a "legitimate domain of the miraculous." Any inconsistencies between natural and spiritual laws would eventually be worked out by rational thought and new discoveries. As ex-

28 *Ibid.*, 164–68.
29 *Ibid.*, 168–71.

plained by LeConte, the theory of evolution was itself in the process of evolving, and with each new stage it came closer to a true conception of religion.[30]

In the decade 1873–1883, LeConte published well over a dozen papers, notes, and book reviews dealing directly with the theory of evolution. Encouraged by the reception of these endeavors, he began to think of writing a comprehensive treatise on the subject, especially after 1883 when Henry Ward Beecher urged him to write a book on evolution and religion. LeConte responded with a letter of thanks to Beecher for his "kind expressions" toward his articles on evolution and his "suggestion concerning a popular work on Evolution." Said LeConte, "I have often thought of undertaking such a work." He explained, however, that "the press of other work" and "the seriousness of the undertaking" had thus far prevented him from following up on his plan. On a later occasion, LeConte stated that he "feared the church was not yet ready to be profited by such a book." By 1885, however, the "reluctant evolutionist" had "commenced the work." He claimed that Beecher's letter "determined" him to undertake the task.[31]

The book was not completed and published until 1888, however, for LeConte was extremely busy with other writing tasks, and he had not yet resolved all of the questions in his own mind regarding evolution and its relation to Christianity. Nevertheless, he continued to work on the manuscript, and in 1887 he wrote to the well-known naturalist Alpheus Hyatt that he was "trying to clear up my own views on the subject [of evolution]. . . . Thoughts on this great subject are still seething in my mind but the elements are so numerous & complex that they are slow to take definite form." In spite of his indefiniteness, LeConte had pushed on with his writing, aiming it primarily at a "popular" audience but hoping also to make it "interesting & even profitable to the specialist."[32]

30 *Ibid.*, 171–74.

31 Joseph LeConte to Henry Ward Beecher, October 16, 1883, in Beecher Family Papers, Yale University Library, New Haven. In his *Autobiography* (pp. 288–89), LeConte stated that Beecher came to California in "the spring of 1885" to lecture on evolution and that, while there, Beecher was given copies of LeConte's papers on the subject by Sherman Day, son of Yale University's president, Jeremiah Day. LeConte claimed that he received the letter from Beecher after that. He further stated that he began writing his *Evolution* "in the fall of 1885."

32 Joseph LeConte to Alpheus Hyatt, May 19 and June 9, 1887, both in Alpheus Hyatt Correspondence, Ms. 1007, Maryland Historical Society, Baltimore.

Early in 1888, *Evolution and Its Relation to Religious Thought* came off the press. "Its success was far greater than my expectations," said LeConte, and he claimed that the "intelligent public seemed to have been waiting for such a book." He noted that he "received letters from many clergymen, of every denomination . . . thanking me for the boldness yet temperateness of the book." Praise also came from many scientists who, in LeConte's words, wrote to thank him for saving them from "blank materialism." Likewise, LeConte received encomiums from "men of the highest distinction in England, France, and Italy." As he reflected on the success of *Evolution* in his autobiography, LeConte claimed that "the book was timely and has done much good, which, of course, greatly gratified me."[33]

Indeed, *Evolution and Its Relation to Religious Thought* was a huge success, and it quickly became a leading work on the subject. Various magazines and journals hailed LeConte's new work, among them E. L. Godkin's *The Nation*, which published a lengthy summary of *Evolution*, praising LeConte as "a man in whom reverence and imagination have not become desiccated by a scientific atmosphere, but flourish, in due subordination and control, to embellish and vivify his writings." The reviewer observed that readers could expect the same "peculiar alertness of mind and freshness of method" that always characterized LeConte's publications, and he lauded the author for avoiding "any repetition of the nauseating incongruities which have been put forward under the guise of a 'reconciliation of science and religion.'" In the same vein, E. L. Youmans' *Popular Science Monthly* called the book "a notable contribution to a discussion perennial in its interest." With respect to LeConte's section on the evidences of evolution, *Popular Science* said that the "presentation of its proofs, though rapid, is masterly, and brought down to date," and it declared that LeConte "presents the factors of evolution tersely and concisely."[34]

Other magazines also offered favorable reviews of *Evolution*, among them the prestigious British journal *Nature*. The *Nature* reviewer said that the "exposition of evolution is well-planned" in LeConte's book, and he praised the writer not only for his "well-selected comparative

33 LeConte, *Autobiography*, 290.
34 S. B. Christy, "Biographical Notice of Joseph LeConte," *Transactions of the American Institute of Mining Engineers*, XXXI (1902), 779–80; *Nation*, July 12, 1888, pp. 34–35; *Popular Science Monthly*, XXXIII (May, 1888), 129–31.

figures" but also for the accuracy of his statements. In addition, the reviewer commended LeConte for his candor regarding the relation of religion to evolution, but he likewise criticized him for so freely accepting the British biologist George Romanes' theory of physiological selection among species because it was a hypothesis "still requiring proof." Moreover, the reviewer found fault with the author's generous position on paroxysmal stages of evolution and the "undesirably broad statements" that characterized the work. *Mind*, on the other hand, praised LeConte for his paroxysmal theory, calling it "one of the distinctive points of the author's view." This magazine also lauded LeConte for his treatment of the evidence of evolution as "not only a good exposition" but also one that contained "some distinctive points that claim the attention of biologists."[35]

A reviewer for the *Critic* applauded *Evolution* as "probably the best book that has been published on the subject, as it is concise, clear and accurate." He also extolled the book for its "popular language" and its "illustrations," which "are such as can be understood by the reader who is not a scientific student." Although the *Critic* reviewer maintained that *Evolution* was likely not to be "the last word on the subject" of the relation of evolution to Christianity, he praised LeConte for being "judicious and yet outspoken." When the second edition of *Evolution* appeared, the *Critic* again offered a very favorable review, but it also criticized LeConte on several points. Although the reviewer conceded that the second edition was "a success" because of "the general interest in the subject" and also because of "the author's earnestness of spirit and clearness of style," he believed that "so elaborate an argument seems at this day unnecessary, for that species have been derived is now generally admitted." That charge may have been accurate, of course, if applied to intellectuals, but it was certainly off base regarding orthodox religionists. But the additional criticism that evolutionary theory still failed to explain "how the derivation takes place" was most pertinent, for LeConte's book shed no further light upon that question.[36]

In addition to these and other American reviews, *Evolution* also drew critical praise from abroad. The French *Revue Scientifique* noted

---

35 *Nature*, XXXVIII (May 31, 1888), 100; *Mind*, XIII (October, 1888), 610–11.
36 *Critic*, n.s., XI (March 9, 1889), 118; *ibid.*, n.s., XVI (November 7, 1891), 248–49.

the clear style and the excellent quality of the book. High commendations also came from a reviewer in the Italian *Rassegna di scienze sociale e politiche*. Moreover, the recognized Italian novelist Antonio Fogazzaro experienced a sort of conversion as a consequence of his reading of *Evolution*. A devout Christian and recognized man of letters, Fogazzaro had been wrestling for some time with the conflicts between Christianity and science, and when he read LeConte's book, he was struck "as though a new revelation were at hand." Fogazzaro told LeConte that he had "read and reread your admirable book." Explaining that he had "always been fascinated by the theory of evolution," Fogazzaro confessed that he had feared to "adhere to it entirely" because it might bring a "deadly shock" to his Catholic faith. He informed LeConte that the third part of *Evolution* had shown him "the grandeur and sublime beauty of the plan of the Creation, teaching me that the reconciliation of science and faith in a magnificent ensemble is not a dream." The book, he said, "dazzled" him "beyond expression."[37]

Such were the typical responses to LeConte's *Evolution and Its Relation to Religious Thought*. At age sixty-five, LeConte had published his sixth book, and although those on geology and vision had a large impact upon a selected segment of the scientific population, neither of them had so widespread an influence as *Evolution*. This was in large measure due to the timeliness of the work, which had appeared during a period of intense interest in the implications of evolutionism for religious, theological, and philosophical beliefs, but it was also due to LeConte's superior ability as a writer and thinker. True, the theoretical explanations offered in *Evolution* contained many flaws, and eventually these would negate the usefulness of the book. But this is

37 John Bascom, "Books on Evolution and Life," *Dial*, IX (July, 1888), 59; Williston S. Hough, "Some Recent Discussions of Religion and Philosophy," *Dial*, XIII (July, 1892), 81; *Outlook*, LXVIII (July 13, 1901), 613; *Revue Scientifique*, XXV (July–December, 1888), 613–14; *Rassegna di scienze sociale e politiche*, VI (September 15, 1888), 99–100; Antonio Fogazzaro to Joseph LeConte, June 2, 1889, and LeConte to Fogazzaro, June 26, 1889, in "Cinque Lettere Inedite del Fogazzaro al Professor LeConte," *La Cultura*, VIII (1934), 82–84. The originals of these and other letters in the exchange have recently been released from restrictions on one box of material in the LeConte Family Papers, UCB. See also Antonio Fogazzaro, "For the Beauty of an Ideal," *Contemporary Review*, LXVII (May, 1895), 671–94; Tommaso Gallarati-Scotti, *The Life of Antonio Fogazzaro*, trans. Mary Prichard Agnetti (London, n.d.), 117–20.

inevitably the fate of a work that, as LeConte understood, must be superseded by new scientific discoveries. In any case, LeConte was certainly justified in his response to a congratulatory letter from Daniel Coit Gilman that "my heart was deeply in the work and it is pleasant to know I have not wholly failed."[38]

38 Joseph LeConte to Daniel C. Gilman, April 23, 1888, in Gilman Papers.

X

"Woe Is Me, If I Preach Not the Gospel"

During the 1880s and 1890s, the theory of evolution became a subject of bitter controversy among American evolutionists and orthodox Christians. As a leading proponent of the theory, Joseph LeConte often found himself under attack by antievolutionists. Not easily driven to anger, the mild-mannered scientist took most of the criticisms in good stride, avoiding both acrimonious debate and personal attacks on his critics. But his commitment to the theory was so complete that he felt almost a missionary sense in propagating it. In fact, the appellation pinned upon him by a San Francisco newspaper writer as the "gentle prophet of evolution" was most befitting.[1]

As early as 1884, LeConte's papers on evolution and religion had come under fire from the "Virginia Presbyterians," according to the Reverend W. R. Atkinson, editor of the *South Atlantic Presbyterian*. Atkinson wrote to Means Davis that he was himself under attack for his "liberal views on science and religion," but he noted that Davis' father-in-law was a chief target of the orthodox critics. If such attacks bothered LeConte, he did not admit it. He told Emma, for example, that he was "quite sure that there is coming about an approximation if not a reconciliation between thinkers on the scientific and theological side[s]." Shortly afterward, he informed Emma that he had "recently delivered a lecture before an association of so-called 'Liberal Christians'—mostly Unitarians—which has created quite a stir. It was on the 'Relation of Evolution to Christian Thought.'" On the one hand, the lecture need not have worried Christians, for it was a positive reaf-

---

1 San Francisco *Examiner*, May 26, 1895, p. 12.

firmation of faith in God. On the other hand, however, it presented a serious challenge to orthodox beliefs.[2]

Although the address was mainly a reiteration of LeConte's previously expressed views, it raised several vital questions. The principal issue at hand, said LeConte, is: "'Either God operates in nature in a more direct way than we have been accustomed recently to think: or else nature operates itself and wants no God at all.' There is no middle ground tenable." Again, however, LeConte rejected the deistic conception of God as "a great master-mechanic far away above us, who once upon a time, long ago and once for all, worked, created matter and endowed it with necessary properties and powers ... and then rested." He added that "the hand [of God] must be introduced from time to time to repair, to rectify, to improve and especially to originate new parts, such as new organic forms." But he also rejected materialism because it "pushed [God] further and further away," leaving "origins and causes" to mere "resident forces and natural law." The proper position, he continued, requires a "return to the old idea of direct divine agency, but in a new, more rational and non-anthropomorphic form." In other words, man must understand God as "immanent, resident in nature, at all times and in all places directing every event and determining every phenomenon."[3] Although this view offered a workable alternative, it certainly undermined traditional notions of God as Divine Intercessor and thus posed a threat to the spiritual security of orthodox Christians.

In his view of man as a spiritual being, LeConte further agitated the orthodox, especially in his statement that, as "a primate," man "shares his primacy with apes." Indeed, he argued, "the structural differences between man and the anthropoid apes are probably not so great as between the sheep and the deer family." But LeConte sought to play down this association by stressing that man has "an immortal spirit," a spirit that differs in degree and kind from the lower animals. Human beings thus possess "consciousness, will, intelligence, memory, love,

2 W. R. Atkinson to R. Means Davis, September 8, 1884, in Davis Papers; Joseph Le-Conte to Emma LeConte Furman, November 25, 1886, in LeConte-Furman Papers, SHC/UNC.

3 Joseph LeConte, The Relation of Evolution to Religious Thought, pamphlet published for the Pacific Coast Conference of Unitarian and Other Christian Churches (San Francisco: C. A. Murdock & Co., 1887), 5–7.

hate, [and] fear" in greater degrees than animals, and these are all of
a different nature. Although evolution is the process of development,
it has a terminal point, which is the unification of man's spirit with
the Divine Spirit.

LeConte likewise endeavored to establish the continual viability of
revelation. That was a difficult task, of course, because his emphasis
on the importance of rational thought and the operation of natural
laws would seem to make revelation a "palpable violation" of reason,
but LeConte explained away the ostensible contradiction by arguing
that revelation must come through the test of reason. Revelations filter
through man's mind like water through the earth and thus contain
imperfections. "Such filtrate," he concluded, "must be redistilled in the
alembic of reason, to separate the divine truth from the earthly im-
purities."[4] Clearly, however, such abstract arguments could hardly satisfy ordinary Christians who for ages had believed that God worked
more directly in the affairs of man.

Bessie LeConte mailed a copy of the address to Sallie, saying, "I wish
I had a thousand copies to scatter among the benighted clergy South
[*sic*]." She added that "it seems to me coming from a scientist and a
good man who is also a philosophical thinker it might produce some
effect, or at least present the subject from a stand point new to them.
I cannot but think they have never thought of it except from the special theories of Evolutionists." Joseph had also written to Sallie on the
subject of evolution and its impact upon orthodoxy. "The whole orthodox mode of looking at religion & the Bible is fundamentally different
from mine," he said. "I sometimes feel as if I were in a false position—
and yet I must think the false position theirs," he added. LeConte admitted to Sallie that he had thought of leaving the church, and he
expressed his opinion that "unless there is a fundamental reconstruction of religious thought, the church will die." But he believed that, in
the meanwhile, "it is our duty to stick by the Church." His own plan
was to "take a humble position—a back seat. Sustain the church
heartily but make no secret of my very divergence from what people
call orthodox. . . . To quit would be to set up my view as orthodoxy."[5]

The evolutionary theory fomented much wrathful controversy in

4 *Ibid.*, 9–15.
5 Bessie LeConte to Sallie LeConte Davis, January 15 [1887], and Joseph LeConte to
Sallie LeConte Davis, June 17, 1886, in Davis Papers.

the South, and LeConte could thank his lucky stars that he had not returned to teach in his native land. One of the liberal southern theologians who suffered most from the controversy was the Reverend James Woodrow, a professor in the Columbia Theological Seminary in South Carolina. Woodrow had followed LeConte as a student of Agassiz, and he had also taught at Oglethorpe University and the University of South Carolina; he had even been offered LeConte's old post at the University of Georgia in 1856 but had declined. As chief of the Confederacy's chemical laboratory in Columbia and editor of the *Southern Presbyterian*, Woodrow was well-acquainted with LeConte, whose views on theistic evolutionism he had adopted. When Woodrow openly espoused the theory in 1884, however, he was critically attacked by orthodox Christians, forcing him to resign his post at the seminary in 1886. He was eventually tried for heresy, but the South Carolina Synod relented and left him in good standing as a Presbyterian minister. LeConte's works were frequently cited as authoritative pronouncements during the course of this controversy, but in the end the conventional position prevailed.[6]

In spite of the esteem that LeConte enjoyed in his beloved South, his revolutionary teachings were unacceptable in the bastion of religious conservatism. LeConte wrote to Sallie, who was now living in Columbia, that he was "sorry of this Woodrow trouble." "The position of the Church is foolish in the extreme," he continued. "What conceivable difference is there between Evolution and any other law of Nature that there should be so much fuss about it?"[7] But LeConte knew that was a rhetorical question, for he was fully aware of the conservative nature of the South. Yet he had hoped that reason would prevail.

6 Clement Eaton, "Professor James Woodrow and the Freedom of Teaching in the South," *Journal of Southern History*, XXVIII (February, 1962), 3–17. The nature of Woodrow's teachings on evolution and religion can be found in Marion W. Woodrow (ed.), *Dr. James Woodrow: Character Sketches by His Former Pupils, Colleagues, and Associates* (Columbia, S.C., 1909), *passim*. The term *controversy* rather than *warfare* seems much more apposite in describing the conflict between religionists and evolutionists, for, as Moore has demonstrated in his excellent work, *Post-Darwinian Controversies*, many compromises were made and, indeed, many (even most) of the American evolutionists were very religious men. Certainly, the conflict was bitter at times, as the case with Woodrow indicates, but it was hardly tantamount to "war."

7 Joseph LeConte to Sallie LeConte Davis, January 11, 1886, in Davis Papers. On the reception of LeConte's address on evolution, presented to the Atlanta Philosophical Society, see the Atlanta *Constitution*, May 1, 1888, p. 4, July 29, 1888, p. 12, and July 31, 1888, p. 8.

After all, he had come a long way in his own mental development. Why, then, should supposedly intelligent people fail to grow likewise!

The evolutionist controversy touched closer to home, too. It hit the University of California briefly in 1888 when Francis Horton, a Presbyterian minister from Oakland, criticized a local Unitarian minister for his praise of the novel *Robert Elsmere*, the story of a man who became disenchanted with the church and gave up his faith. Horton also took the occasion to criticize a university regent, Horatio Stebbins, who pastored a Unitarian church. The charge was likewise aimed at the university president, Horace Davis, who belonged to Stebbins' congregation, but it was also a more general charge that the University of California was inculcating its youth with "a skeptical philosophy." Since Joseph LeConte was certainly among those who criticized "revealed religion," Horton's attack was also aimed at him.[8]

An even more severe attack was launched soon afterward by the Reverend C. H. Hobart, a young minister whose pastorate was the well-established First Baptist Church of Oakland. Hobart censured the university administration for appointing "certain professors" who were guilty of leading youth astray by teaching infidelity. Chief among them, charged Hobart, was Joseph LeConte. He criticized the professor for touting evolutionism as a fact and for saying that all great thinkers accepted it. "These sentiments," he alleged, "are ruining the young people who attend the university." The same newspaper reported that Hobart's sermon had "created the greatest excitement, for Professor Joseph LeConte is the most popular professor in the University. He also ranks among the greatest scientists in the world, and this very work criticized ... is considered a standard throughout the world." Another newspaper flatly accused the Baptist minister of being unfamiliar with "Prof. LeConte's theological belief" and charged that Hobart was ignorant of the widespread acceptance of the theory of evolution. As "one of the most profound of scholars and investigators of nature," Professor LeConte, claimed the newspaper, could better "reconcile nature's teachings with the teachings of the world's savior, Christ ... than the exhortation of a preacher who has to deny the facts which science makes clear in order to keep his flocks from straying from his fold."[9]

8 Stadtman, *University of California*, 100–101.
9 San Francisco *Bulletin*, clipping in Nancy Smith Collection; unidentified newspaper clipping in McMillan Collection.

Other newspapers also boldly defended LeConte, chief among them the *Berkeleyan*. Affirming its great respect for Joseph LeConte, the student newspaper said: "The attacks on the teachings of our beloved Professor LeConte by certain barbarians in the world of thought, deserve no more than a passing reference. An articulate object who calls himself Hobart . . . has made a desperate effort to uplift himself from his native obscurity by raising the mediaeval cry of heresy against evolution and modern science." The *Berkeleyan* went on to compare Hobart with a primate, and if this irony were insufficient, the paper continued: "We desire to nominate 'Brother Hobart' to the first vacancy occasioned by the unchristian zeal of the misguided heathen in the list of missionaries to the cannibal isles. For verily we believe he would effectually convince them of the error of their cannibalistic appetites." The vitriolic slur against Hobart ended with the suggestion that he "be forcibly dragged from his den in the home of the feeble-minded and placed on exhibition in the Anthropological Museum of the Midwinter Fair." [10] Such was the esteem in which LeConte was held, and such were the rejoinders of those who defended him against his critics.

Indeed, LeConte became even more popular, at least in California, and he was invited more and more to lecture on evolution. "Since this miserable controversy about Evolution in the papers," he wrote to Emma, "I am called on every occasion to say something." He noted that he was soon expected to speak on "Evolution & Human Progress" and on the "Effect of Scientific Method on Religious Thought." Moreover, he told his daughter, he had been invited to speak at the Jewish temple in San Francisco on evolution as an element in fundamental religious thought. "The Jews have been such good friends & admirers of mine," reported Joseph, "that I cannot refuse." He noted that many other groups in San Francisco likewise wished to hear him. He did deliver his address before the Congregation Emanu-El, and it was stenographically reported for the *Jewish Progress*.[11]

Assuring the congregation that evolution is not necessarily synonymous with materialism, LeConte declared that the charge of religionists that evolution is incompatible with religion was like the cry of wolf. It was an old cry, he said, and he stressed that it was "time we should recognize it is a false alarm." First it came in response to the

10 *Berkeleyan*, October 6, 1893, p. 72.

11 Joseph LeConte to Emma LeConte Furman, March 7, 1894, in LeConte-Furman Papers, SHC/UNC.

heliocentric theory, then to the gravitational theory, and now to the evolutionary theory, he averred. In the first two instances, the wolf had been found to be "a very noble beast" who lay down "in peace with the lambs." But, he continued, many say that evolution is "the *real* wolf." It may be that the real wolf has come at last, said LeConte, but, if so, it is not "in scientific form, but . . . in sheep's clothing"—that is, from the traditional religionists. Any "great new idea," he concluded, always requires severe mental adjustments. "But do not misunderstand me," he said, "I deeply sympathize with a reasonable conservatism." Conservatism at the expense of established truth, however, is wrong, stressed LeConte. Evolution is true, and it takes nothing from God; indeed, it makes him not the *"absentee landlord* managing His estates indirectly" but a Being "immanent, indwelling, resident in nature . . . in every molecule and atom, and *directly* determining every phenomena [*sic*] and every event."[12]

In part, LeConte's argument for theistic evolutionism represented a delayed response to the objections of several able Christian theologians who had previously attacked the Darwinian theory. One of the most outspoken of these critics was the Reverend Doctor Charles Hodge of Princeton University, who bitterly denounced the theory as atheistic. Hodge declared that Christians had a duty "to protest against the arraying of probabilities against the clear evidence of the Scripture." Hodge's colleague, the Reverend Doctor John T. Duffield, was quite as adamant, and he openly attacked the reconciliationists for their liberal interpretations of the Scriptures. In fact, said both Duffield and Hodge, there can be no compromise; either the Bible is the true account or it is a mere fable. But if the arguments against evolutionism were not all so clearly theological in nature, most continued to be rooted in a fear of materialism, and LeConte ultimately found himself also the target of supposedly more secular critics, one of whom was Josiah Keep. The "doctrines of evolution," charged Keep, "are creeping into textbooks [and] are openly advocated in public lectures." They are especially influential in California, he claimed, because of the lucid explanations presented "by Professor Joseph Le-

12 Joseph LeConte, "Evolu[t]ion, An Element in Fundamental Religious Thought," *Jewish Progress*, March 23, 1894, in a bound volume labeled "Miscellaneous Writings," in Joseph LeConte, Miscellaneous Writings, Bancroft Library Archives, UCB.

Conte, a man whom we all hold in the highest esteem." In fact, Keep averred, LeConte was so effective that, like King Agrippa, one could say, "Almost thou persuadest me to be an Evolutionist."[13]

Keep then proceeded to enumerate his doubts regarding evolution, citing what he considered illustrations that refuted the theory. Although characterized by a rational approach, Keep's essay manifested that its author lacked a thorough knowledge of the evolutionary theory. Regarding the transmutation of botanical species, Keep averred that an apple is always an apple, and a potato always a potato, no matter what their earliest forms may have been. True, they have changed, he admitted, but only in variety, not as transmuted species. The same was true of animals, he opined, and he argued that no missing links had been discovered to verify the hypothesis of transmutation. Moreover, he declared that any so-called intermediate species could not have survived because they could not have functioned in a transitional stage. Thus, although winged insects may appear to bear a relation to purely terrestrial types, they were in fact created by God at the moment when the environment could sustain them in a new form. In effect, Keep was attempting to resurrect the old Agassizian view of special acts of creation to account for the myriad types of species.[14]

Such arguments were common among those who resisted evolutionism, and LeConte was intimately familiar with them. In fact, as he confessed in his reply to Keep, he had himself wrestled with such doubts for many years before he became an evolutionist—that is, before he was "driven to my present position." In response to his critic, LeConte admitted that "every attempt to generate life *de novo* has failed," and he believed that such attempts would "always fail." This was "not sufficient to convince us of its impossibility," however. But his point was that evolution made any attempt to recreate the origin of life unnecessary anyway, because evolution never returns to a beginning point; it always moves on in "some direction." The first spark of life may indeed have been created by supernatural power, said Le-

---

13 Boller, *American Thought in Transition*, 22–29; Andrew D. White, *A History of the Warfare of Science and Theology in Christendom* (2 vols; New York, 1896), I, 78–86; Josiah Keep, "Doubts Concerning Evolution," *Overland Monthly*, 2nd ser., XVIII (August, 1891), 190–98.

14 Keep, "Doubts Concerning Evolution."

Conte, but that is a philosophical, not a scientific, question. Science is concerned only with secondary causes, and the question of the origin of life in no way invalidates evolutionary theory.[15]

LeConte also replied to Keep's view that the idea of special acts of creation made more sense than the theory of evolution by arguing that the case was just the contrary. The notion of gradual transformation by natural means was much more reasonable than the conception that "there has been a continued succession of the most stupendous miracles, in the form of exterminations and re-creations." In the field of geology, for example, said LeConte, one can see with his own eyes that mountains and valleys are in the process of evolution right at this very moment. The same is true with respect to species, he continued. Though the process of evolution among species is infinitely slow, still "there are some few examples of species changing from one form to another, quite distinct, under our very eyes." LeConte freely admitted to gaps in knowledge regarding transformations of species, but he countered with the question of whether "the theory of miraculous creation" was any better. Indeed, he told Keep, the evolutionists are not offering "an *easy solution*"; rather it is the special-creationists who proffer the simplest solution to the "knottiest of problems."[16]

Certainly, LeConte survived virtually unscathed the attacks upon his theistic evolutionism; at least he felt he did. Yet, as he had already noted, the problem of reconciling theology and evolutionism continued to "seethe" in his mind. Eventually, many of his thoughts on the subjects were published, but some of them were too personal to set to print. He thus confined them to a private document that he called "A Brief Confession of Faith."[17] The confession also included some ideas that LeConte had put forward publicly, but it likewise contained others that he kept private.

Altogether the confession consisted of ten topics and a summary statement. The first of these was the "Relation of God to Nature." Ever the theist, LeConte stated that "belief in a Power behind Nature, and the cause of all its phenomena is the underlying condition of Reason,

15 Joseph LeConte, "Origin of Organic Forms—Is It by Natural or Supernatural Process?" *Overland Monthly*, 2nd ser., XVIII (August, 1891), 198–203.

16 *Ibid.*

17 Joseph LeConte, MS labeled "A Brief Confession of Faith, written in 1890,—slightly revised and added to in 1897," in LeConte Family Papers, UCB.

and like our existence is more certain than anything can be made by *reasoning* or which we call *proof*. This power we call God." In short, LeConte could never accept an atheistic position, nor could he even entertain agnostic thoughts. The only thing he had to do was rationalize the existence of God in natural form. "Natural forces," he wrote, "are but different forms of the omnipresent Divine Energy,—natural laws of operation of that Energy,—and natural objects, or what we call Nature itself, but objectification of the Divine Thought." This bordered closely on Pantheism, of course, but LeConte insisted that "God is a Person." The paradoxical nature of this position LeConte clarified by explaining that God's "personality consists not of bodily form or visible appearance, nor in local habitations; but in self-consciousness, free-will, and thought."[18]

What, then, was the nature of the human soul? "Man's spirit," LeConte declared, "is naught else than the Divine Energy . . . individuated more and more until by complete individuation it finally separates itself from nature." The spirit or soul of man thus "preexisted . . . in the womb of nature. First [the soul is] potential in inorganic nature, then as germ in life force of plants, then as quickened embryo in animals, and finally comes to birth into a new spiritual and immortal world in man." Thus, "man's soul is of the same substance as God, and like him immortal." That point made, LeConte turned to the question of the "Origin of Soul," about which he maintained there were "two fundamentally different views." He rejected the "orthodox view" because it put soul into man "from outside the body"—that is, either at the moment of conception, birth, the dawn of self-consciousness, or the "so-called conversion or second birth." These theological notions LeConte called "wholly irrational and unphilosophic." Instead, he defended the view "that soul was not made at once, but pre-existed in all time."[19]

The fifth point in LeConte's "Confession of Faith" was addressed to the question of the validity of the Bible as the source of spiritual authority. "The Bible," LeConte noted, "must be regarded like other habits, our sacred scriptures like other sacred scriptures, only, as we believe, much higher and more sacred than others." But despite its higher nature, the Bible was "not written down in words whispered

18 *Ibid.*
19 *Ibid.*

into the ears of divinely chosen emanuences [*sic*], who thus acted like a type-writing machine." The Bible was "not a direct revelation"; it came "through the spirit of man" and was "therefore subject to all the imperfections of human work." Rather than being "a divine word-book," the Bible is "a record of the growth of the divine idea in the mind of man." Like all "human works," the great book was "deeply affected by the spirit of the age, and the character of the writers. It is full of mistakes and misconceptions characteristic of the age and of the writer." But still it is "a great agent of civilization" and should be used in the schools as "a great power for good."[20]

Such views clearly suggested that LeConte also harbored reservations about the divine nature of Jesus, and indeed he did. The "Ideal Man—the Christ—need not be perfect in knowledge, or omniscient; not infinite in power, omnipotent," he observed. The Christ need be "only perfect in character," for it is character that is "the divine part of us all." It is "essential spirit—it is the attitude of the human spirit to the divine spirit." LeConte confessed that although Jesus was the Christ, he was "not infinite either in knowledge nor [*sic*] in power"; he was only perfect in character. That was sufficient for LeConte, and he declared, "I therefore accept him as the Christ, and worship him as such."[21]

The subject of the Church served as the seventh point of the confession. "The Church is a divine institution," averred LeConte, "only in the sense in which the State is divine, or society is divine, although in higher degree." It is a "human institution," and as such "it must change, and is subject to evolution." Like any other human agency, the Church has experienced both progress and regress, though, on the whole, it has moved steadily forward. The "Gospels are on a higher plane of truth than the Epistles, for these latter are mixed with Judaism . . . [and] the Epistles are higher than the Catholic Church, for this latter is evident mixture with Paganism [*sic*]." LeConte believed these stages in the evolution of Christianity were "the necessary result of environment." The Reformation brought about some "purification" of the Church, "but a hard mechanical theology inherited from the Catholic Church and made still harder by Calvin, still continued, and still continues." A "bondage of blind authority and dogmas" was sim-

20 *Ibid.*
21 *Ibid.*

ply "transferred from a person—the Pope—to a book—the Bible" as a consequence of Luther's effort. What is needed, asserted LeConte, is a better understanding of Jesus' teaching that the Christ is "the way of the Father." Thus, although Jesus came to save, he did so not by "vicarious blood sacrifice" but "by revealing the Father, by showing us the idea of human character, and the right way of life." In LeConte's opinion, Jesus was "the Ideal Man, and therefore the Divine Man." The proper function of the Church, then, is to exemplify ideal character among human beings by following the teachings of Jesus.[22]

LeConte's obvious rejection of the literal and verbal inspiration of the Scriptures became even more pronounced in his views on the Devil, the doctrine of repentance, and the idea of Immaculate Conception. "It is hardly necessary," he observed, "to say there is no such being as the Devil." The whole notion of sin has been misconstrued, argued LeConte. Sin is simply "the discordance between our ideal and our conduct, the willful violation of our best nature." There is therefore no literal Hell; it is a "condition," not a "place." LeConte agreed with the teaching that man had "fallen," but he rejected the orthodox meaning of the word. For him, man had fallen in the sense that he was out of harmony with natural laws. The "whole of this elaborately construed scheme [of Original Sin] is a figment of the theological brain." On the doctrine of the Immaculate Conception, he likewise demurred from orthodoxy, deploring the idea because it implied that all other conceptions were unclean. In fact, he argued that the idea insulted the marital relationship. Sexual union is not impure, said LeConte; although it was originally a "purely animal" act, it has become "the purest and holiest" kind of relationship between husband and wife. LeConte noted that since the idea of "supernatural conception" was "not mentioned by Mark and John, nor alluded to in any of the Epistles," it was "evidently an afterthought."[23]

LeConte added a "Summary" to his confession, in which he stated that "the essential . . . articles of faith are, 1. A belief in God the father of our spirits, 2. In Christ, the Ideal Man, the Divine Man, our Brother, 3. The Freedom of man; and 4. Personal Immortality." But, he declared, "If any of these is less dear to me than the others, it may be the last." He claimed that he would rather give up the idea of immortality than

22 *Ibid.*
23 *Ibid.*

any of the other essential beliefs, for if he ceased to have "conscious existence" he would simply be reabsorbed into the Infinite.[24] Despite his disclaimer regarding the relative importance of the idea of immortality, LeConte commented upon it at length in several articles, and he included a briefer discussion of it in his book on evolution. Not only was it more important to him than he admitted (or realized), it was also crucial to his philosophy of theistic evolutionism. When all was said and done regarding the reconciliation of evolutionism and Christianity, nothing was as essential to the rejection of the materialistic view as the idea of immortality.

A full account of LeConte's views on evolutionism and immortality appeared first in his "The Natural Grounds of Belief in a Personal Immortality" and later in a paper on "Plato's Doctrine of the Soul." In order to dispense with the "physiological objection"—that is, with the purely material approach—to the question of immortality, LeConte posed the possibility of looking inside a person's brain. If this could be done, said he, one would see "nothing but molecular motions, physical and chemical, molecular vibrations or agitations, chemical decompositions and recompositions." Yet, he observed, while this physical and chemical process is occurring, the individual is totally unaware of it; the person being observed is engaging in conscious thought and volitional acts of mind. This is really the "psychical" nature of man, or the "inside set" as opposed to the "outside set" of mental phenomena. The former is a "*microcosm*"; the latter, a "*macrocosm.*" Only man, not other animals, has the "*inside view* of brain-phenomena"; therefore, only humans live in both worlds. If animals possessed that quality, they, like man, could engage in voluntary acts of progress: "they, too, would have their religion, science, and philoso-

---

24 *Ibid.* Comments relevant to these "confessions of faith" were often jotted down by LeConte in his pocket notebooks. In the one labeled "Calico 1887," for example, he penned a number of thoughts on such matters, including the rough sketch of a letter to a Myron [?] Adams "about Feb./89," in which he said, "If [miracles] ever were of use—they are so no longer. Whether actual occurrences or not, [it is] no matter for Christian religion or Christian ethics." In his notebook labeled "Misc. 1891–93 & 94," he called himself a "monist" and sketched some thoughts on why he was not a "dualist." In a letter to Lewis Janes, March [?], 1890, LeConte rejected the use of "the word agnosticism in connection with God & immortality" and refused to accept the use of "the word metagnosticism" as a useful substitute, suggested by his correspondent to explain knowledge of supernatural phenomena. This letter and the notebooks are in LeConte Family Papers, UCB.

phy." In brief, therefore, the self-conscious, "self-active" nature of man constitutes an "immortal spirit."[25]

To bolster his argument, LeConte resorted to another "*indirect*" proof, namely, the ancient argument that mortality of man's spirit was a *reductio ad absurdum*. "Without immortality," said LeConte, "there would be no conceivable *meaning* in human life, nor indeed, in the complex structure and elaborate evolution of the Cosmos itself." He also argued that the concept of immortality must necessarily flow from our essential belief that there is a supreme power behind the "phenomena of nature." Man as a self-conscious being apprehends through his psychical consciousness that there is a power greater than he, and therefore man partakes "of the divine nature." God as "spiritual Father" could not "make laws of nature which eventuate necessarily in delusive hopes of immortality." Moreover, continued LeConte, man by his very nature wishes for "the True and the Good." That is "a necessary *condition of human improvement*." Thus, the "universality and the persistence" of such a hope necessarily constitute the validity of the idea of immortality.[26]

In connection with this point, LeConte argued that the positivists Auguste Comte, George Eliot, and Frederick Harrison implicitly supported the notion because their "apotheosis" of man, or "humanity worship," is really the goal of "rational Christianity." In other words, if the elevation of humanity leads to the making of the ideal man, then it ultimately leads to the development of the "*divine* man." However, rational Christianity, not Positivism, is the true way to the "*Infinite personal* God," and hence to immortality.[27] From all of these arguments, it is evident that LeConte genuinely believed in immortality, but it is equally certain that he had to muster a host of "proofs" to convince his readers—and himself. He had embraced a theory that depreciated the old dogmatic religious views on the origin, nature, and destiny of man, and reconciliation proved to be a hard mental task.

Plato's doctrine of the soul fascinated LeConte, and he gladly accepted an invitation to speak on the topic before the university's Philosophical Union. Admitting that he was uninformed regarding critical

---

25 Joseph LeConte, "The Natural Grounds of Belief in a Personal Immortality," *Andover Review*, XIV (July, 1890), 1–13; Joseph LeConte, "Grounds of My Belief in Immortality," *Christian Register*, April 18, 1889, pp. 247–48.

26 LeConte, "The Natural Grounds of Belief in a Personal Immortality," 6–8.

27 *Ibid.*, 8–13.

interpretations of Plato, LeConte confined his remarks to a critique based upon his own personal reading of the *Phaedo*. As he understood Plato on the subject, the soul or spirit of man "is pre-existent from all eternity," taking up "its abode in an earthly tabernacle—a material body—for a brief space, only to be released again by death." LeConte also interpreted Plato's theory of knowledge as the "recalling of forgotten *experience*, . . . an awakening of an essential property—a reassertion of an essential attribute." Knowledge thus forgotten could be recalled by the method of dialectic. Moreover, he argued that Plato's philosophy was "intensely ASCETIC" and led to "a *despising* of nature and the human body, and of all the senses, appetites, desires, affections, associated with the body." The last, along with Plato's "false theory of knowledge," had resulted in a "baleful influence" upon mankind, stated LeConte, because it denigrated the study of nature and, through the doctrine of "innate ideas," sought "*to impose*" deductive laws upon nature.[28]

This doctrine had been all the more harmful, declared LeConte, because it had been picked up by early Christian thinkers and "is still preached in a modified form from nearly every pulpit in Christendom." Thus, the "unscientific condition" of Plato's approach—the stress on deductive reasoning, the ascetic philosophy of life, and the confining perception of the soul—had deterred man's rational progress. A solution to scientific mysteries must come first through the gathering of facts, followed by the formulation of laws and generalizations, and the extension of those laws by logic. In other words, deductive reasoning must follow inductive reasoning, explained LeConte. Both are necessary; but their order is crucial to intellectual progress. That, of course, was what LeConte felt he was doing. One cannot deduce the origin and destiny of the soul from pure thought. One must first gather the facts of nature, then formulate laws. Then, and only then, can one employ deductive reasoning. The law of evolution is a necessary step to the law of immortality, concluded LeConte.[29]

In 1895 LeConte again had occasion to elaborate upon his concep-

28 Joseph LeConte, "Plato's Doctrine of the Soul, and Argument for Immortality, in Comparison with the Doctrine and Argument Derived from the Study of Nature," *University of California Philosophical Union Bulletin*, VIII (1891), 1–19 (originally delivered as an address before the Union on June 20 and 24, 1890).

29 *Ibid.*

tion of God and immortality of spirit. The event was another meeting of the Philosophical Union and involved several of America's leading philosophers. Presenting addresses along with LeConte were Josiah Royce, George H. Howison, and Sidney E. Mezes. Royce was a former student and later a colleague of LeConte at the university. He openly admired his former teacher and freely admitted that LeConte's teachings had had a profound impact upon his own mental development. Howison, who had been a colleague of LeConte since the early 1880s, was a noted authority on Kant and Hegel. Although he and LeConte differed greatly on a number of philosophical issues, the two men were strong intellectual companions. Mezes was a well-known educator from the University of Texas. Royce, a philosophical idealist, presented the main paper, titled "The Conception of God." His address was a modification of his work *The Religious Aspect of Philosophy*, which LeConte had reviewed shortly after its publication in 1885. In his review, LeConte had called Royce "one of the acutest and most independent of American thinkers" and praised Royce's book as "stimulating in the highest degree."[30]

The purpose of Royce's address was to deal with three conceptions: the reality of God, the nature of His being, and the relation of the idea of God to autonomous moral actions. LeConte offered no criticism of Royce's proof of the existence of God, but on the question of God's nature, he differed somewhat, offering a mediating hypothesis between Royce's notion of God as "Thought" and Howison's idea of "Immanent Being." Yet LeConte differed principally on grounds that Royce's God as Thought was too "passive, powerless [and] passionless." Otherwise, he accepted the idea that the true nature of God is "Omniscience." In a related sense, LeConte held that Royce's God was far removed from the natural world, thereby too remote from scientific scrutiny. God is *in* nature, argued LeConte, and He gives order and meaning both to natural processes and to moral standards. Nevertheless, said LeConte, man is free to choose between right and wrong, but, according to evolutionary principle, man must "struggle with an apparently inimical environment" until he achieves the state of being

---

30 Josiah Royce, *et al.*, *The Conception of God* (New York, 1898; reprint ed., St. Clair Shores, Mich.: Scholarly Press, 1971); Joseph LeConte, "Review of Royce's *The Religious Aspect of Philosophy*," *Overland Monthly*, 2nd ser., V (May, 1885), 542–44. A longer and slightly different manuscript draft of the review (probably the original) is in the possession of Joseph LeConte Smith, Macon, Ga.

ideal man. LeConte also rejected Royce's argument that God is a Divine Self-Consciousness, countering with his theory of the evolutionary individuation of man into a state of self-consciousness. Man's goal, his ideal, is to complete the process of evolution by rational moral thought until he has once again achieved union with the Divine—that is, by attaining immortality through reabsorption into the Divine Energy.[31]

In truth, LeConte ever strove to personalize God, but his theistic evolutionism, despite its strong emphasis on immortality, had the effect of making God as Divine Energy remote from the personal lives of individuals. Even Royce's God as Thought and certainly Howison's God as Immanent Being were more personal than LeConte's Divine Energy, no matter how hard LeConte pressed to make man's spirit a part of the Divine Spirit. LeConte *qua* scientist had to answer metaphysical questions in light of organic laws. He could not compartmentalize physical and metaphysical problems; he must see things as a whole, constantly seeking a reconciliation of conflicting truths. He devoted most of his later life to the effort, and at the end of his autobiography he said, "Woe is me, if I preach not the Gospel."[32] It was the Gospel of Evolutionary Theism.

From the days of his early childhood to the end of his life, Joseph LeConte maintained a gentle and enduring spirit of piety, never wa-

31 Royce, *et al., The Conception of God,* 4, 67–68, 81–83, 90–91, 112–19, 130–32, 348–54.

32 LeConte, *Autobiography,* 336; Joseph LeConte to William "Willie" Louis LeConte, September 23, 1894, in possession of Mrs. Joseph Nisbet LeConte. In addition to his numerous addresses and papers on evolution and religion, LeConte also reviewed several books on the subject. They included: G. Frederick Wright's *Studies in Science and Religion,* in *Science,* I (June 15, 1883), 543–45; John Fiske's *Excursions of An Evolutionist,* in *Overland Monthly,* 2nd ser., III (March, 1884), 329–31; Arnold Guyot's *Creation,* in *Science,* III (May 16, 1884), 599–601; Josiah Royce's *The Religious Aspect of Philosophy,* in *Overland Monthly,* 2nd ser., V (May, 1885), 542–44; John Fiske's *Darwinism and Other Essays,* in *Overland Monthly,* 2nd ser., VII (March, 1886), 334; Max Hark's *The Unity of Truth,* in *Popular Science Monthly,* XXXIII (September, 1888), 699; and Howard Mac-Queary's *Evolution of Man and Christianity,* in *Overland Monthly,* 2nd ser., XVI (July, 1890), 110–12. He also wrote reviews of Samuel Laing's *Modern Science and Modern Thought* (1885) and Sir J. W. Dawson's *Egypt and Syria: Their Physical Features in Relation to Bible History* (1885), but neither of these reviews seems to have been published (MSS in possession of Joseph LeConte Smith). LeConte was extremely critical of several of these works for failing to present an accurate picture of scientific findings and for what he considered a superficial analysis of evolutionism as a theistic conception of natural phenomena.

vering from the firm faith in God imbued in him by his father but continually making adjustments as necessary to reconcile his beliefs with scientific discoveries. Indeed, the influence of his father profoundly shaped his religious views, providing him with an unshakable faith in Christian precepts, a liberal attitude toward the interpretation of Deity, a capacious ability for accommodation, and, above all, an unrelenting commitment to rational thought and the value of scientific inquiry. Over the decades, he never parted from the fundamental faith of his father, but he ultimately modified his early teachings to conform to the doctrine of evolutionary theism, and of that doctrine he played the role of ardent apostle. Certainly, he was not consumed by the duties of discipleship, and the mission of converting others to the dogma always remained an effort of gentle persuasion through the power of reasoned argument. In light of his background, training, and experiences, and in keeping with the tenor of his time, LeConte must be judged as a constant truth-seeker who loosened the fetters of convention but could not completely free himself from the bonds of traditional belief. Although his efforts to reconcile conflicting views often represented skillful but unwitting contortion, they nevertheless reflected a genuine attempt at free inquiry. Hardly more could be expected.

XI

A Widening Reputation, 1885–1894

In 1885 Joseph LeConte celebrated his sixty-second birthday and his sixteenth year in California. He was in good health, and serious thoughts of retirement had not yet entered his mind. Teaching remained as fresh a challenge to him as it had more than thirty years before, and each morning as he took a brisk walk for exercise and reflection upon the daily lectures, the dedicated professor looked forward to his classes in physiology and geology. When these were over, he turned his attention to any number of topics on which he desired to set his thoughts on paper. If the tasks seemed heavy, the great lover of Nature could contemplate the renewal of spirit that would come during his excursions into the mountains. Life, he felt, was to be lived in the pleasures of creation and recreation, not in the doldrums of despair over its brevity.[1]

As the summer of 1885 approached, LeConte was invited to accompany his old friend Clarence Dutton on a trip to northern California and Oregon to examine the geological structure of that region. Unfortunately, the funds for the trip, sponsored by the U.S. Geological Survey, were delayed, and the two geologists had to spend half of their ten-week excursion in camp. During their wait, LeConte and his companion "took daily rides to explore the country" or sat in camp discussing a variety of topics, of which the contractional theory of mountain formation served as the center. By LeConte's own admission, Dutton had written "an able paper . . . combating every form of the so-called 'contractional theory.'" The geologists also disagreed on other points regarding diastrophic changes. But they did not confine their debates to geology, for they both held strong views on other subjects,

---

1 LeConte, *Autobiography*, 297.

including the theory of evolution, religion, music, and philosophy.[2]

On August 6, LeConte and Dutton were finally able to leave for Oregon. During the next two weeks LeConte recorded many of his findings in his notebook, devoting much of his attention to the evidence of "the great lava flow," a subject on which he had by now become an authority. He and Dutton also examined the structure of the Klamath Falls and the surrounding region. Eventually, they arrived at Crater Lake. LeConte later claimed that he and Dutton were "the first scientific party that ever visited the lake." Dutton revisited the lake the following summer, and later it was explored by J. S. Diller. LeConte at once recognized the geological significance of the region, and he soon set to work writing a paper on the post-Tertiary elevation of the Sierra, an idea he had dealt with "imperfectly" in a previous paper on the old riverbeds of California. He completed the paper later that year, and in April, 1886, he read it before the National Academy of Sciences. Unfortunately, publication of the paper was delayed, and Diller, who had in the meantime begun work on the same topic, published his paper a month before LeConte's appeared. In his autobiography, LeConte asked, "Who should claim the credit?" To his own question, he replied, "I neither know nor care." It seems rather obvious that he did care, however, for his explanation of the circumstances surrounding the publication of his paper reveals a desire to be credited with the discovery. At any rate, the idea was important to geologists, and it was one of the many LeConte studies of landform geology often cited by contemporary specialists on the subject.[3]

LeConte's next field and camping excursion came in 1887 when he and his son joined the Reverend George Wharton James, a liberal Methodist minister, on a trip into northwestern California (Modoc County) and the northwestern section of Nevada, mostly in the vicinity of Pyramid, Winnemucca, and Blue Lakes. For the seventeen-year-

2 *Ibid.*, 278–82; Joseph LeConte, notebook labeled "Mt. Shasta 1885 & Modoc 1888," in LeConte Family Papers, UCB.

3 LeConte, *Autobiography*, 282–85; Joseph LeConte, "The Old River-Beds of California," *American Journal of Science*, 3rd ser., XIX (March, 1880), 176–90; Joseph LeConte, "A Post-Tertiary Elevation of the Sierra Nevada Shown by the River-Beds," *ibid.*, 3rd. ser., XXXII (September, 1886), 167–81. See Richard Chorley, Anthony J. Dunn, and Robert P. Beckinsdale, *The History of the Study of Landforms* (2 vols.; London, 1964 and 1973), for a comprehensive treatment of the history of landform geology and an assessment of LeConte's contributions to that field of earth science.

old Joseph Nisbet, this was the first lengthy camping trip with his father.

The trip to Modoc County and Nevada began on July 9 and lasted for a month. LeConte and his son spent many pleasurable hours reading, swimming, bird hunting, and mountain climbing. In addition, LeConte found time for a number of geological investigations. The region had already been extensively surveyed and written about by the geologist I. C. Russell, but LeConte "greatly enjoyed verifying his [Russel's] results by personal examination." He was especially intrigued by the geological formation of Surprise Valley, "an example of mountain making by block tilting."[4] Later, he would use some of the information gained here in a paper titled "The Origin of Traverse Mountain-Valleys." As usual, LeConte combined his pleasure trips with field investigations, though often, as in this instance, the latter was secondary to the former.

In 1889, LeConte again spent a portion of his summer vacation in camp, this time returning to his favorite spot, the High Sierra. By the time the six-weeks' trip was over, Joseph and a group of university students, including little Joe, had traveled nearly five hundred miles on horseback. The beginning of the trip, especially across the San Joaquin plains, proved extremely arduous for the sixty-six-year-old LeConte. On the second night out, he was awakened by an "awful spell of cramps" in one leg, and the aching limb had to be rubbed by one of the students for a quarter of an hour "before it was relieved." As they moved on during the next two days, Joseph recorded in his notebook, "hot, hot, inconceivably hot," and again, "*Oh*, the heat almost unbearable . . . wind blowing as out of a furnace." He began to feel that "for the first time" he was "losing [his] physical endurance." He lost his appetite and was "greatly fatigued." His feeling of despair was no doubt greatly increased by the rapidly deteriorating health of his brother John. This pressed hard upon LeConte's mind as he trekked

4 LeConte, *Autobiography*, 291–92; LeConte, Autobiographical MS.; LeConte, notebook "Mt. Shasta, 1885 & Modoc 1887"; George Wharton James, "Camping with LeConte," *Sunset*, XI (October, 1903), 563–66, and *ibid.*, XVI (December, 1905), 197–201; Joseph N. LeConte, "Recollections"; Joseph N. LeConte to James Augustus LeConte, January 27, 1891, in possession of Mrs. Joseph Nisbet LeConte; Joseph N. LeConte to Carolyn "Carrie" LeConte, February 23, 1885, in LeConte Family Papers, UCB; Joseph N. LeConte to Elizabeth "Bess" Furman, March 2, 1891, in Nancy Smith Collection.

across the fiercely hot plains of San Joaquin. Would this be his last trip?[5]

As the party reached Hetch Hetchy, high in the mountain air, Joseph's spirits revived, and he put away thoughts of death. For ten days the group camped in the valley that has sometimes been called "Little Yosemite." Joseph explored the valley at length, looking especially for evidence of glaciation. He did not climb to the top of the falls there with Joe but contented himself with his "usual swim in [the] river." The students of the party often asked their esteemed professor for explanations of various geological phenomena, and Joseph readily responded, giving at one point a very long lecture on "the glacial phen[omen]a of Hetchhetchy." From this region the party pushed on toward Yosemite, arriving there on the first day of August. Once in the great valley, LeConte "was as strong as ever and enjoyed life as much as the youngest of the party." After more than two weeks in the Yosemite region, the party set out for home.[6] LeConte had done precious little geological investigation this time, but the trip had rejuvenated his spirits.

Meanwhile, Joseph kept busy with his teaching, writing, correspondence, and popular lectures. He had become such a sought-after speaker that he had to turn down some of the requests lest they eat away at his time for writing. He also had extracurricular duties at the university, such as the development of a geological and zoological museum. In addition to these activities, he entertained many prominent visitors, including Alfred Russel Wallace and E. D. Cope. Moreover, correspondence flowed between him and other scientists. Among geologists, for example, he exchanged letters with Alexander Winchell, Lester Frank Ward, O. C. Marsh, and Alexander Agassiz. On his evolutionary views, he corresponded with such men as John Thomas Gulick (on divergent evolution among varieties of species) and George J. Romanes (on physiological selection as a factor in evolution). The busy man also found time to revise his books on geology and evolution and to write

5 Joseph LeConte, notebook labeled "Hetchhetchy, Yosemite and High Sierra 1889," in LeConte Family Papers, UCB; Joseph N. LeConte, "Recollections"; Charles Palache, "Six Weeks in the Saddle," MS in Charles Palache Collection, Manuscripts Division, Bancroft Library, University of California, Berkeley; LeConte, *Autobiography*, 294–95.

6 LeConte, notebook "Hetchhetchy"; Joseph LeConte to Matilda "Tillie" Harden Stevens, September 20, 1889, in LeConte Family Papers, APS.

many major articles, as well as a great variety of minor pieces on such topics as the application of the principles of art to the novel, shad propagation, inherited polydactylism, glycogenic functions of the liver, the germ of hydrophobia, "sound blindness," and bird-flight as a basis for a flying-machine.[7] Altogether, LeConte published well over sixty articles and completed revisions of his *Elements of Geology* and *Evolution* during the decade 1885–1894.

By 1887, thoughts of retirement had finally begun to enter LeConte's mind, but he told Emma that it would be another four or five years before he terminated his active work. Shortly thereafter, he told Sallie that her recent letter had "brought up all my old yearnings for the South." He then informed her that "my plan is to continue to work

7 Joseph LeConte to Spencer Baird, February 22 and April 28, 1886, and Baird to R. E. C. Stearns, March 2, 1886, all in Record Unit 29, Box 3, Folder 17, SIA; Henry Fairfield Osborn, *Cope: Master Naturalist* (Princeton, N. J., 1931), 260; Alfred Russel Wallace, *My Life* (2 vols.; New York, 1906), II, 158; San Francisco *Daily Examiner*, May 26, 1887, p. 2; Joseph LeConte to George Davidson, January 16, 1888, and June 25, 1891, both in George Davidson Papers, Manuscripts Division, Bancroft Library, University of California, Berkeley; LeConte to Alexander Winchell, December 10, 1887, and December 24, 1888, both in Alexander Winchell Correspondence, Michigan Historical Collections, University of Michigan, Ann Arbor; LeConte to Lester Frank Ward, April 12, 1886, and September 13, 1889, both in Lester Frank Ward Papers, Brown University Library, Providence, R.I.; LeConte to O. C. Marsh, January 6, 1885, in Marsh Papers; LeConte to Alexander Agassiz, in Agassiz Papers. Among other prominent persons to whom LeConte wrote during this period were Stephen Bowers, September 23, 1885, in Stephen Bowers Papers, American Philosophical Society Library, Philadelphia; Sereno Watson, February 27, 1891, and Mrs. Asa Gray, November 29, 1888, both in Asa Gray Letters, Gray Herbarium Library, Harvard University; David Starr Jordan, May 18, 1891, in David Starr Jordan Papers, Stanford University Library, Stanford, Calif.; and W. A. Spalding, March 13, 1893, in W. A. Spalding Collection, UCLA Library, University of California at Los Angeles. In addition, he wrote in his notebook labeled "Misc. 1891–1892 & /93" that he sent a letter to Herbert Spencer on May 27, 1892 (in LeConte Family Papers, UCB). See also Addison Gulick, *Evolutionist and Missionary: John Thomas Gulick* (Chicago, 1932), 304; John Thomas Gulick, *Evolution, Racial and Habitudinal* (Washington, D.C., 1905), 214–15; George J. Romanes, *The Life and Letters of George John Romanes* (4th ed.; New York, 1898), 237–43. LeConte's articles included "The General Principles of Art and Their Application to the 'Novel,'" *Overland Monthly*, 2nd ser., V (April, 1885), 337–47; "The Result of Shad Propagation on the Atlantic Coast," *Science*, VI (December 11, 1885), 520; "A Case of Inherited Polydactylism," *Science*, VIII (August 20, 1886), 166; "Glycogenic Functions of the Liver," *American Naturalist*, XX, pt. 1 (May, 1886), 473–74; "Pharyngeal Respiratory Movements of Adult Amphibia Under Water," *Science*, VII (May 21, 1886), 462; "Germ of Hydrophobia," *Science*, VIII (July 30, 1886), 102; "Sound-Blindness," *Science*, X (December, 1887), 312; "The Problem of a Flying-Machine," *Popular Science Monthly*, XXXIV (November, 1888), 69–76; and "Ptomaines and Leucomaines, and Their Relation to Disease," *Pacific Medical Journal*, XXXII (September, 1889), 529–32.

here for 4 or 5 years more, at least until Joe graduates, and then ease up my labors by continuing my professorship only as a lectureship in Geology alone—and during only 6 months of the year." He added that he would spend the other six months in the South. Having already consulted one of the university regents on the matter, LeConte learned that "there *was no doubt that anything I wanted could be done.*" Under the proposed arrangement, LeConte would receive a yearly salary of $2,000–2,500 from the university, and that, coupled with his royalties, should suffice, he thought. At last he had begun to think of slowing down, but the actual deceleration was yet to be several years away.[8]

In spite of his nostalgia for South Carolina and Georgia, LeConte was wedded to California. His love of the West Coast, his warm friendships, and his strong attachment to the university were bonds not easily broken. In fact, LeConte enjoyed a warm social life that, although it did not excel, in memory at least, the grand relationships of old Columbia, was very meaningful to him—and to Bessie, though apparently to a lesser degree. Among Joseph's and Bessie's closest friends, apart from John and Josephine, were the E. W. Hilgards. Hilgard, a respected geologist and an authority on soils, had for many years lived in the South. In 1875 he was appointed as professor of agriculture at the University of California. Sharing much in common, Hilgard and LeConte soon developed a close relationship. Indeed, it was Hilgard more than anyone else who had "strongly urged" LeConte to publish his *Elements of Geology.*[9]

The LeContes were also intimate friends of William and Mary Keith. A fairly well-known and prolific artist, Keith was a close friend of John Muir as well. In addition to numerous sketches and paintings of Yosemite and mountain scenes, the capable but not outstanding artist painted numerous portraits of prominent Californians, among them Joseph LeConte. Joseph, Bessie, Carrie, and Joe were often entertained in the Keith home, where the social play usually involved cha-

8 Joseph LeConte to Emma LeConte Furman, December 10, 1887, in LeConte-Furman Papers, SHC/UNC; LeConte to Sallie LeConte Davis, January 23, 1888, in Davis Papers.

9 Joseph N. LeConte, "Recollections"; Hans Jenny, *E. W. Hilgard and the Birth of Modern Soil Science* (Pisa, Italy, 1961), 116; Hilgard, "Memoir of Joseph LeConte," 176; E. W. Hilgard, "An Estimate of the Life Work of Dr. Joseph LeConte," *University of California Magazine*, VII (September, 1901), 231–34.

rades and the "recitation of nonsense rhymes." The relationship between the LeContes and the Keiths seems not to have been strained
by two important occurrences, the first of which involved a sharp difference between Joseph and Mary Keith especially. The latter was an
ardent advocate of woman's rights, particularly of woman suffrage—a
point of sore contention with Joseph.[10] The other involved Carrie's affection for William Keith.

This matter was deeply rooted in other problems faced by Joseph's
youngest daughter. As Joseph stated in his autobiographical manuscript: "Carrie was a slender & apparently very delicate child." Her
illness and lame leg had "prevented her from joining fully in the
sports of other children [and] threw her on her own resources for
amusement." As a result, wrote Joseph, "she became reflective, imaginative & very fond of books." Carrie was especially interested in literature and drawing, and the latter partly attracted her to Keith. But,
continued Joseph, "her health easily breaks down with continuous
work." He also referred to her "delicate fancy-fine imagination." Carrie
overcame the handicaps of a crippled leg insofar as physical activity
was concerned, and, as Joseph observed, "she became a great walker
& mountain climber." He concluded in his assessment of his youngest
daughter that Carrie "is physically sound & enjoys good health when
not overstrained."[11]

The young woman had already begun to show signs of what Joseph
called "nervous prostration," and the psychological illness increased
over the years. The real nature of the problem is unknown, but it was
most certainly related to her physical handicap and the severe deprivations suffered during childhood. Eventually, at least in lay terms, it
became a spiritual crisis for Carrie. She seems to have fallen in love
with Keith when she was eighteen, and her unrequited love for the
happily married artist who was twenty-five years her senior caused
her severe suffering for well over a decade.[12]

Keith showered fatherly affection upon his young protégée, and
Carrie returned it with total adoration, a fact known to both Mary

10 Joseph N. LeConte, "Recollections"; Joseph N. LeConte to Elizabeth "Bess" Furman, February 13, 1895, in Nancy Smith Collection; Brother Cornelius, *Keith, Old Master
of California* (2 vols.; New York, 1942), I, 157.

11 LeConte, Autobiographical MS.

12 Susanna B. Dakin, "Introduction to 'C,'" in Caroline E. LeConte, *Yosemite, 1878:
Adventures of N. & C.* (San Francisco, 1964), vi–viii.

Keith and the LeContes. When Keith departed in 1883 for three years of study in Munich, Carrie went into a fit of depression that severely tested Joseph and Bessie. The young woman herself later admitted that Keith's departure threw her into "a great shock of horror and pain." After Keith returned to San Francisco, Carrie improved, and she resumed her study under the man she called her foster father. Keith in turn called her "Fostie." The lonely spirited woman wrote a number of verses for her love-idol; for example, there is one in which she spoke of having been found not as a baby in a garden paradise by her father but in the "rough and barren hills" by her foster father. She also wrote to her "Goodle" of how she loved his "queerness." "You are so wild and yet so mild," she penned in one of her poems of love. Keith played the game with her, as a father might play with a small child, and he heaped attention upon Carrie, encouraging her to write fanciful stories and imaginative interpretations of his art.[13]

This play-affection continued on into the 1890s, and Joseph and Bessie often tried to force Carrie into the world of reality. She only plunged deeper into her work of drawing and writing. Her parents sent her on trips to the South to visit her sisters as a means of encouraging her independence. Joseph also paid his daughter to draw illustrations for the revised edition of *Evolution*, which she did with considerable skill. In part, of course, he did this to get the drawings he needed, but in larger measure, it seems to have been directed toward giving her a sense of professional and personal importance. At times, however, both he and Bessie became exasperated. Referring once to "Carrie's trouble," Joseph averred that it was because his daughter pursued one line of activity too diligently. He also encouraged her to get "plenty of exercise and bathing" as a cure for her illness. Bessie lamented that "scolding does no good" and that she and Joseph had to treat Carrie gently and bear with her lack of patience.[14]

Carrie's "nervous prostration" continued for many years, and in 1898 she became virtually incapacitated, requiring so much attention from her devoted father that he gave over most of the year to her care. Carrie was indeed a bright woman, but her imagination simply ran

13 Cornelius, *Keith, Old Master*, I, 188–93.

14 Bessie LeConte to Sallie LeConte Davis, January 15 [1887], and Caroline LeConte to Sallie LeConte Davis, March 10 [1888], both in Davis Papers; Joseph LeConte to Daniel C. Gilman, April 23, 1888, in Gilman Papers; Joseph LeConte to Emma LeConte Furman, January 30, 1888, in LeConte-Furman Papers, SHC/UNC.

wild. Nowhere was this more evident than in the little book she wrote, titled *The Statue in the Air*. Published in 1898, the work was reviewed in the *Dial*, whose critic said: "The book presents in myth-like phantasietta the doings in some apocryphal pseudo-Hellenic fairy-land. It is a tale of shepherds and prophets, of hollow echoes and flushing apple-blossoms, of gods and harpies." Continued the *Dial* reviewer, the book seems to deal "mystically" with the "ultimate victory over Evil of Love through Art." Carrie's mystical story related the creation of "a marble child" out of "the cold air," and how this "sacred statue was enshrined in the cave of Love." Said the reviewer, "We have not wholly mastered the story" of the book, but he added that the work "has doubtless a significance beyond the details of the story."[15]

Actually, the book might never have been published had not Josiah Royce used his influence with the Macmillan Company. Royce had, of course, known Carrie since she was a young girl, and Joseph LeConte pressed his old student and friend to persuade Macmillan to accept the manuscript. In 1897 the LeContes (Joseph, Bessie, and Carrie) visited Royce in Cambridge, where he was a professor of philosophy. As Royce told a friend, Carrie was "slight, eccentric—a lonely, yet on the whole, despite moods, sensitiveness and imperfect constitution, a happy nature—a not uninstructed lover of art, a thoughtful dreamer of airy visions, without poses, and unlike, in some respects, the other dreamers whom we see." Later, in reference to Carrie's manuscript, Royce wrote that he agreed "that many things in the work of Miss LeConte are open to criticism." He cited "the oddities of style, etc.," but he believed these simply reflected "the inevitable oddity of this person." Royce concluded, however, that the oddities could not be altered "without destroying it [the book] altogether," and he secured the impartial opinion of a colleague who agreed with him.[16] Thus, the little book was published much as Carrie submitted it.

Carrie's mystical expression in *The Statue in the Air* offered no relief to her psychological dilemma, however, and Joseph had to attend to his daughter constantly, especially between 1898 and 1901. As his own

15 Carrie referred to her "seven years of a terrible nervous prostration" and suffering in a letter to Katherine "Kate" Furman Smith, July 2, 1930, in Nancy Smith Collection. This is confirmed in various LeConte correspondence and in LeConte, *Autobiography*. Carrie's little book was reviewed in *Dial*, XXIV (April 16, 1898), 266.

16 Josiah Royce to Mary Gray Ward Dorr, November 5 and 14, 1896, both in *The Letters of Josiah Royce*, ed. John Clendenning (Chicago, 1970), 348–49.

life drew near the end, he became more encouraged because Carrie seemed to be finding an answer to her problem in the Catholic faith. She had often mentioned to Keith that she was attracted to Catholicism, finding its forms and dogmas more suited to her tastes than those of the Episcopal Church, which she had joined as a teenager. In 1901, "in the midst of a long nervous prostration" after her father's death, she sought out a priest and told him of her desire to become a Catholic, which she did officially in 1903. Joseph had been well aware of her inclinations, and he had not discouraged her.[17]

Meanwhile, more grief was in store for Joseph. The health of his brother John continued to decline, while that of Josie grew worse also. Indeed, the health of the older LeConte was so bad that he had to request a year's leave of absence in May, 1889, which the university regents readily granted "with continuance of salary." John had become so feeble that one student complained that the study of physics "was made repellent by [his] senility," a point confirmed by Joseph in a letter to Tillie. John planned to take a recreational trip with Josie to Europe, but when Josie became seriously ill, the trip had to be canceled. The aging man therefore spent the academic year of 1889–1890 nursing his wife back to health, and he had to resume his duties at the university without achieving the rest he had sought. In late April, 1891, John took to his bed with a severe attack of bronchitis. Joseph was not surprised at John's rapid deterioration, for John had come to his room a few weeks before, "looking very weary," to inform his brother that "he felt his life work was done." He told Joseph that he wanted to retire at the end of the term. Joseph had "immediately consulted the regents" and arranged for John to be relieved of all but minor duties. But John could not ward off the illness that struck him only three weeks prior to his "promised rest," and he died on April 29, 1891, just a little over four months past his seventy-third birthday and two months short of his golden wedding anniversary.[18]

17 Cornelius, *Keith, Old Master*, II, 126–27.
18 LeConte, *Autobiography*, 295–97; Minutes of the Board of Regents, University of California, Vol. VI, May 13, 1889; Charles Palache, "Recollections of Studies at the University of California" (Typescript in Palache Collection); Joseph LeConte to Matilda "Tillie" Harden Stevens, June 23, 1889, March 16, 1891, and May 11, 1891, all in LeConte Family Papers, APS; Joseph LeConte to Emma LeConte Furman, April 30 and May 19, 1891, both in LeConte-Furman Papers, SHC/UNC; LeConte, "Memoir of John LeConte," 371–89; New York *Evening Post*, May 16, 1891. Joseph told one of his former students, who later became his attorney, that, in regard to John's death, "I feel as though half of my very self

Shortly after John's death, Joseph wrote to Emma of the details of her uncle's final illness, and then he added: "the generation to which I belong is practically gone—I am the last of my father's family. . . . But I know it is useless to think that way. I feel well and strong. I believe there is work in me yet, and I hope a good deal." The loss was nevertheless "inconceivable," as Joseph said in 1901. "With but brief interruptions, we had been companions all our lives; as children on the old plantation, as fellow-students in college and professional school, and as colleagues in Athens, Columbia, and Berkeley." Joseph's recollections were not quite accurate, of course, because he and John were not fellow students in college and professional school, but then, they had been together for so long that, when he reflected many years later upon the death of his brother, it must have seemed to Joseph that they were rarely apart from one another. John's widow was prostrated by the loss of her husband, and she never fully recovered, living out her final days in sadness and illness. Josephine lingered on in her melancholy mood until mid-December, 1894, when she met a tragic end. While she was dozing in her rocking chair, a newspaper fell from her lap into the fireplace, and the resulting blaze set her gown afire. The seventy-year-old lady, whom a California newspaper described as "one of the most magnificent appearing women in the state," died from the injuries she received.[19]

Although the year 1891 brought tragedy to Joseph with the death of his dear brother, it also brought rewards. During that year and the previous one he had published numerous papers, including several important contributions to geology. By this time, many of his fellow scientists had taken note of LeConte's extensive contributions to scholarship, and some of them thought it high time to consider their aging peer for the presidency of the American Association for the Advancement of Science. LeConte had, of course, first become a member of AAAS in 1850 and had served as its general secretary in 1861. As

---

had gone" (Joseph LeConte memorial address by William R. Davis, February 26, 1912, in LeConte Family Papers, UCB).

19 LeConte, *Autobiography*, 296; Joseph LeConte to Emma LeConte Furman, May 19 and April 30, 1891, both in LeConte-Furman Papers, SHC/UNC; Joseph LeConte to William "Willie" Louis LeConte, September 23, 1894, in possession of Mrs. Joseph Nisbet LeConte; Sallie LeConte Davis to Emma LeConte Furman, January 14, 1895, in Nancy Smith Collection; Louis Julian LeConte to Richard LeConte Anderson, June 6, 1972, in possession of Richard LeConte Anderson.

LeConte observed in his autobiography, "I was on the road to advancment and would doubtless in a few years more have been made president." But the war intervened and AAAS meetings were suspended. "When they were resumed in 1866," said LeConte, "I did not attend; the embittered feelings engendered by the war had not wholly abated, and moreover I was too poor to afford the expense." His membership thus "lapsed by default." LeConte noted with a hint of acrimony that after he moved to California, he was "practically cut off from intercourse with Eastern scientific men and had to work alone."[20]

LeConte's statement certainly contained an element of truth, but, on the other hand, the expatriated scientist had made no special effort to resume the "intercourse." Pride played heavily in his decision; he would not genuflect before the eastern establishment. He did not have to, for the AAAS came to him. After being elected a fellow in 1881, LeConte regularly paid his annual dues, but he did not attend the meetings, principally because of the expense involved. Thirty years after his last attendance at an AAAS meeting, he "received letters from some of the most prominent members" of the organization, informing him that he would be elected president if he attended the meeting in Washington that year.[21]

LeConte did attend, and the promise was kept; he was elected head of the largest scientific body in the United States. The belated honor was well deserved, for LeConte was certainly among the most capable and gifted scientists in the country. True, his accomplishments did not match those of Agassiz, Gray, or several other outstanding scientists of the nineteenth century, but those men were exceptional. On the other hand, LeConte was most surely qualified as a leader among that second echelon of capable scientists who contributed significantly to the advancement of their respective disciplines. Had he been less the universalist in a rapidly changing time of specialization, the old southerner might have achieved the top rank. At any rate, LeConte was accorded not only the honor of the presidency of the AAAS but also, only a few days later, was elected to one of the three vice-presidencies of the United States division of the International Congress of Geologists, which also met in Washington during August, 1891. The other two American vice-presidents were the noted geologists J. W.

20 LeConte, *Autobiography*, 298.
21 *Ibid.*, 299.

Powell and Raphael Pumpelly. The honor was enhanced when Le-
Conte was chosen to serve as acting president, a decision made necessary because the president, J. S. Newberry, could not attend on account of illness and advanced age. The occasion provided LeConte
with an excellent opportunity to become personally acquainted with
most of the leading geologists of Europe.[22]

Having received no hint that he might be elected to the vice-presidency of the ICG until two days prior to the meeting, LeConte was
most pleasantly surprised. He had attended a meeting of the American Committee of the ICG back in 1888, when his active participation
drew the attention of his confreres. The notice of his election also
came as something of a shock, since he had not prepared an address.
"I chose as my subject," he related in his autobiography, "The American Continent as a Geological Field and compared it in this respect
with Europe." The speech came off well, and LeConte presided over
several sessions of the congress with great skill and composure. Of all
the participants in the congress, he was certainly among the most
exuberant. In the short space of a few days, two crowns had been
placed upon his aging head after forty years of labor in the cause of
science.[23]

When the conference ended, LeConte left Washington for New York
to oversee the fourth edition of his *Elements of Geology* and to enroll
his son in the graduate program of electrical engineering at Cornell
University. Young Joe was the first recipient of the LeConte Fellowship
established by the Alumni Association of the University of California
in honor of his father and uncle. Meanwhile, Bessie and Carrie had
come to New York in September. They and Joseph left shortly thereafter to visit Sallie and Emma. Joseph stayed with his daughters for
four months, and on November 8 he wrote to his young colleague

22 *Ibid.*, 299–300.

23 Report of the International Congress of Geologists, *American Geologist*, II (September, 1888), 141–284; *Congrès Geologique International, Compte Rendu de la 5ᵐᵉ Session, Washington, 1891*, published in *Miscellaneous Documents of the House of Representatives*, 53rd Cong., 2nd Sess. (Washington, D.C., 1894), XIII, 83–89; *Report of the
Proceedings of the Fifth International Congress of Geologists, August 27–31, 1891* (Washington, D.C., 1891); International Congress of Geologists, *Circular of Information, No. V*
(Washington, D.C., 1891); International Congress of Geologists, *Members of the Committee of Organization and List of Members and Delegates Present* (Washington, D.C., 1891);
excerpts from newspapers reporting on the conference, in a volume of circulars—all
located in the United States Geological Survey Library, Reston, Va.

Andrew Lawson, "I am in perfect health—never better in my life." During his leisurely stay, LeConte began to prepare a paper on the "Race Problem in the South," which he would later deliver before the Brooklyn Ethical Association. He was also invited to present a paper on "The Relation of Philosophy to Psychology and Physiology" to the Philosophical Society in Washington.[24] Both of these papers, given in early 1892, represented important extensions of LeConte's evolutionism.

For Joseph LeConte, the year 1891 had indeed brought many rewards, and the following year was to bring even more. For a long time he had desired to visit Europe, and in March, 1891, he asked the university regents for a leave of absence. "After a continuous connection with the University of 22 years," he wrote, "I now for the first time respectfully ask for a leave of absence for one year commencing August 1891." He added in his request to the regents that he hoped to "make my absence profitable not only to myself in the way of health and recreation but also to the University in the wider knowledge of University methods which I hope to acquire." The very next day, the regents acted on his request, granting him leave "with full pay." Joseph and Bessie viewed the trip as not only beneficial to themselves but also to Carrie, thinking it might help to relieve their daughter's depression. So to Europe they went, departing on February 27, 1892, only one day after Joseph's sixty-ninth birthday. Two weeks later they landed at Genoa. Their stay in Europe lasted for six months, and LeConte later wrote that the trip was "an important epoch in my life."[25]

The LeContes spent two weeks in Rome. From there, they traveled to Naples, which particularly fascinated Joseph because of its beautiful natural scenery. They also visited Pompeii and ascended Vesuvius. After a week in Naples, the family made the rounds of the famous art galleries in Florence, Venice, and Milan, all of which evoked great appreciation from Joseph, who was, as he called himself, "an intense lover of art." Moving on to Switzerland, they spent a week in Zurich because Joseph wanted to study the program of the city's famous old university and that of its reputable Polytechnicum. Afterward, they

24 LeConte, *Autobiography*, 300–301; Joseph N. LeConte, "Recollections"; Joseph LeConte to Andrew Lawson, October 5 and 19, and November 8, 1891, all in Andrew Lawson Papers, Manuscripts Division, Bancroft Library, University of California, Berkeley; letter from the Brooklyn Ethical Association to LeConte, May 19, 1891, in LeConte-Furman Papers, SHC/UNC.

25 Minutes of the Board of Regents, University of California, Vol. VIII, March 9 and 10, 1891; LeConte, *Autobiography*, 302.

visited the German cities of Heidelberg and Cologne, but apparently Joseph made no effort to go to Berlin and visit with Helmholtz, the famous authority on physiological optics.[26]

LeConte expected to be accorded greater recognition in France anyway, so he spent the entire month of May in Paris. There he was especially delighted over his frequent visits with the geologists Albert Gaudry, Pierre Boule, and Gabriel Auguste Daubrée. LeConte had met Gaudry a few months before in Washington at the International Congress of Geologists, and at the same time he probably became acquainted with Boule, an authority on the mountains of central France. Daubrée, a lithologist and mineralogist, was president of the French Academy of Science. LeConte also visited Émile Javal, the French authority on physiological optics and the man who had translated Helmholtz's volumes on that subject into French. Invited by the Sorbonne professor to a luncheon in his home, LeConte quickly accepted. Upon entering Javal's study, he spied a copy of his *Sight* lying on the Frenchman's desk. Javal informed LeConte that "he used it in his teaching of physiological optics." LeConte took the opportunity to note that he differed with Helmholtz on a number of fundamental points, to which Javal replied that he considered LeConte "right in every case." Fortunately for LeConte, Javal and his family all spoke English, and, as LeConte put it, "I spent a very delightful day."[27]

In June the LeContes crossed the Channel to England. "Ah! the delight of hearing my mother tongue again!" exclaimed Joseph. "It was like returning home." The three happy travelers remained in London until early July, finding there a number of old friends. But LeConte also made new friends, among them Joseph Prestwich, well-known geologist and retired Oxford professor. Almost eighty years old at the time of LeConte's visit, Prestwich (later knighted) was "still full of life and of interest in science." The "genial and kindly" old fellow invited the LeContes to stay for several days in his home. LeConte was particularly intrigued by Prestwich's collection of palaeolithic instruments, of which the generous Englishman gave his guest a number of specimens. Another geologist of whom LeConte saw much was the Scotsman Sir Archibald Geikie, one of Britain's most famous earth scientists.[28]

26 LeConte, *Autobiography*, 307–308.
27 *Ibid.*
28 *Ibid.*, 308–10.

He also had the pleasure of meeting other scientists of repute, including Sir John Lubbock, the famous astronomer and mathematician, and the geologists S. H. Woodward, Wilfrid Huddleston, and John W. Judd. Moreover, he was introduced to Sir Andrew Clarke, the British colonial administrator and engineer, who greeted LeConte with the question, "Are you the author of this book?" as he held a copy of *Evolution* up for his guest to see. Sir Andrew showed LeConte "how carefully [he] had read it" by thumbing through page after page of "marginal annotations." LeConte was "of course greatly gratified," as he always said when someone praised his work. He also met the Scotsman George Croom Robertson, founder and editor of *Mind*, the noted British philosophical journal.[29]

Rejoicing in such cordial receptions, LeConte felt that he had gained something of an international reputation—and justifiably so. At Cambridge he stayed in the home of T. McKenney Hughes, the respected geologist with whom LeConte had become well acquainted at the Washington meeting of the ICG. Through Hughes's auspices, LeConte was introduced to other Cambridge dons. Even more satisfying perhaps was LeConte's stay at Oxford in the home of George Romanes. The English biologist and physiologist, who worked out a modification of his friend Charles Darwin's theory in the form of "physiological selection," had often corresponded with LeConte. He praised LeConte's book on evolution not only for its treatment of the theory but also for its discussion of the relation of the theory to religion. In fact, Romanes considered it the best book written on the latter topic. He had also lauded LeConte for his paper on the evolution of flora in coastal California and adjacent islands, primarily because Romanes felt that it bolstered his own hypothesis regarding physiological selection in isolated groups of species. But he had likewise disagreed with LeConte, specifically on the matter of sexual selectivity among species, because he believed the American evolutionist had overemphasized the tendency of species to propagate "for the sake of better results in the offspring." That, argued Romanes, makes "natural selection act prophetically," a point that gave LeConte considerable difficulty.[30]

29 *Ibid.*, 310–12.
30 Romanes, *Life and Letters of George John Romanes*, 237–38, 240–42, 295; George J. Romanes, *Darwin and After Darwin* (3 vols.; Chicago, 1901–1906), I, 412; Romanes to LeConte, April 20, 1889, in LeConte Family Papers, UCB.

Nevertheless, Romanes was impressed with LeConte's writings on evolution, and he invited many people to meet his guest at a dinner in LeConte's honor, among them the Anglican prelate Charles Gore, later to be elevated to the office of Canon of Westminster. Gore, an able essayist, took LeConte on a tour of the Oxford grounds after dinner, and, as the delighted American put it, Gore "took my breath away by telling me that he thought so highly of my book, Evolution and Its Relation to Religious Thought, to which Professor Romanes had drawn his attention." Gore also informed LeConte that he used the book in his classes on evolution and religion.[31]

After a brief trip to several cities in Ireland, the LeContes set sail for America at the end of July, arriving in New York in early August. Joseph was in high spirits: he had received greater recognition abroad than he had anticipated; he was president of the AAAS and vice-president of the ICG; his son had received his master's degree in electrical engineering and an appointment to the faculty of the University of California; and his own health was excellent. In addition, the revised editions of his *Elements of Geology* and *Evolution* were off the press and doing well. Upon returning to California, LeConte found himself still much in demand as a speaker, and during the winter of 1892–1893 he was called upon to give lectures in Los Angeles and San Diego on the glacial epoch in California and on "The Relation of Organic Evolution to Human Progress" at the California Teachers' Convention in Fresno. On the occasion of his seventieth birthday in February, 1893, the university's Academic Senate honored him at a dinner in the Palace Hotel of San Francisco. It was a grand affair attended by a large number of guests, "including several distinguished visitors." University President Martin Kellogg and others paid tribute to their most illustrious teacher, their beloved "Professor Joe." It was only the first of what became annual tributes to the esteemed teacher.[32]

The University of California regents had long recognized the excellent services of their honored faculty member. They had, of course,

31 LeConte, *Autobiography*, 311–12.

32 *Ibid.*, 312–18; LeConte, "The Relation of Organic Evolution to Human Progress," *Pacific Coast Teacher*, II (February, 1893), 131–39; *Berkeleyan*, I (February 3, March 3, 17, and 24, and May 17, 1893), 11, 59, 81, 106, and 202, respectively; Joseph LeConte to Elizabeth "Bess" Furman, February 15, 1893, in LeConte-Furman Papers, SHC/UNC. On September 24, 1892, the Alumni Association of the University of California had given a grand banquet and testimonial tribute to him; see *Proceedings at the Banquet Given to Professor Joseph LeConte*, in McMillan Collection.

readily granted his request for a leave of absence with "full pay," and they would soon do so again. As early as 1884, the regents had bestowed upon him an honorary professorship in biology in the dental department of the university, resolving that he deserved the honor "in consideration for his purity of character as a citizen: eminent ability as a scholar and Lecturer:... [and] the long period of usefulness in the University of California as a teacher." In 1892 the regents unanimously agreed to increase LeConte's salary "to $4,000 per annum." The new salary was the highest paid to any faculty member at the university, and it compared favorably with that paid by other leading universities. The regents were fully aware of the prestige of their oldest faculty member, and the praise and honors heaped upon him could not be ignored.[33]

The year 1893 was indeed a high point in the long career of Joseph LeConte, but it also brought some problems. The first of these came during the summer, when LeConte decided once again to visit the High Sierra. Leaving home on June 22, LeConte planned to spend three weeks in the area of Yosemite. But the vacation turned into an unpleasant trip, both physically and mentally, for the seventy-year-old man. Suffering from a bad cold, LeConte was wracked with fever and sinus pain most of the time, and he had to take morphine, ginger, and quinine to relieve his aching body, especially at night. It may have been largely the physical pain that precipitated a mood of morbidity in LeConte, but in part it was also the result of his realization that he had now lived the three-score-and-ten years allotted to man. Perhaps he was also thinking of his deceased brother John. Whatever the case, the aging fellow was sure that he was on the last visit to his beloved Yosemite.[34]

During a ride by himself out to Nevada Fall, LeConte "turned about and gazed ... in delight and yet in grief." He recorded in his notebook that he felt he "was taking leave of [the] grand falls perhaps *forever*. Associated with so much happiness—to leave it seemed like leaving my life." To this is added, "I feel I am at last an old man," and later, at the foot of Vernal Fall, he noted, "I gazed ... with passionate grief &

33 Minutes of the Board of Regents, University of California, Vol. V, November 5, 1884, and Vol. IX, October 11, 1892; David Starr Jordan, *The Days of a Man* (2 vols.; Yonkers-on-Hudson, N.Y., 1922), I, 217, 450.

34 LeConte, *Autobiography*, 318–19; Joseph LeConte, notebook labeled "Yosemite 1893," in LeConte Family Papers, UCB.

even tears." Yet LeConte would not completely concede that all was over, and he wrote, "But if physically I am old & must acknowledge it—I yet have y[ea]rs to do much intellectual work & I will do it." He concluded his morose entry with a note that he had nothing to regret and that he was "willing to go" because he had "perfect faith in the Good Father of all." [35]

In spite of his febrile condition, LeConte could not resist further speculation on the origins of Yosemite, and he recorded some of his thoughts on the geological phenomena of the region. By the fourth day of July, his condition had improved somewhat, and he sat around the campfire with others as they played games and sang, "winding up with [the] patriotic 'Star Spangled' and 'God Save America.'" He departed for home by stagecoach the next day, feeling slightly better in spite of the "hard ride" and longing to be with his beloved Bessie. On the seventh of July, he arrived at home, "took a good supper," and then went "to bed with the dearest & best of women." [36]

LeConte soon forgot his depression and resumed his normal activities. His health returned, and he plunged back into his work again. Indeed, as he had said, there was still much intellectual vigor left in him, and he proved it by writing more articles, drafting a memoir on his brother John for the National Academy of Sciences, setting to work on a revision of his *Sight*, and staying active in teaching and in professional organizations. In fact, the rest of the year was most pleasant— until the very end of the month of December, when he received a highly disturbing letter from Alexander Agassiz, his old friend and son of his great teacher. He and the younger Agassiz had corresponded on a number of occasions, and they had exchanged copies of their publications. LeConte thought so highly of the accomplished marine zoologist that he had consulted Agassiz for advice on the development of "a marine station" and an accompanying chair of biology at the University of California. He also invited Agassiz to visit him and stay in his home in Berkeley. Only a day or so before Christmas in 1893, however, LeConte received what was apparently a very unkind letter from Agassiz, and it upset the old scientist greatly. [37]

"I confess it [the letter] was not a very pleasant Christmas gift," re-

35 LeConte, "Yosemite 1893."

36 *Ibid.*

37 Joseph LeConte to Alexander Agassiz, October 8, 1888, and April 21, 1893, both in Agassiz Papers.

plied LeConte to Agassiz. "I confess I was both surprised & pained that you should think my reclamation unwarranted and especially that you should think it unjust to your father." LeConte added that he held Louis Agassiz "in such veneration & love that I take every occasion, in season & sometimes perhaps out of season, to do honor to his memory." What was at issue was LeConte's claim that he had been the first scientist to show how the Florida peninsula was built up by action of sediments deposited by the Gulf Stream. Just why Alexander Agassiz should choose that time to take issue with LeConte over his claim is unknown. LeConte had published his original paper on the topic in 1857 and had written brief modifications of it in 1880 and again in 1883, but, for some reason, Agassiz did not dispute LeConte's position until 1893. This is strange, for Agassiz was surely well acquainted with the literature on the subject of coral formation, one of his own special areas of interest. Perhaps his reaction was triggered by a brief address that LeConte delivered to the Harvard Club of San Francisco. The address dealt with the "Work of Agassiz," as LeConte titled the short speech later when he noted it in his "Index Rerum."[38]

Mainly a reiteration of LeConte's previously published praise of Louis Agassiz for his role in developing the comparative method of studying species, the little talk briefly referred to LeConte's work with Agassiz on the Florida reefs. LeConte claimed that "subsequent reflection on the observations then made gave rise to my first and, I think, one of my most important scientific papers, viz: 'The Agency of the Gulf Stream in the Formation of the Peninsula, the Keys and the Reefs of Florida.'" It was no more than LeConte had previously claimed, and it hardly seemed immodest, nor did it depreciate Louis Agassiz's work on the topic.[39] At any rate, for whatever reason, the younger Agassiz must have written a fairly caustic letter, and LeConte was much dis-

38 LeConte to Agassiz, December 26, 1893, in Agassiz Papers. In addition to his "On the Agency of the Gulf Stream" (1857), LeConte's published papers and letters on coral reefs included: "Rate of Growth of Corals," *American Journal of Science*, 3rd ser., X (March, 1875), 159–64; "Coral Reefs and Islands," *Nature*, XXII (October 14, 1880), 558–59; and "The Reefs, Keys, and Peninsula of Florida," *Science*, II (December 14, 1883), 764. In the last of these, he gave special acknowledgment to the work of both Louis and Alexander Agassiz.

39 LeConte, "Work of Agassiz," in Joseph LeConte, Miscellaneous Writings, Bancroft Library Archives, UCB. In a brief "Note on the Florida Reef," *American Journal of Science*, 3rd ser., XLIX (1895), 154–55, Agassiz offered a correction of some of his father's conclusions but did not mention LeConte.

tressed when he received it. It was especially painful for LeConte because he did indeed venerate his former mentor, and he never tried to claim credit for Agassiz's work.

Despite these two low points encountered by LeConte in 1893, the year was, on the whole, a time of great pleasure for him, and he continued to work, think, write, and enjoy life as though he had many fruitful years left. Indeed, he did, and for the next eight years he remained as active as he could. His fame increased, as did his ever-optimistic outlook on life.

—••❮ XII ❯••—
Evolutionism and a "Science of Society"

A strong wind of optimism swept across the United States during the 1890s as American intellectuals touted the inevitability of social, cultural, and political progress. Their firm faith was grounded in the Neo-Lamarckian theory of evolution by physical adaptation. As the budding science of sociology began to flower, social theorists found a ready model for the explication of cultural development. Like the majority of his intellectual contemporaries, Joseph LeConte embraced the Neo-Lamarckian view of the inheritance of acquired characteristics and the role of environmental adaptation in organic evolution, and, also like his peers, he firmly believed in the direct applicability of the theory to social and cultural progress. Indeed, the 1890s marked the high point of LeConte's evolutionist writings, and in paper after paper he freely related Neo-Lamarckism to every topic he considered.[1]

As a leading spokesman for the theory, LeConte articulated and popularized the peculiarly American brand of evolutionism. Perennially optimistic, he held an almost undeviating faith in the idea of progress, which served as a key term in his philosophy of organic and social life. In 1890, for example, he resolutely declared that "all human progress is a process of expansion of the mental horizon,—an expansion from the narrow limitations of the animal to the universality of the spiritual." He continued: "Now science is the type of human progress in the field of thought. This law of expansion, of generalization, and finally of universalization is not only conspicuous in her own

---

1 Pfeifer, "The Genesis of American Neo-Lamarckism," 156–67; Pfeifer, "United States," in Glick (ed.), *The Comparative Reception of Darwinism*, 198–203; Ray Allen Billington, *Frederick Jackson Turner: Historian, Scholar, Teacher* (New York, 1973), 112–15.

domain, but by a sort of intellectual contagion she has gradually in-
fected every other department with her own spirit and methods."
There are, he noted, two kinds of evolution: "organic" and "human."
The first is effected through various factors: the "pressure of a changing environment," changes through "inheritance" of modified structure (*i.e.* the "use and disuse of organs"), "natural selection," "sexual
selection," and, finally, "physiological selection." In the case of human
evolution or "human progress," however, "there is introduced a new
and higher factor which immediately takes precedence of all others."
This he identified as "the *conscious, voluntary coöperation of the human spirit in the work of its own evolution.*" In actuality, explained
LeConte, the last factor of evolution "cannot come until the *end*"—that
is, until the progress of organic evolution is complete.[2]

But such an extension of the Neo-Lamarckian theory was no better
than the original theory itself, and that had come under serious scrutiny by a number of thinkers whose criticisms threatened to wreck the
fundamental tenets of environmentalism held by LeConte, Lester
Frank Ward, and other intellectuals who sought to apply the concept
of physical adaptation to the social order. LeConte could not be
moved, however, and he summoned forth his staunchest defense of
evolutionary social theory. A good example of this effort appeared in
his 1896 address on "The Relation of Biology to Philosophy," the focal
point of which was a critical review of Professor John Watson's book
Comte, Mill, and Spencer. Commencing with the statement that he
hoped to "clear up some points left obscure in [his] previous writings,"
LeConte proceeded to identify two "fundamental mistakes" in Watson's argument against evolution. The first error, he contended, was
the synonymous usage of "Darwinism" and "natural selection," and the
second, "the limitation of Darwinism to natural selection." Claiming

---

2 LeConte, "The Relation of Organic Evolution to Human Progress," 131–39; Joseph
LeConte, "The Theory of Evolution and Social Progress," *Monist*, V (July, 1895), 481–500;
Joseph LeConte, "The Relation of the Church to Modern Scientific Thought," *Andover
Review*, XVI (July, 1891), 1–11. LeConte's previous statements on the topic are found in
his "The Relation of Organic Science to Sociology," *Southern Presbyterian Review*, XIII
(April, 1860), 39–77 (revised and reprinted as "Scientific Relation of Sociology to Biology,"
Popular Science Monthly, XIV [January, 1879], 325–36); and "Relation of Biology to Sociology," *Berkeleyan*, XXIII (May, 1887), 123–31. For a discussion of the early sociological
theories of LeConte, see Theodore Dwight Bozeman, "Joseph LeConte: Organic Science
and a 'Sociology for the South,'" *Journal of Southern History*, XXXIX (November, 1973),
565–82.

that Watson's interpretation was grounded in August Weissmann's and Alfred Russel Wallace's "extreme view" on Darwin's idea of natural selection as the sole, or at least the major, cause of evolutionary change, LeConte reiterated his belief in the five factors of organic evolution. He insisted not only that "selection of the fittest" was "the most wasteful and unintelligent, though perhaps most effective, of all possible methods of improvement" among species but also that the idea failed to account for "purposism" in the process. In his estimation, the fortuitous nature of natural selection thus made a plaything of life. As a Neo-Lamarckian proponent, LeConte therefore argued diligently for evolutionism as a "process of having an end, a goal, a purpose." In the process of evolution, he said, species become more complex and subsequently their "functions are separated and localized, each in its own organ, and therefore proportionately perfected, and their purpose plainly declared."[3]

Following his long-held belief in the correlation of chemical, physical, and vital forces, LeConte asserted once more that "evolution is far wider than organic evolution." Indeed, he argued, "there are several different kinds and grades of evolution, of which organic evolution is only one." The four kinds he identified as "1st, Physical, or Cosmic; 2d, Chemical; 3d, Organic; and 4th, Human, or Rational,—each determined by a different force and carried forward by a different process, to reach a different goal." In the beginning, through "physical forces, mainly by gravitative attraction," a "universal cosmic order" evolved "out of primal chaos." That process, he explained, "preceded all," but it continues in the present. While physical forces were reaching their peak of cosmic evolution, chemical affinities began to form as the earth's surface cooled from the "intensity of primal heat." Compounds formed into "combinations and re-combinations" until protoplasm was "achieved," after which organic evolution commenced. In this third stage of evolution, "higher and higher forms of plants and animals" evolved over the eons, finally culminating in man, "the highest

3 George Daniels (ed.), *Darwinism Comes to America* (Waltham, Mass., 1968), 75–76; Stocking, *Race, Culture, and Evolution*, 253–54; Mrs. Ole Bull, "The Cambridge Conferences," *Outlook*, LVI (August 7, 1897), 845–49; Joseph LeConte, "The Relation of Biology to Philosophy," *Arena*, XVII (March, 1897), 549–67. LeConte stated that he had read the paper on several previous occasions and complained that it was published "without my permission." He called the matter "a grievous wrong to me, particularly as the sense of the article was marred by some bad typographical errors," but he absolved the editor of blame.

possible animal." Once this process was complete, man himself continued to evolve, and the stage of "Human Evolution—Social Progress" began.[4]

From this broad, speculative, philosophical base, LeConte ventured into the ramifications of his theory for many areas of social activity, one of the most prominent of which was the educational system. Thus, when he was invited to address the National Education Association in Denver, Colorado, the old professor chose to speak on "The Effect of the Theory of Evolution on Education," a topic that represented the culmination of many years of thought on the training of American youth. For many years, American educators had been advocating reform of the educational system, and if some of them disagreed with LeConte's emphasis on evolution as a guide to change, few of them rejected his proposal for pedagogical modifications.[5]

Ever since Herbert Spencer's *Education* was published in 1860, a number of educators had worked hard to reform the traditional program of American schools and colleges. Tied closely to pre-Darwinian evolutionary theory, Spencer's ideas had served as a launching point for reformist efforts. By the late 1880s, however, the Englishman's proposals had ceased to hold center stage, giving way to other social theories more suitable to American pedagogues. Actually, American social reformers had split into two camps, the Social or Conservative Darwinists and the Reform Darwinists. The first group believed that since the process of evolution was now complete, current institutions represented the highest stage of evolutionary progress. The latter group, on the other hand, supported the idea that the process of evolution was incomplete, and therefore social institutions must also continue to develop. The reformers thus wanted to control the environment in harmony with evolutionary laws. In reality, the eclectic thinker LeConte borrowed from the opposing theories, synthesizing and adding to them to provide what he considered a more plausible explanation of the role of science and evolution in the function and development of human knowledge.[6]

4 LeConte, "The Relation of Biology to Philosophy," 554–65.

5 Joseph LeConte, "The Effect of the Theory of Evolution on Education," *Proceedings of the National Education Association, 1895,* 149–61; Lawrence A. Cremin, *The Transformation of the School* (New York, 1961), 90–126; Boller, *American Thought in Transition, passim*; Hofstadter, *Social Darwinism in American Thought,* 31–84.

6 Eric Goldman, *Rendezvous with Destiny* (New York, 1952), 85–97.

Some of LeConte's ideas were absorbed into the sociological theories of his friend Lester Frank Ward and thereby indirectly influenced the thought of a generation of social theorists. As a leading sociologist in the latter decades of the nineteenth century, Ward exerted a strong influence upon the social and educational thought of his time. Ward also subscribed to the idea of progress, but, unlike some of his contemporaries, he maintained that man had to work actively to achieve the perfect, or democratic, society. Like LeConte, he firmly believed it possible to develop a genuine *science* of society, and he considered education the means of rectifying all social ills, achieving progress, and reaching the goal of happiness. A staunch egalitarian, Ward rejected the elitist conception inherent in Social Darwinism, or the idea of "natural selection," favoring instead the Neo-Lamarckian theory of evolutionary progress.[7]

LeConte also opposed the idea that the intense struggle for survival of the fittest was the chief factor at work in the evolutionary process, stressing instead the altruistic principle of love and mutual dependence as the most characteristic operational force in human progress. It was particularly on that point, of course, that LeConte differed from Spencer. Although he acknowledged Spencer as the individual who had done the most to apply the theory of evolution to sociology, LeConte argued that Spencer's ideas had had little practical effect upon social progress. That was the case for two reasons, said LeConte: first because the subject was extremely complex and difficult, and, second, because "the impulse" of Spencer's work had "taken a *wrong* direction." The Spencerian school, he insisted, had fallen victim to a "materialistic or mechanic philosophy" that tended "to identify the social organism with the animal organism."[8]

In short, LeConte charged that Spencer and his disciples had ignored the "many *kinds* of evolution under guidance of *different forces*,

---

7 Ward cited LeConte in most of his volumes on sociology, and the two men corresponded with each other several times between 1886 and 1898. The letters are in the Ward Papers. For a discussion of Ward's sociological and educational theories, see Boller, *American Thought in Transition*, 64–69; Hofstadter, *Social Darwinism in American Thought*, 67–85; and, especially regarding Ward's ideas on education, Samuel Chugerman, *Lester F. Ward: The American Aristotle* (Durham, N.C., 1939), 459–96.

8 LeConte, "The Theory of Evolution and Social Progress," 487–91; LeConte, *Evolution*, 26–27. Ward's estimate of LeConte is in his *Outlines of Sociology* (New York, 1923), 79, and examples of his reliance upon LeConte's theories can be found in Lester Frank Ward, *The Psychic Factors of Civilization* (Boston, 1901), 221, 240, 292, 314.

operating by *different laws* and on *different planes*." He further argued
that Spencer disregarded the higher forces by confusing organic evolution with social progress. "Besides the unconscious evolution by
natural laws, *inherited from below*," said LeConte, "there is a higher
evolution, *inherited from above*." The higher evolution was voluntary
rather than spontaneous as in the lower or natural evolution. Moreover, the higher evolution he viewed as a conscious effort "of the best
members of the social aggregate" to attain "a recognized *ideal*." In
other words, it had moved from mere unconscious, material *evolution*
to a deliberate endeavor to achieve social *progress*.[9]

One of LeConte's key arguments was that the idea of evolution was
itself evolutionary. He contended that Lamarck had imbued the idea
with life; that Darwin had accomplished the task of convincing the
scientific world of the validity of organic evolution; that Spencer had
"extended the law of evolution to embrace every department of nature"; and that T. H. Huxley had fought the necessary battles to solidify
scientific gains in the theory of evolution. But now it was the responsibility of American evolutionists "to complete the evidence from paleontology . . . [and] to apply it fearlessly . . . to religious and social
thought."[10]

LeConte believed that scientific thought had progressed from philosophic idea to semiscientific theory to true scientific theory. Under the
first, evidence was derived by speculation, but ideas were removed
from practical application and progress was provisional or empirical.
In the second stage, which was transitional or critical, philosophy still
prevailed over true science, and although new ideas had been entertained by Georges Cuvier early in the nineteenth century and by Robert Chambers (later acknowledged as author of *Vestiges*) in the 1840s,
the general approach was characterized by misconceptions, poor reasoning, and sometimes specious evidence. LeConte contended that
the third stage of genuine scientific theory had been inaugurated in
the work of Darwin, who not only established evolution as a fact but
also demonstrated "*how* it could and did take place." Progress in this
stage was permanent because rational or scientific thought had
triumphed. At last, maintained LeConte, a law of causation or law of

9 LeConte, "The Theory of Evolution and Social Progress," 487–91.
10 LeConte, "The Effect of the Theory of Evolution on Education," 150; LeConte, "The
Theory of Evolution and Social Progress," 484–87; Joseph LeConte, "Illustrations of a Law
of Evolution of Thought," *Princeton Review*, 4th ser., VIII (November, 1881), 392–93.

derivation of successive forms had been firmly established. Now, he continued, the law of "the derivative origin of all things ... is certain and applicable to all nature and therefore to human society." Hence, LeConte felt that society and education could proceed to advance in accordance with the firmly established laws of organic evolution applied through scientific sociology and pedagogy.[11]

As early as 1861, LeConte had called for reform of the college curriculum, saying: "A course of education is an *organized* system. It may be, therefore, compared with other *organisms*, and is subject to their laws." But the basic goal, he later declared in agreement with Spencer, is to construct an educational scheme whose "aim and end ... is preparation for a worthy life." To attain that goal, the curriculum must cultivate the major aspects of our intellect by including studies in science, art, and philosophy—each structured in evolutionary progression. The science course would commence with the study of mathematics, proceed through mechanics, physics, chemistry, biology, and geology; then it would progress finally to sociology. The art course would begin with the study of language, progressing thence to literature, fine art, history, and ultimately to philosophical history. It would likewise culminate in the study of sociology. The study of logic would initiate the third course, philosophy. Then it would progress to the study of mental and moral philosophy, connecting finally with the other two courses in the study of sociology—the "all-embracing subject." Each of these three courses, LeConte continued, has a special "correlative function." The science course is intended to cultivate "an intense love of truth for its own sake." The function of the language course is to cultivate "the power of expression," while the philosophy course is to provide "rational grounds of the validity of all knowledge."[12]

Before embarking upon such a college program, however, each student should be properly trained in a lower school. LeConte reasoned that since every normal human being is acted upon in two ways—

11 LeConte, "The Effect of the Theory of Evolution on Education," 150–51.

12 Joseph LeConte, "Natural History as a Branch of School Education; and the School, the College, and the University, in Relation to One Another and to Active Life," *Southern Presbyterian Review*, XIV (July, 1861), 186; Joseph LeConte, "The True Idea of a University," *University Chronicle*, I (February, 1898), 3–6; Joseph LeConte, "The Essential Characteristics and Mutual Relations of the School, the College, and the University," *Princeton Review*, 4th ser., V (March, 1880), 177–204.

through the senses and by fellow-beings—the lower school should provide a "nature-culture" and a "language-culture." The first of those two cultures was to be achieved through the study of natural history and then later, in higher education, through the study of natural science. A nature-culture course should thus include inorganics (physics and chemistry) taught in laboratories, and it must include the study of "living things" taught by field work. Pure book work, he exclaimed, serves to destroy initiative and spontaneity. Finally, the lower school should stress "hand-work." But that, LeConte quickly pointed out, was not the same as manual training for practical utility as taught in trade schools. He was not opposed to such technical education, but he objected to its connection with the public schools. The true aim of handwork was to train the mind. Both nature-culture and handwork were "natural methods," and a return to them was essential, for, as civilization progresses, "man separates himself from nature."[13]

Stressing the mutual relationship of the school, the college, and the university, LeConte further outlined what he considered the ideal curriculum. He stated that culture consisted of two grades: natural history and science, the former suited for "lower education," and the latter, for "higher education." Natural history in the schools would include chemistry (names and properties of elements and how compounds are formed); *Erdkunde* (forms of the physical phenomena of the earth); biology (forms, habits, structures, and classifications of organic life); history (detailed accounts of past events); languages (grammatical rules and lexicon); and mathematics (arithmetic and algebraic operations). These studies, he averred, would cultivate the memory and perceptive powers and would develop skills of "rapid comparison."[14]

Both the nature and the language courses were to be continued in the college curriculum, but mathematics would be transferred to the nature course, which now became the scientific course. The language course would be enlarged to cover "all modes of expressing or embodying thought and feelings, such as art, literature, etc.," becoming thereby a "language-art course." And a third area would be added:

13 Joseph LeConte, "Sense-Training and Hand-Training in the Public Schools," *Pacific Educational Journal*, III (March, 1888), 41–52.
14 LeConte, "Essential Characteristics," 180–96; LeConte, "True Idea of a University," 3–6.

philosophical or metaphysical studies, which established the bases and certainty of all knowledge acquired in the scientific and language-art courses. Thus, the college program would fulfill two laws: the truth of receiving and the truth of distributing. In the end, therefore, a perfect culture would be achieved, modelled as a three-roomed edifice—the center, the most imperfect, represented as philosophy; one wing, the most perfect, represented as science; and the other wing, language-art. Moreover, each wing of LeConte's tripartite structure contained several floors, signifying the varying levels of study within each of the three courses.[15]

An important element of LeConte's social philosophy was that all educational schemes were tools of environmental, hence individual and consequently social, improvement. Like Ward, LeConte believed that natural selection still operated in the evolutionary process, "*but not without limit*." As a Neo-Lamarckian, he strongly believed that all improvements in one generation were transferred to the next. The principle did not apply necessarily to individuals, but to the whole race, and although the "residuum" that carried over might be ever so slight, it was there nonetheless. The "fit" could not be selected but had to be "made"; otherwise, our only alternative would be to destroy "the sick, the helpless, the old ... with Spartan firmness," a thought that LeConte considered absolutely repulsive. He argued instead that we must be guided by reason in *making* the fit in order to elevate society. He also believed that "knowledge and character" were acquired through operation of the law of heredity. Both evolutionary factors—environmental adaptations and hereditary acquisitions—could be passed on from phylogeny to ontogeny, or from the social aggregate to the individual. In keeping with his optimistic views on social progress, LeConte declared that rapid improvement could be expected since the hereditary principle operated also by the "law of acceleration." Consequently, "each generation repeats the form and structure of the previous ... [and] there is a tendency for each successively-appearing character to appear a little earlier in each successive generation." The

15 LeConte, "Essential Characteristics," 196–204; LeConte, "True Idea of a University," 6–19. See also Joseph LeConte, "Science and Literature," *Overland Monthly*, 2nd ser., I (February, 1883, supplement), 8–10; Joseph LeConte, "A Note on the Religious Significance of Science," *Monist*, X (January, 1900), 161–66; LeConte, "Illustrations of a Law of Evolution of Thought," 388–91; Joseph LeConte, "The Factors of Evolution," *Monist*, I (April, 1891), 321–35.

acceleration of inheritance thus left time "for the introduction of still higher *new* characters."[16]

LeConte did not ignore the relationship of the individual to collective society in his social philosophy. He typically argued that social thinkers had succumbed to extremism on the issue of individual rights. On the one hand, he criticized the extremists who "apotheosized" the individual, while on the other hand, he spoke against the opposite group for their socialist views. As always, LeConte found extremism unacceptable, and he called for a reconciliation, or a middle view. Thus, he concluded that "individual interests must be subordinated to social interests . . . chiefly because society is the only means of achieving the ideal," but, he also insisted, "subordination is not sacrifice." A higher principle must prevail: commitment to achieving the goal of the "divine ideal [of] humanity," based upon the operation of evolutionary laws.[17]

The first of these he called the law of differentiation, which signified for society an increasing diversification of individual activities and functions. But LeConte firmly believed that the consequence was an increasing merger of the independent life into the common social life, and humans thus became more and more mutually dependent. To fail to realize this and to persist in the prevailing position of "selfish antagonism and competitive struggle," could, in LeConte's view, cause a reversion to the old fierce contest that characterized the animal kingdom in the evolutionary process. LeConte identified the second major law of evolution as the law of progress, whereby life forms "advance to successively higher points along every line." Again this applied to the organism as a whole and not to individual parts. It logically followed for LeConte, then, that the same law of progress applied to the social body: some individuals advanced to "philosophic thinkers, the teachers of the race," while some sank to "street-sweepers and sewer-cleaners." Overall, however, the entire social organism continues to progress.[18] By implication, LeConte supported an elitist philosophy of life

16 LeConte, "Relation of Sociology to Biology," 336; LeConte, "The Theory of Evolution and Social Progress," 492–500; LeConte, *Evolution*, 178–79.

17 LeConte, "Relation of Sociology to Biology," 331–32; Boller, *American Thought in Transition*, 65.

18 LeConte, "Relation of Sociology to Biology," 331–32; LeConte, "Natural History," 218. In his *Evolution* (1891, p. 25), LeConte rephrased the sociophysiological comparison as follows: "Some members of the social aggregate advance *upward* to the dignity of statesmen, philosophers, and poets; some advance *downward* to the position of scavengers

and education, which in reality partially reflected his own personal background of experience but also mirrored the views of a sizable body of American intellectuals who were acutely conscious of social class distinctions.

Not unlike other influential intellectuals of the age, LeConte magnified the role of science in the educational process. But he also believed that there would be no "speedy advent of a society regenerated by science [and] . . . cut loose from the past." Man must realize that he was placed in the world "to conquer and reduce it to rational method—to attain, by Reason, a complete knowledge of Deity, as revealed in the Book of Nature." On the other hand, the incessant march of evolution and progress naturally required periodic readjustments of man's ideas or, occasionally, a total rearrangement of his "whole mental furniture to a new and higher order." Too many people, opined LeConte, have their mental furniture "so screwed to the floor that it is impossible to readjust without tearing up the whole mental flooring." Progress depends upon a truly rational spirit, which can be achieved only through "the clear and thorough apprehension of the idea of Evolution," and American education offered the best hope for attaining such an apprehension. Yet he maintained that the American educational system had not kept pace with the advancement of knowledge: it was still somewhat unrealistic, tending toward a formalism that dealt "too much with words and ideas, and so little with things." Educational reform, he concluded, was essential if society was to progress to a higher stage.[19]

LeConte did not confine his social theories solely to education, of course. For him, science or, later, more specifically, evolutionism, was

and sewer-cleaners." In a footnote, he added: "Of course I mean downward in *social function*. Individually the scavenger may be nobler than the statesman."

19 LeConte, "True Idea of a University," 11–18. LeConte had made a similar statement in his first article on "The True Idea of a University," *Berkeleyan*, August, 1876, p. 6; he repeated it in his "Essential Characteristics," 199–203, and in his "Address on Behalf of the Faculties," *Addresses at the Inauguration of Martin Kellogg, LL.D., President of the University of California* (Berkeley, March 23, 1893), 22. See also his "Sense-Training," 41–43; "Instinct and Intelligence," 653–64; "From Animal to Man," 356–81; "Address" (on James Dwight Dana), in Daniel C. Gilman, *The Life of James Dwight Dana* (New York, 1899), 250–51; and *Outlines of the Comparative Physiology and Morphology of Animals* (New York, 1900), 82–83. LeConte's views on education are discussed at greater length in Lester D. Stephens, "Joseph LeConte on Evolution, Education, and the Structure of Knowledge," *Journal of the History of the Behavioral Sciences*, XII (April, 1976), 103–19.

simply *the* basis of all advanced thought; progress depended upon its application to a broad spectrum of intellectual activity—indeed, to virtually every facet of human endeavor. The "reluctant evolutionist" thus frequently expressed his ideas on the whole course of human development or, in brief, on history. Although he never drafted a paper on the philosophy of history, he often commented upon the course of man's past in societies and civilizations, systematically applying evolutionary concepts to the interpretation of cultural development.

The idea of the evolutionary development of civilization first appeared in the writings of the eighteenth-century Italian philosopher Giambattista Vico, who postulated that human societies go through specific cyclical stages of growth and decay. Later, in the nineteenth century, Georg W. F. Hegel also formulated a theory of the evolutionary development of civilizations in a dialectical progression of societies in which one phase declined as another (and higher) form was born. Afterward, the French positivist Auguste Comte presented a theory of the historical progress of human thought in three determinate stages. Like Vico, Comte found evolutionary cycles in history, but, unlike the Italian philosopher, he subscribed explicitly to the idea of evolutionary progress; thus his theory also bore a relationship to Hegel's. All of these evolutionary theories appeared before the comprehensive theory of organic evolution was set forth by Charles Darwin in 1859, however, and so were based upon metaphorical comparisons of societies with the cycle of human life and the development of thought rather than upon the concept of the organic evolution of species. LeConte was certainly one of the first to construct a philosophy of history based upon the idea of organic evolutionism.[20]

Although he argued consistently that evolution was "continuous progressive change," LeConte believed that the process occurred "not at uniform rate but in a succession of cycles by the rise, culmination, and decline, of higher and still higher dominant functions, principles, ideas, etc." He also believed that "every system of correlated parts" consisted of both internal change and "equilibrium and stability." Every

---

20 A brief but useful statement of the development of evolutionary theories can be found in Robert Scoon, "The Rise and Impact of Evolutionary Ideas," in Persons (ed.), *Evolutionary Thought in America*, 4–29. LeConte was intimately familiar with Comte's theory of the evolution of thought; see his "Comte's Classification of the Sciences," *Berkeley Quarterly*, II (April, 1881), 97–117. In this article, he objected to Comte's scheme and presented his own system of hierarchical classification of the disciplines.

system was thus characterized by "progressive movement" toward a higher form and by "constant slight disturbance and readjustment of parts" in order to remain stable. The first of these was governed by the "laws of evolution" and the second, by the "laws of sustentation of the system." Societies must constantly readjust in order "to produce social equilibrium," and they must move forward to avoid "social stagnation and decay." In the organic kingdom, the rise and decline of cycles had resulted in the evolution of one-celled organisms into higher forms of life and subsequently into the development of man as the most complex of all living matter. LeConte also noted that certain physical functions "rise, culminate, and decline" within individual humans, but he likewise believed that this "law of progress" or "law of development" did not signify that all parts of the human organism "advance to successively higher points along every line." Instead, "the highest parts become successively higher, and the whole becomes successively greater."[21]

The concept of evolutionary gradation also applied to society, wherein "progress is accomplished by a successive rise, culmination, and decline, of higher and higher dominant ideas or principles, determining different phases of civilization." LeConte therefore declared that the law of cyclical progress governed past and present societies and would naturally determine the nature of future civilizations. Societies advanced, "not uniformly, but by successive waves, each higher than the last . . . and [each] embodying a new and higher phase." Quite clearly, LeConte was influenced by Comte's conception of the evolution of society, but unlike the French positivist, he grounded his explanation in the theory of organic evolution. Although his idea of cyclical progress resembled the spiral theory suggested by Vico, it seems unlikely that the American evolutionist was familiar with the Neapolitan's *New Science*. LeConte was admittedly familiar with the writings of Comte and Hegel, however, and thus indirectly exposed to the train of Vico's thought inasmuch as both of those men had observed some of Vico's ideas in their explanatory schemes.[22]

21 LeConte, "Scientific Relation of Sociology to Biology," 331–32; LeConte, *Evolution*, 4–8; LeConte, "The Theory of Evolution and Social Progress," 481.

22 LeConte, *Evolution*, 26; LeConte, "Scientific Relation of Sociology to Biology," 333; LeConte, "The Relation of Organic Science to Sociology," 50. In his *Autobiography* (p. 287), LeConte assessed the contribution of these philosophers to his own thought, acknowledging a special indebtedness to his University of California colleague G. H. Howison,

In his cyclical theory of evolutionary progress, LeConte posited the idea that "there is always a residuum, which accumulating throughout geological times, goes to form the cycle of the earth's life and development." The earth, he stated, is "a complex organism which, under the operation of . . . circulating air and water, has been through all time developing into higher and higher life." All organic matter changed through the process of circulation, very gradually at first, but increasingly as it became "more diverse in form and complex in organization." As each cycle culminated, "a small, almost infinitesimal residuum" was retained in the subsequent cycle. The residua thus gradually accumulated and ultimately developed into still higher stages of life. Therefore, argued LeConte, species evolved into higher forms; man himself developed into a more complex organism, and society and civilization progressed ever upward. Moreover, since the spirit of man was "entangled in matter," it was also "conditioned by its laws." As humankind was gradually freeing itself "from slavish bondage to [its] lower, sensuous and material nature," it was beginning to "approximate steady, onward, rectilinear progress." Ultimately, man would realize the nature of evolutionary cyclical development, and by following social laws he would be able to break out of the cyclical pattern and march toward the goal of the Ideal.[23]

Following the notion of residual accumulations, LeConte posed that, when the life cycle of individual men ends, "there is always a small residuum of influence, good or bad, which accumulating through successive generations, enters into the composition of the cycle of the national life." Thus, specific individuals would experience little direct impact from their immediate forbears, for the residuum was too small to alter their personal nature. But the *accumulation* of tiny amounts of good or evil would ultimately shape the race or civilization as a whole. Likewise, the residual influences of individual nations were too small to affect their immediate successors significantly. As those influences were passed on in minute quantities through the ages, however, their cumulative mass would appreciably transform

---

whose system of "personal idealism" was grounded in Kantian theory. In a prefatory note to LeConte's essay on "Evolutionary Idealism," Howison stated that LeConte's view was a variation of the idealist philosophy and claimed that it served as a "counter-system" to his own theory.

23 LeConte, "The Natural Law of Circulation," 62–66; LeConte, "Relation of Organic Science to Sociology," 50; LeConte, "Scientific Relation of Sociology to Biology," 333.

later societies. "The cycle of nations and civilizations closes," LeConte contended, "but there is always a residuum of ideas and principles which, accumulating through successive ages, enters into the progress of the race." He therefore concluded that it is the "duty" of both individuals and nations to see that the most "cultivated" aspects of man's higher nature were carried forward in order to move "the race in the right direction."[24]

As the law of evolution or development proceeded on its ineluctable course, it operated under another phenomenon, namely, "a law of derivation." The very nature of the evolutionary process was *change,* stated LeConte, but it was change of some preexisting form, not the creation of a totally new form. Every new form in the organic world thus represented the transformation or modification of an older form. The same was true of human society. While some societies "die out and leave no progeny . . . [others] are transformed into the higher new." In the related realm of ideas, the same law likewise applied: "old theories" were never "utterly *exploded*"; some of them certainly died, but all new theories contained bits of the old. This was a process of transmutation. Hence, by transformation and transmutation, new and higher ideas and new and more advanced civilizations appeared. LeConte claimed that civilizations had become "successively higher and higher social forms, embodying higher and higher social forces." It was certain, he contended, that each civilization passed through the stages of rise, culmination, and decline, "but as each decays, its characteristic dominant forces and principles are not lost, but incorporated into the next higher form, and subordinated to the next higher dominant forces."[25]

The process of growth or decline of civilization was tied to a "dominant idea, principle, or social force," and as long as a nation could "accept the incoming principle"—that is, respond to new challenges—it could maintain its vitality. Every civilization experienced "periods of comparative quiet and prosperity, during which the forces of change are gathering strength." Likewise, every civilization experienced paroxysmal periods of revolution "when these forces show themselves in

24 LeConte, "The Natural Law of Circulation," 66.
25 LeConte, "The Theory of Evolution and Social Progress," 487; LeConte, "The True Idea of A University," 18; LeConte, "Illustrations of a Law of Evolution of Thought," 387; Joseph LeConte, "Mutual Relations of Intellectual and Moral Culture," *Overland Monthly,* 2nd ser., I (January, 1883), 15.

conspicuous effects." But neither a civilization nor its dominant force could "continue in its pristine power," and each must accept new dominant principles or forces or it would be "left behind" and ultimately decay. Every dominant force of a civilization culminated and declined, but it did not "*perish.*" Instead, it was merely "subordinated to the next coming and higher characteristic."[26]

In LeConte's view, each society organized itself upon "some principle of classification," and no matter how ultimately imperfect, that principle worked for the society at the moment of its birth. Gradually, "the organized form *consolidates* into a more perfect and effective instrument of social activity, then *hardens*, then *petrifies*, and farther [*sic*] growth of humanity is rendered impossible." The same was true of religion: dogmas and creeds were first necessary, but eventually they "*consolidate* first into greater *effectiveness*, then harden into *protectiveness*, then petrify into obstructiveness." Whenever societies, or religions, or intellectual movements of any sort, reach the stage of obstructiveness, then, like Crustacea, they must throw off the shell that was once protective if they are to continue to grow. Many civilizations, he declared, have ceased to grow. These are "*arrested civilizations*, of which nearly all barbarous and semi-civilized races are examples, but the Chinese and Japanese are the most conspicuous." Other civilizations have died. Of these, "the Greek and the Roman are the most conspicuous." Such nations had gradually accepted "fixed customs and habits," which, enforced over time, led ultimately to petrification of "*national character.*" In part, their vitality was decreased because they never allowed "cross-breeding, or mixing of varieties." Such mixing conferred "plasticity" and prevented "the formation of fixed national character, and the consequent arrest of progress by petrification."[27]

"Organic evolution," said LeConte, "is by a *vis a tergo*, a *pushing* upward and forward from below and behind." Thus, in the animal kingdom, the "fittest are those most in harmony with the physical environment." In social evolution (*i.e.*, progress), however, growth occurred "by a *vis a fronte*, a *drawing* upward and forward from above and in front by an aspiration, an attraction toward an ideal." In the latter form

---

26 LeConte, "Scientific Relation of Sociology to Biology," 333–34; Joseph LeConte, *Compend of Geology* (New York, 1884), 242.

27 LeConte, "Illustrations of a Law of Evolution of Thought," 389–90; LeConte, "Instinct and Intelligence," 664.

of evolution, therefore, "the fittest are those most *in harmony with the ideal.*" Or to phrase the idea differently, organic evolution operated under a "law of *force,*" whereas social progress was determined by a "law of *love.*" Also, in the former, the object was survival of the fittest, but in the latter, the aim was "to *make* as many as possible fit to *survive.*" LeConte offered a caveat, however, for he feared that man could also "perpetuate weakness by inheritance." This was, for him, the most serious social problem created by man's ability to affect his own evolution. On the one hand, then, modern societies were faced with the question of how they could observe "the higher, spiritual law of love," while on the other, they were beset by the problem of how to avoid "weakening the blood of the race." Since he correlated the success of a civilization with its racial character, LeConte decried any reversion to lower cultural influences because that would result in retrogression. He was thus confronted with a paradox, for he believed that intermarriage among racial types would elevate a culture, but he also feared that it would increase the possibility of debasing the higher culture.[28]

As in organic evolution, the law of universal development among societies resulted in increasing differentiation, since the "lines of progress branch and diverge, apparently *ad infinitum.*" But, in LeConte's view, these lines of divergence are limited, and "at a certain point they begin again to converge, and approximate unity." The law of development was therefore correctly expressed as follows: "first, *simple unity,* or *homogeneity,* then infinite *diversity,* and finally, the coordination of this diversity into *organic unity.*" LeConte warned, however, that "the way of evolution toward the highest—i.e. from Protozoan to man and from lowest man to the Ideal man—is a very 'strait and narrow way,' and few there be that find it." In the case of organic forms, deviation was "fatal to upward movement of the diverging form toward its goal, man." Once the organic form had diverged, moreover, it could never get back on the track, and no living deviation among organic forms was ever able to develop into man. In the case of human evolution, however, such divergence was not necessarily fatal. "If an individual or race gets off from 'the strait and narrow way' to the high-

28 LeConte, "The Theory of Evolution and Social Progress," 490–93; LeConte, "The Relation of Organic Evolution to Human Progress," 136; Joseph LeConte, "The Effect of Mixture of Races on Human Progress," *Berkeley Quarterly,* I (April, 1880), 85–86.

est, the divine ideal, it is hard, very hard to get back," but it was "*not impossible*, because man's own voluntary effort is the chief factor in his own evolution." Thus, any living civilization that had strayed could return through "self-activity." By "the use of reason and co-operation in the work of his own evolution, man is able to rectify an error of direction."[29]

Science provided the method of improving physical, mental, and spiritual conditions, and thereby of improving civilizations. In Le-Conte's judgment, science was "the *image* of Divine Truth—a revelation of the Divine thought." The "distinctive mission of Science . . . is to perfect that image in the human Reason as ideal truth." Now that organic evolution was reaching its culmination, man must apply science and reason to the evolution of society and work toward the goal of its perfect organization. His optimism forever intact, LeConte declared that "when we understand fully the principles of social progress . . . then this science [of society] will build the social structure on rational principles." In other words, he argued, man must consciously carry progress to a higher plane and ultimately achieve the ideal state. Then, at last, mankind could free itself from the "melancholy law of the cyclical movement of society [and] the rise and fall of nations and civilizations." This could be accomplished by "science alone," for science, in LeConte's words, was "the pledge of human progress and its type."[30]

29 LeConte, "Natural History as a Branch of School Education," 216; LeConte, "The Scientific Relation of Sociology to Biology," 332–35; LeConte, "The Relation of Organic Evolution to Human Progress," 136–37; LeConte, "Science and Literature," 9; LeConte, "Address on Behalf of the Faculties," 20; LeConte, "The Theory of Evolution and Social Progress," 489; LeConte, "The Effect of the Theory of Evolution on Education," 159; Le-Conte, *The Higher Utilities of Science*, Chit Chat Club Pamphlet, San Francisco, November 14, 1881, pp. 9–10; LeConte, "The Relation of Organic Science to Sociology," 51.

30 LeConte, *Higher Utilities of Science*, 13; LeConte, "Illustrations of a Law of Evolution of Thought," 384; LeConte, "Natural Law of Circulation," 63. For a fuller discussion of LeConte's ideas on cultural history, see Lester D. Stephens, "Joseph LeConte's Evolutional Idealism: A Lamarckian View of Cultural History," *Journal of the History of Ideas*, XXXIX (July–September, 1978), 465–80.

In Their Proper Place: The "Race Problem" and the "Woman Question"

Riding high on their belief in cultural progress, American intellectuals in the Gilded Age expanded the reach of their interests to the problems of race and the place of women in the perfect society, applying scientific theories freely to meet the demands of their expectations. Certainly, Joseph LeConte, the enthusiastic evolutionist, could not ignore these questions as he developed his theory of the ideal civilization. They had concerned him long before he embraced evolutionism, and when they became white-hot issues, he turned to face them head-on. They too could be resolved when fitted into the framework of the Neo-Lamarckian explanation of evolutionary development. In essence, LeConte merely articulated the views of his contemporaries and enunciated the racial and sexual stereotypes of his generation, but his facile and popular style of explanation enhanced his reputation as a scientific spokesman on these prominent topics in the 1890s.[1]

LeConte rarely spoke publicly on the subject of race before the late 1870s, the first such comments appearing in 1870 in a paper on the origin of sex. He enlarged upon these remarks in three papers published respectively in 1880, 1889, and 1892. Significantly, these papers corresponded with the intensification of interest in racial matters among white American Protestants. The first two papers were based, at least theoretically, upon laws of heredity, while the next two represented a sweeping sociological explanation grounded in empirical

1 One of the most thorough and balanced analyses of LeConte's racial views is given in John S. Haller, *Outcasts from Evolution* (Urbana, Ill., 1971), 150–66. See also George M. Fredrickson, *The Black Image in the White Mind* (New York, 1971), 247.

reasoning. Since the latter papers reflected LeConte's southern sympathies, they reveal the roots of the former and hence deserve prior
description.

In 1888, LeConte read his "The South Revisited" before the California
Historical Society, and he had it published the following year in the
Overland Monthly. He began his paper by stating that he believed the
antebellum South had erred in remaining "stationary, while the rest of
the civilized world rushed on." This "arrest of development" was a direct result of the South's failure to realize that the "peculiar institution" of slavery was "a mere survival from an extinct world, and therefore more and more an anomaly in the modern world." In short,
southerners had not kept pace with the process of evolutionary development. LeConte admitted that slavery might have been ended by
some means other than war, but he concluded that it probably "could
not have gone in any other way." That was now a dead issue, however,
and nothing was to be gained from rehashing the alternatives. Instead,
southerners must accept the loss of their economic system of bonded
labor and, without any shame for their past, move forward.[2]

Then, he asked, "how goes the putting on of the new" in the South?
The answer was, by "mental readjustment," which the optimistic
LeConte said "is already well nigh complete." He contended that with
changing attitudes would come two other adjustments: industrial,
and social and political. Like the "New South" spokesman Henry
Grady, LeConte insisted upon "an increased diversification of pursuits" in the southern economy, especially in the production of farm
crops. He also called for a good public school system and a liberali-
zation of the southern college curriculum. The southern system of
education, especially in its literature and art, represented an "intellectual isolation" that could not "spur . . . ambition." But the most serious
adjustment would have to be made in the "social and political" realm,
a particularly difficult problem because it was "complicated with the
dread *race problem*." The matter could be resolved, said LeConte, only
by allowing "*the higher race . . . [to] assume control and determine the
policy of the community*." He observed that "elsewhere . . . negroes are
treated much the same as people of other color," but he argued that
"*relative number* is a prime factor in the problem." Where only a few
blacks were present in a population, a community could "afford to

2 LeConte, "The South Revisited," 22–31.

recognize, nay, more to patronize, to pet, or even to lionize them," but where their numbers were large, as in the South, the "laws must be made and the state policy must be determined by the superior race."[3]

Although he took a liberal stance on the modification of some southern institutions, LeConte remained as recalcitrant as ever on the question of racial equality. To objections that racial inequality was unconstitutional, he replied that the postwar constitutional amendments were "in conflict with the laws of nature, and therefore unjust." After all, he said, the "law of self-preservation" applies to communities as well as to individuals. He charged that critics of southern society ignored the genuine commitment of whites to "the real best interests of the negro." Indeed, he asserted that white southerners "earnestly desire" to elevate blacks "by education and by acquisition of property." Strongly denying any prejudice, LeConte flatly stated that "there is less race repulsion at the South than elsewhere." "Race repulsion," he argued, " . . . is an instinct necessary for the preservation of the purity of the blood of the higher race."[4]

Although he stressed that education alone could "not solve the problem in this generation nor in many generations," LeConte nevertheless contended that it offered the best hope for racial improvement. Under the operation of evolution (i.e., by inheritance of acquired traits), however, "education of the race" would require "centuries, perhaps millenniums." The process could be accelerated through racial mixing, of course, but LeConte held reservations about that kind of approach. In the meantime, the solution to the political problem must come through "a limitation of the ballot, by a qualification both of education and property." That such qualifications would limit "many whites who ought not to vote" bothered LeConte not at all. He told his audience that he had proposed such a solution during the early days of Reconstruction in South Carolina: "I urged on many friends who were members [of the constitutional convention] the extreme importance of opening the franchise to all without distinction of color." But, he lamented, the convention "had not the backbone even to propose it because it would disfranchise many whites also, as well as nearly all the blacks."[5]

3 *Ibid.*, 24–27.
4 *Ibid.*, 28.
5 *Ibid.*, 28–30; LeConte, *Autobiography*, 235–36.

In 1892, LeConte was invited to read a paper before the Brooklyn Ethical Association on "The Race Problem in the South." Insisting that he was approaching the question of race from a rational point of view, he stressed the importance of applying the scientific method to the problem. He admitted, however, that since the problem also involved social and political questions, the scientific method was not wholly applicable; its "application must be made with the greatest caution and modesty, and in strict subordination to a wise empiricism." Thus, a two-pronged approach was necessary: scientific method should be used insofar as possible, but otherwise a strictly rational or logical analysis was necessary as the question extended further into social and political matters. In brief, LeConte had to find a *reasoned* justification for denying full equality to blacks, since he was convinced that they were an inferior race.[6]

LeConte continued by arguing that the subject could be understood "in the light of its history," or by the "historic method," which was, as always, no more or no less than "the evolution method." At one point in time, he argued, slavery was the "normal, and indeed the necessary, result of the close contact of civilized with savage races." As such, it was the "natural" outgrowth of the operation of the law of survival of the strongest among races. Since it was therefore natural, it could not be "wholly wrong." In the South, however, "slave labor became more and more profitable" to plantation owners "by the increasing culture of cotton and rice." The possibilities of emancipation thus grew less likely, partly because of the economic value of slaves but more "especially on account of the extremely grave social question involved." As the prospects of manumission decreased and as critics of slavery intensified their attacks upon the South, southerners suffered from "a painful sense of isolation from the rest of the civilized world" and drifted out of "the current of progress." But, LeConte insisted, the South could have soon reentered the mainstream if it had realized that emancipation had not ruined its economy but only brought about a change in the "form of labor." In fact, if the freed Negroes had been willing to work for wages, the southern economy would have continued to prosper. But, unfortunately, these "plastic,

6 LeConte, "The Race Problem in the South," 349–82 and 383–402 ("Abstract of the Discussion").

docile, imitative" people are "unwilling to work for wages," and they must thus be guided by civilized whites.[7]

Such prejorative ascriptions necessarily entailed many assumptions about race, and LeConte had carefully worked them out to his own satisfaction. He did so by egregious extrapolation from studies on selective breeding—that is, from the somewhat tenuous findings of scientists engaged in horticultural and livestock experiments. In addition, he fitted the whole idea into his theory of organic evolution and molded it to conform to his sociological scheme. The result was a thoroughly systematic theory of race. Quite naturally, of course, the validity of the theory depended upon the strength of its premises, the empirical testing of its hypotheses, and the applicability of natural and physical science to culturally acquired habits of behavior. On most of these, LeConte's theory was seriously deficient, but it was not viewed as such in his own time. Indeed, to many well-read Americans, it represented one of the most "scientific" and sophisticated explanations of racial differences, but it likewise reinforced the deeply entrenched racial stereotypes of the age.[8]

As early as 1866, LeConte had shown an interest in the effects of artificial breeding. He resumed his interest in the late 1870s under the effects of his acceptance of the theory of organic evolution. In his "The Genesis of Sex," LeConte first presented a history of the development of modes of sexual reproduction and then took up the matter of the "crossing of varieties." He outlined a theoretical scheme on the effects of interbreeding between what he designated as "slight varieties," "strong varieties," and "races" of mankind. A few years later, he elaborated upon this idea in a lengthy paper on "The Effect of Mixture of Races on Human Progress," which became his major "scientific" explanation of interracial crossing.[9]

Crucial to LeConte's theory of racial evolution was the commonly held belief that mankind consisted of races and varieties that had not

7 *Ibid.*, 354–59.

8 See Haller, *Outcasts from Evolution*, for a general discussion of racial theories grounded in evolutionism, and compare with the older theories discussed in William Stanton, *The Leopard's Spots: Scientific Attitudes Toward Race in America, 1815–59* (Chicago, 1960).

9 Joseph LeConte, "The Genesis of Sex," *Popular Science Monthly*, XVI (December, 1879), 167–79.

achieved the true status of *Homo sapiens*. Also important to his theory was a conception of two laws of reproduction: heredity and "divergent variation of offspring." The first law resulted in the more or less permanent fixing of racial types, while the latter led to constantly changing variations among the human types. LeConte thus argued that there were "many grades of variation short of species." The first of these was represented as *"small individual differences,"* the second, as *"varieties* produced by continued selection which may be slight or strong, but are not fixed," and the third, as *"races,* or permanent varieties." By "artificial," that is, selective, breeding, changes among varieties could be achieved, but not on a permanent basis.[10]

The results of selective breeding, said LeConte, would be "unstable" or "plastic" because, by the "Law of Reversion," the inherited traits would eventually revert beyond the immediate parents to "the whole line of ancestry." In other words, selective breeding would culminate in "the integrated sum of a diminishing series," thus allowing reversion to "ancestral forms." On the contrary, the tendency to revert was negligible in the case of natural breeding because the process of "natural selection" gradually "eliminated" those "ancestral tendencies." In reality, therefore, selective breeding formed new varieties, but since the "original stock" could not be "pushed" beyond certain limits, permanency was impossible. By natural process, however, "varieties pass into races, races into species, species into genera, genera into families, etc."[11]

LeConte believed that cross-fertilization occurs most frequently among the inconstant varieties and least among the permanent varieties or races. In effect, the latter have become "commencing species," but as their tendency to become permanent increases and interbreeding ceases, they qualify as species. It was by such a process, said LeConte, that varieties and races of human beings have evolved, and this has resulted in the formation of "national varieties" and "primary races." The former "tend to mix and make a common type" unless some extraneous force prevents the process from freely operating. The second, however, tend to shun mixing, but their ability to cross-fertilize is "not destroyed." From these more permanent types early man originated, and by the process of evolution great differences

---

10 LeConte, "The Effect of Mixture of Races on Human Progress," 81–82.
11 *Ibid.*

emerged. Now, noted LeConte, civilized man has reached the point where he is trying "to reverse this process of differentiation and to assimilate the races again." This reversal is occurring "partly by the control of physical conditions, and partly by mixing of blood and by commerce."[12] What, then, would be the long-range results of such assimilation? This was the question that most bothered LeConte and many of his fellow intellectuals.

To answer the question, LeConte began by citing once more the effects of controlled breeding in horticulture and animal husbandry. "It is well known," he observed, "that judicious cross-breeding of different varieties of the same species strengthens, while too close and long-continued consanguineous breeding enfeebles the stock." After a lengthy discussion of the results of selective breeding among plants and animals, he applied the findings to "the human race." But he added that, although the same laws of cross-breeding applied to man, they were also "modified by the higher nature of man and the higher conditions of human life." Thus he claimed that "close and consanguineous breeding" among humans not only weakened physical strength but also "especially the mental constitution." In other words, "too true and too close breeding, by the accumulation of limited inherited qualities generation after generation ... tends to fix the type too rigidly." Such breeding leads to petrifaction and loss of "plasticity," and, as such races are "brought into contact by commerce," the only outcome can be "an extinction ... or an assimilation."[13]

LeConte admitted that the crossing of varieties and races would avoid the bad effects of rigidity of types and the petrifaction of societies and cultures, but only within limits. Too much crossing, especially of extreme types, would reduce evolutionary progress "below the mean" of desirable physical and "psychical" development—a phenomenon that LeConte believed could best be observed "in the Southern United States." Although he admitted that such observations were yet unscientific and not "sufficiently careful" to warrant strongly reliable generalizations, he nevertheless asserted that "some facts are too obvious to be doubted," adducing the "mulatto" as a noteworthy example of evolutionary regression. Although he viewed the mulatto as more intelligent than blacks, LeConte believed that this mixed breed

12 *Ibid.*, 83–86.
13 *Ibid.*, 86–99.

possessed less physical strength and was subject to more disease than "either the white or negro race." Mulattoes, he said, are frequent victims of "hereditary diseases, especially various forms of scrofula," and their offspring are "even still feebler." The "worst effects" are seen in the crossing of extreme types, namely "the light-haired blue-eyed Teutonic and the negro." Less disastrous is the crossing of Mexican and South American Indians with Spaniards, a mixture that "has produced a physically hardy and prolific race" but discouraging results for "social progress."[14]

LeConte recognized that contact between races would increase with the "advance of commerce." Since he could not accept the idea of exterminating "inferior types," he had to concede the "alternative" of "absorption of mixing"; but, in order to avoid the deleterious effects of crossing extreme races, this "absorption" must be a "judicious crossing . . . of the slight varieties." By this means, the inferior type would be elevated, and the superior type would not degenerate too rapidly, being able by its strength to lift the lower to "a still higher plane." Careful crossing would thus "produce a generalized human type capable of universal progress in *all* directions." The eventual result would not be an "Anglo-Saxon, nor Teutonic, nor European, nor Aryan, but *human*" civilization. If judicious mixing did not occur, warned LeConte, "then are the lower races indeed doomed." Just how such desirable mixing could be attained, however, he did not specify.[15]

Meanwhile, contact between the races could only increase and problems therefore intensify. What was to be done in the interim, particularly in the South? LeConte reiterated his belief in the long-range benefits of education and the restriction of the franchise to educated property holders. He likewise supported the poll tax, and he favored segregation of public facilities, a policy that he viewed as an ethical problem. Important decisions of this sort had to be made in the best interests of "groups of classes," not individuals. As in all matters dealing with Nature, a system of classification, no matter how tentative,

14 *Ibid.,* 100–101.
15 *Ibid.,* 102. See also his "'The Race Problem in the South" for additional comments upon what he labeled "The Laws of the Effects of Race Contact" (pp. 359–75). Theodore Dwight Bozeman has explored LeConte's evolutionist views of the crossbreeding of races in his "Joseph LeConte: Evolutionary Science and the Sociology of Miscegenation" (unpublished paper).

had to be followed. "Now, race-classes not only come under the same head, but are more natural and rational than many others, because founded on a real natural difference—i.e., a difference in the grade of evolution." Thus, at least for the time being, the "natural caste-line" should be continued "until we understand better than we do now the laws of the effects of race-mixture." The evolutionary process must be allowed to run its course, but beneficial intervention was necessary. LeConte could justify this ostensible inconsistency by appealing to his belief that man could now work out his own evolution.[16]

Shortly after his formulation of a systematic theory of race, LeConte began to develop a "scientific" explanation of sexual differences, an effort precipitated by the woman's rights movement in the 1890s. Over the years he had offered some comments on woman's rights, but these were scattered about in various papers only peripheral to the topic. A chance to elaborate at length came in 1895 when the woman's rights movement reached a high point of excitement in California, where an effort was under way to strike the word *male* from the suffrage clause of the state constitution. To support the movement, a group of woman suffragists, including Dr. Anna Shaw and Susan B. Anthony, had planned a Congress of Women to debate the proposed amendment and organize the efforts of an increasing number of prominent Californians who favored extension of the franchise to women. Joseph LeConte was invited to address the congress at its final session on May 25, 1895.[17]

Attendance at the conference, which began on May 20, swelled each evening, and by mid-week the crowds could no longer be accommodated in San Francisco's Golden Gate Hall. More than 2,500 persons gathered for the final session in the First Congregational Church, but many of them had to be turned away. The sanctuary was filled to overflowing, and over 500 people had to stand throughout the long session. As the distinguished old professor rose to speak, the eager audience poised to hear what the "gentle prophet of evolution" had to say about woman's rights. To their surprise, LeConte spoke not at all

16 LeConte, "The Race Problem in the South," 375–82.
17 Ida Husted Harper, *The Life and Work of Susan B. Anthony* (3 vols.; Indianapolis, 1898), II, 819–38; San Francisco *Chronicle* and San Francisco *Examiner*, May 19–26, 1895, *passim*.

upon the general topic of "The Home" as announced but instead repeated his old address on "The Psychical Relation of Man to Animals."
For almost two hours he rambled on, none of his comments seeming
to be directly related to the question at hand. Apparently LeConte
intended to provide a theoretical base for showing why women were
different from men and therefore not suited to cast the ballot. Not
unexpectedly, some of the audience grew restive, and finally some of
the most agitated suffragists began to hiss the venerable scientist before he finished his lengthy discourse.[18]

Following the meeting, LeConte's friend and candid critic, Mary
Keith, remarked to the otherwise popular speaker that he was lucky
the audience had not walked out on him. In fact, she bluntly told
LeConte that only their respect for him and their determination to
hear the suffragist leaders had saved him from a more embarrassing
response. Exactly what prompted LeConte to take such uncharacter-
istic liberty with his audience is unknown, but it seems likely that he
intended to show how evolutionism was important to consideration
of the role of women in society—and he thought his audience would
see his point. In addition, he had apparently failed to work out a systematic response in his own mind.[19]

LeConte had touched upon the subject of woman's rights in 1879 in
"The Genesis of Sex," claiming even then that the "form of woman's
rights which would assimilate as much as possible of the two sexes is
certainly in direct conflict with the law of evolution." Moreover, although he had published some stray thoughts on the "woman question" as early as 1861, he apparently had not completely clarified his
ideas on woman suffrage. After his speech to the congress, LeConte
penned some notes on "Biology and the Woman Question," but he
seems to have set them aside for other work. Then, in May, 1896, he
attended another woman's congress in San Francisco. This meeting,
held on the eve of the amendment ballot, again featured Susan Anthony. LeConte later noted that he had been impressed with Miss Anthony's "simple earnestness." Afterwards, he returned to his notes and
wrote out his defense of antisuffragism. It was never published, how-

18 San Francisco *Chronicle*, May 22, 1895, p. 14, and May 26, 1895, p. 19; San Francisco
Examiner, May 26, 1895, p. 12; Cornelius, *Keith, Old Master*, I, 356. LeConte's address had
been previously published in the *Princeton Review*, 4th ser., XIII (May, 1884), 236–61; it
was revised and published as "From Animal to Man," *Monist*, VI (April, 1896), 356–81.

19 Cornelius, *Keith, Old Master*, I, 356.

ever, probably because LeConte was not totally satisfied with the statement and intended to refine it later.[20]

As a Neo-Lamarckian believing strongly in the purposive nature of evolution, LeConte sought to demonstrate the progressive nature of man and society and thereby to lay the grounds for explaining the ideal roles of male and female in the perfect society. He could not construct his argument solely on the basis of organic evolution, however, and he thus found it necessary to change the title of his initial notes on the topic from "Biology and the Woman Question" to "The Relation of Biology and Sociology to the Woman Question" when he wrote out his complete explanation.[21]

The heart of LeConte's thesis on the role of women in society was his statement that "the most fundamental law of evolution is the Law of differentiation of form & specialization & limitation of function." Sex was not excepted from this biological law. LeConte argued that in the lowest forms of life the sex organs are united in the same organism, but in the ascending scale of animal development those organs are separated and are most fully perfected in the human male and female. As the separation of the sex organs occurred, "sexual *feeling*" developed, and the intensity of that sense was increased in the higher animals. Thus, he declared, "the tendency of evolution is to make men more & more manly & women more & more womanly," thereby sharply differentiating their characteristics.[22]

Since LeConte viewed society as progressive, he found the biological differentiation of sex compatible with the idea of separate social functions for men and women. If those functions or roles were ignored or if any attempt were made to unify them, mankind would be taking "a step backward toward savagery." Although he added that such reversion could not really occur because it violated natural law,

20 LeConte, "The Genesis of Sex," 177; LeConte, "Female Education," 60–91; LeConte's notes on the "Woman Question," in LeConte Family Papers, UCB.

21 LeConte, notes and MS on "The Woman Question," in LeConte Family Papers, UCB. An excellent summary of the biological and the sociological arguments against woman suffrage is given in Aileen S. Kraditor, *The Ideas of the Woman Suffrage Movement, 1890–1920* (New York, 1965), 18–42. LeConte's fellow scientist Edward D. Cope had previously published some thoughts on the subject in his article, "The Relation of the Sexes to Government," *Popular Science Monthly*, XXXIII (October, 1888), 728. LeConte may have been partially influenced by that article, though he deemphasized Cope's stronger view on the thoroughly retrogressive consequences of woman suffrage.

22 LeConte, notes and MS on "The Woman Question," in LeConte Family Papers, UCB.

the very nature of his argument reflected a concern that society might in fact regress. Current efforts to violate these "ineradicable" principles, he averred, only created "restlessness, discontent & incalculable misery." Singling out the extremists in the woman's rights movement, LeConte criticized their efforts to remove all restraints on competition because it would create a struggle that could only "*brutalize* the *sex-relations* & tend to reduce it to the animal plane."[23]

LeConte did not, of course, concur fully in the Darwinian view of the function of natural selection in the evolutionary process, contending instead that the theory of the survival of the fittest applied mostly to organic, not human, evolution. Human progress superseded the fierce, animalistic struggle by replacing it with the principle of altruism. Yet the application of the latter demanded rational, thoughtful action. Any homogenization of the social roles of male and female would negate the new principle and thereby weaken civilization. Evolution had intensified sexual desire in the male, and this function was "more *outside* of the essential physiological being" than in woman, whose weaker sexual desire "lies at the center of the whole bodily organization." To ignore these basic differences, said LeConte, would be to vitiate evolutionary development by placing the less fit in a position contrary to the facts of progression.[24]

Believing not only in the evolution of species but also in the evolution of functions of both the human body and the brain, LeConte stated that the "lower" functions of man and woman were to be found correspondingly in the lower parts of the anatomy. Sex, then, was still within the realm of animal instinct, and the brain, as the highest organ of the body, had progressed to the psychical realm. But even though the brain was equally developed in both male and female, different biological bodies had created dissimilar roles for each sex. The human male was therefore more suited for "external" activities, or the world of business, security and protection, and political affairs. The human female, on the other hand, was adapted for domestic activities. LeConte contended that these special functions did not diminish woman's importance; on the contrary, they enhanced her role in pre-

23 *Ibid.*
24 *Ibid.* See also his "Science and Mental Improvement," 96–101; "The Factors of Evolution," 321–35; and *Evolution*, Ch. 3.

serving civilization. Nor, LeConte insisted, did they depreciate her mental faculties, for woman indeed equaled man in intelligence.[25]

"The male and female standards are equal, absolutely equal," said LeConte. Under the basic law of evolution, however, the two sexes, though equal, were *different*, and this distinctiveness must be recognized. Woman was not inferior to man, and the mistake of the extremists was to ignore the necessity for two standards: one for the female and the other for the male. The ardent advocates of woman suffrage committed the error of "degrading their own sex in setting up the male standard as the *one* standard for both." After all, LeConte argued, the progress of civilization is correlated with the "increasing estimate of the feminine standard." In reality, however, LeConte was not only applying the theory of evolution as he understood it but also revealing his own southern background, which assigned woman a special position in society. The "manly heart," he had previously scribbled in one of his notebooks, "rejoices & glories in woman" because of her "exquisite grace & harmony"; she is the "efflorescence of humanity."[26] Despite its stress upon evolution, LeConte's argument fitted hand-in-glove with the conventional antisuffragist duality of standards; it was really a reaffirmation of the separate-but-equal doctrine.

Actually, LeConte had expressed these views in another form over three decades earlier in his commencement address to the Laurensville Female College in South Carolina. In that address, he had stressed the value of the "ardent and virtuous love of a noble woman" to the preservation and progressive uplifting of civilization, and he had reminded his young listeners of woman's moral obligation as a "moulding influence" on the plastic character of children. The credulous champions of the woman's rights movement, he asserted, aimed to diminish the "distinctive character" of the female by upholding the male standard as their goal. This could only result in an unwholesome uniformity and thereby undermine the distinct "spheres of activity" peculiar to each sex. Education, while equal, must be different because the male mind was unlike the female mind: "In man we have the predominance of the *formal* reason; in woman, of the *intuitive*

25 LeConte, notes and MS on "The Woman Question," in LeConte Family Papers, UCB.

26 *Ibid.*; LeConte's notebook, in possession of Helen Malcolm LeConte.

reason." These views represented the stock antisuffragist argument that women were incapable of dealing rationally with political and other problems that required emotional detachment and clear logic for their solution.[27]

When LeConte wrote out his more elaborate views on woman's rights some thirty-six years later, he perceived a deep-seated problem in current society that had compounded misunderstanding about woman's status and role. Women, he insisted, had been unfavorably affected by the current economic struggle for livelihood and by increasing opportunities to participate in cultural affairs. As a general rule, therefore, women were no longer content to be confined at home with large families. In keeping with the contemporary theme of "race-suicide," he maintained that decreasing family size and negligence of domestic duties among the cultured population would pass responsibilities for propagation and moral instruction to "the ignorant, the unthrifty, and the vicious classes." In their clamor for participation in external affairs and for de-emphasis of the sexual relationship, the extreme suffragists were touting an individualism that endangered the survival of the family. "The integral element of society is *not the individual but the family*," asserted LeConte. To remain unmarried was thus to weaken the social fabric. "The organization of society on the basis of the individual makes society a herd" and contravenes the laws of evolutionary progress.[28]

All of this was not to argue for the satisfactory status of woman's position, however, for LeConte explicitly stated that men must readjust certain of their views toward women. First, it was essential to employ more women in industry, because "the new conditions of modern society" required a greater diversity of occupations for women, including "teaching, typewriting, telegraph & telephone operating, sales . . . , hospital nurses," and perhaps the medical profession. Second, it was imperative that women be provided with better educational opportunities. But their education should differ from that provided for men because women required distinctive training, which could be accomplished best through an elective system of courses. For LeConte, the question was a matter of "equivalency, i.e. equal

27 See Kraditor, *The Ideas of the Woman Suffrage Movement*, 18–42.

28 LeConte, notes and MS on "The Woman Question," in LeConte Family Papers, UCB; LeConte, "Female Education," 60–91; Joseph LeConte to Matilda "Tillie" Harden Stevens, April 25, 1878, in LeConte Family Papers, APS.

value, equal *worth*, equal *dignity*, equal *rank*," and not "equality," because the latter term was tantamount to a single, or a male, standard.[29]

Although he remained as consistent in practice as in theory on this point, LeConte failed to consider the fact that some women were plainly as qualified as some men to enter certain occupations held exclusively by males. Again, however, his vision was obscured by a belief in the evolutionary distinctions between male and female traits of intelligence and by a conservative social philosophy. As an advocate of male dominance, he believed that male teachers should receive higher wages than female teachers because higher salaries attracted men into the teaching profession. Only men could "present the manly ideal to our boys during those plastic periods of childhood & youth." Moreover, the male teacher ought to receive higher pay because he had to support his family, and since the family stood as the pillar of civilization, an inequitable pay scale for the same performance by equally capable women was perforce justified.[30]

LeConte conceded that men had "erred in judgment as to what is best for women," but he labeled the extremists' plea for "emancipation" an absurd argument. As for the accusation of intentional injustice against women, LeConte replied that such was impossible because it contravened the laws of nature. True enough, he admitted, society was undergoing change, and that necessitated modifications of woman's status. But the current suffragist movement had accelerated too quickly, and, unless curtailed, it could disrupt society and hamper genuine social progress. LeConte urged the suffragists to be patient, for in due course men would recognize the need for change and respond favorably as long as "such changes *touch not her* [woman's] *essential womanhood.*" The "passionate women's conventions" only served to exacerbate ill-feeling, he added, calling such meetings a "fad" that had "run to irrational extremes." Evolution had developed the sexes for a good purpose, and it had clearly differentiated between them. Therefore, change as a law of evolution "needs

29 LeConte, notes and MS on "The Woman Question," in LeConte Family Papers, UCB.

30 *Ibid.*; Joseph LeConte to Emma LeConte Furman, June 29, 1871, in LeConte-Furman Papers, SHC/UNC. An indication of LeConte's previous opposition to coeducation is indicated in Sallie LeConte to R. Means Davis, September 10 [1872], in Davis Papers. LeConte's letter of recommendation for an alumna of the University of California, July 20, 1893, reveals a change of attitude toward coeducation (in Phoebe Hearst Papers, Manuscripts Division, Bancroft Library, University of California, Berkeley).

no urging," and tampering with it can only lead to "infinite harm." As
he had stated elsewhere, knowledge will continue to advance, and
changes will occur, but all problems must be solved by rational
thought proceeding through the laws of evolution.[31]

Evolutionism must not be ignored in the progress of society. It had
determined the course of civilization; it was central to the structure of
the educational system; it signified the proper relationship among
races; and it set the standard for sexual roles. In other words, it was,
for LeConte, the "grandest of all modern ideas." On the one hand, he
viewed it as a freely operating process; on the other hand, as a Neo-
Lamarckian, he believed in the necessity of manipulating it along progressive lines. Evolution was both physical and psychical, both natural and spiritual, both organic and social—in short, it was everything
in terms of progressive advancement. It was also synonymous with
"rational" thought, and it was basically only another word for "science." It was, in LeConte's view and in the opinion of many of his
contemporaries, the vehicle through which cultural progress could be
brought to its fullest flower.[32]

31 LeConte, notes and MS on "The Woman Question," in LeConte Family Papers,
UCB.

32 LeConte, "Relation of Biology to Sociology," 123. For a more detailed discussion of
LeConte's ideas on woman's social role, see Lester D. Stephens, "Evolution and Woman's
Rights in the 1890s: The Views of Joseph LeConte," *Historian*, XXXVIII (February, 1976),
239–52.

─••ᢓ XIV ᢞ••─
"The Marks of His Greatness": Honored Years, 1894–1898

Thirty years after he had stood solemnly in the gray ashes of Columbia and forlornly pondered the end of his scientific career, Joseph Le-Conte could but dimly recollect those heavyhearted days of 1865. He had risen from the depths of despair to the heights of international recognition. Few men would have begun anew in their late forties, and even fewer would have persisted so strongly in building a unique career at that stage of life. The rewards had come slowly at first, but they grew into a golden harvest during the last decade of LeConte's life.

As a popular speaker, LeConte was much in demand. Early in 1894, for example, he was invited by President David Starr Jordan to deliver an address to the students of Stanford University. An accomplished biologist and icthyologist and a firm evolutionist, Jordan was especially impressed by LeConte's efforts to reconcile evolutionism and Christianity. Thus, in addition to his opinion of the "beloved" professor as a man of "a singularly sunny disposition, a lucid literary style, and a deep feeling for nature," Jordan also praised LeConte as an "evolutionist of advanced type." As Jordan viewed him, LeConte was a man who always strove to "conciliate rather than to confute opponents" and "did much to reconcile believers" to evolutionism.[1]

LeConte could not accept all of the speaking invitations received. He was still teaching a heavy load, preparing lectures for scientific groups, and actively working in behalf of the university, especially in soliciting funds for research, publication, and support of graduate stu-

---

1 Jordan, *The Days of a Man*, I, 450; Joseph LeConte to David Starr Jordan, January 6, 1894, and April 3, 1895, both in Jordan Papers.

dents in the department of geology. The University of California was now one of the fastest growing institutions in the country, and its graduate programs in the sciences were increasing rapidly in reputation, for which much of the credit certainly belonged to LeConte himself. In addition, the ever-active LeConte could not tear himself away from his writing. He was constantly busy with revisions of his book or drafting articles. Among the latter were two popular papers on bird flight as the basis for the development of a flying machine. The old ornithologist was a better student of avifauna than he was an aeronautical engineer, however, and his thoughts on the topic ultimately made no impact upon the science of manned flight. Still, they revealed the range of his interests and the serious spirit of his inquiries. Unfortunately, it was just this sort of universalism that was outmoding the generation to which he belonged. Specialization of interest was now the order of the day, and LeConte was still operating as a Renaissance Man.[2]

But even though he was basically a universalist, LeConte did pursue some specialized interests, such as the nature of ore deposits. When his studies on metalliferous vein formation had been published in 1882 and 1883, they quickly gained the attention of mining geologists. Over a decade later, he was invited to honorary membership in the American Institute of Mining Engineers. Since membership in this reputable organization was tied to publication in the institute's *Transactions*, the respected geologist was asked to contribute a critical review of Franz Posepny's *The Genesis of Ore Deposits*. Along with many of the leading geologists of the land, he did so and was duly elected to the institute in April, 1895. But LeConte never pursued mining geology as a special field of study, and most of his contributions were

2 Examples of his lectures and activities can be found in the *Berkeleyan*, III (February 22 and October 11, 1894), IV (February 26 and 28, March 7 and 18, April 1, 4, and 16, and May 7, 1895), and V (February 13, 17, 18, 21, and 27, and April 20, 1896). See also LeConte to Arthur Rodgers, February 1, 1894, in Lawson Papers; LeConte to Adolph Sutro, February 15, 1895, in Adolph Sutro Papers, Manuscripts Division, Bancroft Library, University of California, Berkeley; LeConte to William H. Dall, SIA, Record Unit 2, p. 189. In his pocket notebooks, LeConte often jotted down topics that he planned to explore. See also LeConte, "The Problem of a Flying Machine," 69–76; Joseph LeConte, "New Lights on the Problem of Flying," *Popular Science Monthly*, XLIV (April, 1894), 744–57. His views on a flying machine are discussed in Lester D. Stephens, "Birdflight and Aviation: An Unheralded View," *American Aviation Historical Society Journal*, XXIII (Winter, 1978), 267–71.

theoretical rather than practical. As one geologist later noted, LeConte was never a specialist in "the realm of mining, mineral veins, and ore deposits . . . [but] his chapters [in *Elements of Geology*] on veins and ore deposits have been, and probably always will be, standards on such matters." The writer also praised LeConte for his "broad common sense" on the topics: "He saw that in the various disputes about the origin of veins no one side was absolutely right or entirely wrong, but that there was a truth in nearly all, which could be harmonized into a broad general principle."[3]

As the year 1895 rolled around, LeConte was in good spirits. He wrote to Emma that he had never felt better and that he was able to "put off old age." Among his reasons for feeling so good was his recent election as first vice–president of the Geological Society of America. Founded in 1888, the GSA had quickly become the leading organization for geologists, who had formerly utilized the AAAS as their chief professional association. LeConte was an original fellow of the GSA, and he had been initially appointed to its important Committee on Publications. Also active in reading papers before the fledgling association, LeConte had early entertained hopes of serving as an officer. His name had been placed on the ballot as a candidate for the presidency or first vice-presidency every year since 1891, and he was finally elected to the latter in 1895.[4]

With the accolade of the GSA vice-presidency pinned to his vest and the almost certain assurance of election to the highest office in the association waiting in the wing, LeConte felt good as he prepared to celebrate his seventy-second birthday on February 26, 1895. As he entered his classroom that day, he found his lectern bedecked with flowers and his students standing in his honor, ready to present him with a portrait of Louis Agassiz. It was the beginning of annual student tributes to "Professor Joe."[5]

3 LeConte, *Autobiography*, 320; Rossiter W. Raymond to LeConte, April 16, 1895, in LeConte-Furman Papers, SHC/UNC; Joseph LeConte, "Criticisms of Posepny's *Genesis of Ore Deposits*," *Transactions of the American Institute of Mining Engineers*, XXIV (1894), 996–1006; Joseph LeConte, "Don's Work on Auriferous Veins," *ibid.*, XXVII (1898), 993; "Obituary: Joseph LeConte," *Mines and Minerals*, XXII (August, 1901), 24–25.

4 Joseph LeConte to Emma LeConte Furman, January 13, 1895, in LeConte-Furman Papers, SHC/UNC; Minutes of the Society, December 27, 1888, December 27, 1889, December 27, 1892, August 15 and December 26, 1893, August 14 and December 26, 1894, August 29 and December 26, 1895, in Geological Society of America, Boulder, Colorado.

5 LeConte, *Autobiography*, 322; *Berkeleyan*, IV (February 28, 1895), 1.

LeConte was unquestionably a popular teacher. He had always enjoyed a reputation as such since his days at the University of Georgia. As he grew older, the gentle man was virtually venerated by his students. Josiah Royce recalled in his memoir of LeConte how as a freshman he had "looked forward to the time when [he] would reach the level of work where 'Joe's' lectures would form part of [his] task." Continued Royce: "More than once I listened, as others did, near the door of Professor LeConte's lecture room . . . fascinated by the wonderful quality of his voice." He also noted that the first public lecture he had heard from the lips of LeConte gave him "a deeper impression of what the architect of an argument upon a complex subject" could do than any other he had ever listened to. Indeed, so impressed was the future philosopher that he enrolled for six of LeConte's courses. LeConte, said Royce, was the ideal teacher and thinker: "His wealth of knowledge, his instinct for order and lucidity of reflection, have, indeed, always remained my hopelessly distant ideal."[6]

Other former students also praised LeConte as a teacher. S. B. Christy, who later became a colleague of LeConte, recalled his initial impression of the master-teacher: "I well remember the first time I heard him lecture: it was in the basement-hall of one of the local churches. As I wandered down the steps, my ear was touched by a thin, rather high-pitched voice, vibrating like a high-strung but clear and true violin string. . . . The peculiarity of the voice impressed me even before I had seen the speaker. But I stopped to listen; and I forgot the voice in the theme." Christy noted that he had often heard LeConte "repeat an address," but he listened to his great mentor "with new pleasure every time." Although his "notes were never written out in full, and there was no attempt at memorizing," remarked Christy, LeConte had thought out his subject "so thoroughly, so entirely on his own, that the best words naturally suggested themselves to his mind at the right moment." Christy also recalled how, as a student, he had daily observed the "agile figure eagerly stepping off for a brisk morning stroll of half an hour," then going to his study at nine o'clock, where he "shut himself up alone . . . and paced slowly up and down, reviewing those beloved notes of his." At eleven o'clock, LeConte "appeared

6 Josiah Royce, "Joseph LeConte," *International Monthly*, IV (1901), 327–29; Vincent Buranelli, *Josiah Royce* (New York, 1964), 52–53; Clendenning (ed.), *Letters of Josiah Royce*, 15–16, 107, 235.

before his class, full of the inspiration of his great theme." Other students were equally inspired by LeConte: "We leave [his] lecture-room with a feeling that we have been in the presence of something high and worthy," they wrote in the student yearbook for 1892.[7]

In 1901 John M. Stillman, who eventually became a Stanford University professor, declared that he could "safely say that by no other teacher was I so much influenced in my ideals of scholarship and personal character" as by LeConte. He was especially impressed with LeConte's "remarkable generalizing power," and although he acknowledged that he had encountered other outstanding teachers in the University of California, Stillman claimed that none was "so generally admired and so universally loved as Dr. Joseph LeConte." No teacher, said Stillman, had been so "easily approachable," so generous of his time to students, so "lenient to harmless youthful follies," and so appreciative of "a good-natured jest, even if he were himself the subject of it." Concluded Stillman in words that echoed LeConte's own sentiments, "teaching was to him a pleasure."[8]

One of the last students to sit at the feet of the inspired teacher was Ralph T. Fisher, who, as president of the Associated Students of the University of California, delivered a memorial tribute to LeConte in 1901. "One of the marks of his greatness," said Fisher, "lay in the fascinating simplicity with which he explained the most difficult subjects in the class-room." Young Fisher also lauded LeConte for his exemplary devotion to truth, the contagious "earnestness" of his convictions, and the "faithfulness" of his lecture preparations, and he closed with a tribute to LeConte's "love for humanity" and "his Christlike humility."[9]

LeConte also enjoyed a reputation as a teacher in a broader sense,

7 Christy, "Biographical Notice of Joseph LeConte," 765–93; University of California *Blue and Gold*, XVIII (1892), 113.

8 "LeConte Memorial Meeting," *University Chronicle*, IV (October, 1901), 244–49, 254–62; William E. Ritter, "Professor LeConte As Seen Through His Biological Work," *University of California Magazine*, VII (September, 1901), 218–25. The influence of LeConte's evolutionist teaching is evident in the literature of Frank Norris, although perhaps not in a way that would have necessarily pleased the old scientist. See Donald Pizer, "Evolutionary Ethical Dualism in Frank Norris' *Vandover and the Brute* and *McTeague*," *Proceedings of the Modern Language Association*, LXXVI (December, 1961), 552–60; and Malcolm Cowley, "Naturalism in American Literature," in Persons (ed.), *Evolutionary Thought in America*, 315.

9 "LeConte Memorial Meeting," 262–64. See also Charles Woodward Hutson, "The South Carolina College in the Late Fifties," *Sewanee Review*, XVIII (July, 1910), 342.

a reputation that extended beyond his classroom lectures. This was especially notable in his excursions to the mountainous regions of the West Coast. As John Muir recalled his initial meeting with LeConte: "I was at once drawn to him by the charm of his manners, as to a fine lake or mountain." So fascinated was the famous Sierra naturalist with the "sinewy, slender, erect" figure of LeConte that he remembered vividly the latter's "fine poetic appreciation of nature" and how, in the Sierras, "with buoyant, sparkling delight like that of a child, [he kept] up running all-day lectures, as if trying to be the tongue of every object in sight." Muir averred that, as a teacher, LeConte "stood alone on this side of the continent, and his influence no man can measure. He carried his students in his heart, and was the idol of the University." LeConte, concluded the man who was himself a great object of adoration, "had the rare gift of making dim, nebulous things clear and attractive to other minds, and he never lacked listeners."[10]

Frank Soulé, LeConte's colleague and friend for over thirty years and the man who was with LeConte on both the first and the last of his trips into the High Sierra, reiterated Muir's judgment of LeConte as teacher par excellence in the classroom of Nature. Said Soulé of his great friend, "He quoted poetry by the volume, and seemed to have an unlimited store in his memory. The beautifully descriptive and the humorous were usually on these [camping] occasions given his preference. He was full of anecdote seasoned by wit and humor of a high order." It was little wonder, then, that LeConte was so highly honored by his colleagues and former students, and it was not surprising that some of them should attribute to him great qualities of revelation, as did the poet Charles Keeler, who dedicated a volume of poems on evolution and nature to his former mentor. Keeler called LeConte the "dear friend who late unsealed my eyes to lore of Nature's mystic laws." Others named various landmarks in honor of their great teacher.[11]

10 John Muir, "Reminiscences of Joseph LeConte," *University of California Magazine,* VII (September, 1901), 209–13.

11 Frank Soulé, "Joseph LeConte in the Sierra," *Sierra Club Bulletin,* IV (January, 1902), 3–4. Charles Keeler wrote several poems commemorating his teacher and friend (he was frequently a guest along with the LeContes in the William Keith home in San Francisco); see, for example, his "To Joseph LeConte," *Sierra Club Bulletin,* VI (January, 1908), 223–24. Keeler's collection, *A Light Through the Storm* (San Francisco, 1894), contains several poems reflecting LeConte's scientific teachings. See also LeConte to Keeler, November 5, 1894, and March 6, 1895, both in Charles Keeler Correspondence, Huntington

One such honor came in August, 1895, when two mountain-climbing enthusiasts ascended a peak in the Sierra three miles southeast of Mount Whitney and left a monument on the south side of the dome in honor of LeConte. The pair of climbers placed in a small can a photograph of LeConte and the following memorandum: "To-day, the 14th day of August, 1895, we, undersigned, hereby named this mountain LeConte, in honor of the eminent geologist, Professor Joseph LeConte." The name stuck, and the old mountain-lover now had two peaks bearing his name, one on the West Coast, and the other, some two thousand miles across the continent, on the East Coast. Many other landmarks in the Sierra came to bear LeConte's name, some temporarily and some permanently, including a divide and a waterfall. After his death, a memorial lodge bearing his name would be built in Yosemite. These tributes befitted the man who loved the Sierra so dearly and wished to preserve its pristine beauty.[12]

In fact, LeConte was a charter member of the Sierra Club, founded in 1892 "for the purpose of exploring, enjoying, and rendering accessible the mountain regions of the Pacific Coast, and to enlist the support and cooperation of the people and the government in preserving the forests and other features of the Sierra Nevada Mountains." Although LeConte was in Europe at the time of the club's incorporation, he had been an enthusiastic supporter of the idea, and his name was included on the charter. After his return to Berkeley, he was elected to the board of directors and served for a time as a vice-president. In 1898 he relinquished his place on the board to his son, who served for many decades as an officer of the club. A few years earlier, LeConte had been asked to preside over a meeting of the club and to offer

Library, San Marino, Calif.; Keeler to LeConte, September 14, 1896, and February 3, 1900, both in LeConte Family Papers, UCB.

12 A. W. de la Cour Carroll, "The Ascent of Mt. LeConte," *Sierra Club Bulletin*, I (May, 1896), 325–26; *Berkeleyan*, IV (September 16, 1895), 2; Francis P. Farquhar, "Place Names of the High Sierra," *Sierra Club Bulletin*, XII (1924), 52–54; Francis P. Farquhar, *Place Names of the High Sierra* (San Francisco, 1926), 56–57. An elementary school in Berkeley, California, was named for LeConte; see letters from pupils in that school to LeConte, February 23–26, 1897, in LeConte Family Papers, UCB. A former LeConte student named a glacier and a bay in Alaska for LeConte, while city officials in Berkeley, California, and Atlanta and Athens, Georgia, named avenues and streets for him. The universities of Georgia, South Carolina, and California subsequently named buildings in his and his brother John's memory. A number of fossils were also named in honor of Joseph LeConte by his colleague John C. Merriam and others who held him in high esteem as a teacher.

comments upon the topic of "The National Park and Forest Reservations." Said LeConte to the sympathetic group, "I do not believe there
is any interest connected with our country . . . of more pressing ur-
gency than this of the preservation of our timber land." He expressed
alarm over the rapid rate of timber depletion and stressed the importance of the forests to agriculture. Admitting the necessity of "legitimate use" of timberlands, he lamented that "the wasteful disappear-
ance of the timber is simply dreadful." Once more, LeConte saw this
as a consequence of "individualism . . . run mad." Like Muir and other
members of the Sierra Club, he was "perfectly satisfied that nothing
can save our timber land except complete reservation by the Government," and he called for a plan whereby wilderness lands would be
"raised in a thoroughly rational way for legitimate uses only." LeConte
was thus one of the early conservationists whose efforts led eventually
to governmental action by Theodore Roosevelt and Gifford Pinchot.[13]

LeConte did not take an extremely active role in the affairs of the
Sierra Club, but that is understandable in light of the other demands
placed upon his time. Indeed, by 1896 he had decided that he must
give up his undergraduate classes in zoology and geology. In part he
relinquished them because they had "become so large." Enrollment in
his geology class now exceeded four hundred students, and it took
him "nearly a month to look over and grade . . . examination papers."
But he also felt a need for some relief from undergraduate teaching in
order to concentrate on advanced courses in geology and comparative physiology. In addition, the old teacher sensed that his two young
and extremely able colleagues in geology, Andrew Lawson and John
C. Merriam, ought to "have a better chance" to exercise their teaching
skills. Moreover, he believed it was necessary to give up these classes
in order to complete revisions of his *Elements* and *Sight* and to prepare more articles. Thus, as he noted, "the year 1896 was an especially
prolific one with me." Since he was also serving as president of the
GSA, he was occupied with the duties of that office, and he was preparing for a trip to England in the fall to attend the meeting of the
British Association for the Advancement of Science. He had already

13 Holway R. Jones, *John Muir and the Sierra Club: The Battle for Yosemite* (San Fran-
cisco, 1965), 55–63, and "Appendix A," 170–71; Francis P. Farquhar, *History of the Sierra
Nevada* (Berkeley, 1965), 217–18; Ansel Adams, "The Photography of Joseph N. LeConte,"
Sierra Club Bulletin, XXIX (October, 1944), 41–46; "Proceedings of the Meeting of the Sierra
Club, Held November 23, 1895," *Sierra Club Bulletin*, I (January, 1896), 268–71.

secured from the university regents the promise of a leave from August 10, 1896, to January 10, 1897.[14]

Early in September, 1896, LeConte, Bessie, and Carrie set sail for England. Although the visit lasted only a few weeks, it proved to be a rewarding trip for LeConte, especially because he met and "saw much" of the aging Herbert Spencer, whose works had greatly influenced him and whose sociological theories he had modified so much in his own writings on evolutionism. At this point in their life and career, the two old men could find much to discuss on the science of society. Spencer introduced LeConte to Andrew Carnegie, the great "distributor of wealth," who was on one of his visits to the United Kingdom. Carnegie and LeConte apparently hit it off well, for the wealthy steel magnate not only introduced the American evolutionist to the Athenaeum but also had him made a member of that organization of intellectuals.[15]

LeConte also visited old acquaintances, including Sir Archibald Geikie, and he made new friends as well. Among the latter were the Reverend John Watson and Robert Pearsall Smith. With Watson, who had written a number of popular sketches of Scottish rural life under the pseudonym Ian Maclaren and several religious works under his own name, LeConte discussed his views on evolutionism and religion, for which he was perhaps best known in Britain. He also did the same with Smith, an American expatriate who eventually became a naturalized British subject. Smith had become something of a religious skep-

14 LeConte, *Autobiography*, 321–25; LeConte to Andrew Lawson, September 20, 1890, and February 19, 1892, both in Lawson Papers; Minutes of the Board of Regents, University of California, Vol. X, May 12, 1896; S. P. Langley to Joseph LeConte, November 25 and December 13, 1895, and LeConte to Langley, December 3, 1895, all in Secretary's Correspondence, Smithsonian Institution Archives, Washington, D.C.; Bessie LeConte to Emma LeConte Furman, June 4 [1895], in Nancy Smith Collection; Joseph LeConte to Emma LeConte Furman, November 3, 1895, in LeConte-Furman Papers, SHC/UNC.

15 LeConte, *Autobiography*, 324–25; Clarke, *James Hall of Albany*, 540–41; Joseph LeConte to James A. LeConte, July 13, 1896, in possession of Mrs. Joseph Nisbet LeConte; Herbert Spencer to Joseph LeConte, September 28 and 29, 1896, both in LeConte Family Papers, UCB. Founded in 1824, the prestigious Athenaeum included among its members many distinguished men of letters and science. Under its famous "Rule II," nine prominent men were annually selected for membership. The rigorous standards did not apply to the election of "extraordinary members," who (by "Rule XII") simply had to be "certain persons of high official rank." Apparently Carnegie and LeConte were "extraordinary members." See Humphrey Ward, *History of the Athenaeum, 1824–1925* (London, 1926); Bernard Darwin, *British Clubs* (London, 1943), 25–31; and Joseph Frazier Wall, *Andrew Carnegie* (New York, 1970), 428–29.

tic, and LeConte said, "I talked much and earnestly with him, and like a drowning man, he caught at my views with a joy and love that were overpowering." Upon LeConte's departure, Smith hugged the old believer and "almost wept." Such was the overwhelming personal impact of the gentle evolutionist whose religious faith had a contagious effect upon the doubting Thomases of the intellectual world.[16]

But LeConte could not tarry long in England; he was expected back in the United States in October to attend the sesquicentennial celebration at Princeton University, where he would be an honored guest. To highlight the significance of the institution's history, a number of distinguished foreign guests had been summoned to receive the honorary degree of LL.D. LeConte was one of only two Americans selected, and although in his autobiography he claimed the degree had "become so common" that he cared "little for it *per se*," he was nevertheless "under these circumstances" thoroughly delighted to receive such "a distinguished honor." It was at this ceremony that he met another man whose role in United States and world affairs would ascend in years ahead, though long after LeConte's death. Yet LeConte already recognized Woodrow Wilson as a leading historian of the land, and he took especial note of Wilson's speech, particularly because of its criticism of the role of science in education and other affairs.[17]

From Princeton, LeConte, Bessie, and Carrie went to Cambridge to visit old friends, among them Josiah Royce, Mrs. Louis Agassiz, Alexander Agassiz, James Peirce, and Mrs. Asa Gray. It was while he was there that LeConte delivered his paper on "The Relation of Biology to Philosophy." In addition, he "spent one evening discussing evolution and its relation to religion" with the faculty and students of the Harvard Divinity School, and another he spent at a dinner with over two dozen of his former University of California students, known as "the Berkeley Colony." In November, he attended the NAS meeting in New York and then went on to Columbia, South Carolina, to visit with Sallie

16 LeConte, *Autobiography*, 325–26; R[obert] Pearsall Smith to "Mrs. and Miss LeConte," September 30, 1896, and to "Dr. LeConte" on the same date, both in LeConte Family Papers, UCB. See also T. McKenny Hughes to Joseph LeConte, August 25, 1896, and July 8, 1897, and Archibald Geikie to LeConte, October 2 and 4, 1896, all in LeConte Family Papers, UCB.

17 LeConte, *Autobiography*, 326–27; New York *Times*, October 21, 1896, p. 8; *Critic*, October 24, 1896, p. 241; *Berkeleyan*, V (October 30, 1896), 1; LeConte, "The True Idea of a University," 18–19.

and her family. At the end of December, he was in Washington to deliver his GSA presidential address on "Earth Crust Movements and Their Causes."[18]

LeConte hurried back to the South as soon as his schedule would permit, for he was excited over the long-planned celebration of his golden wedding anniversary on January 14, 1897. The affair would be held in Emma's huge home at Scottsboro, only two miles from Bessie's old home, where the couple had been married. "It was a happy occasion for all," noted LeConte, "but most of all for my dear wife and me." A flood of telegrams from colleagues, students, and the regents in California heightened the festive mood. But there would be more. A great celebration was planned for the LeContes when they returned to Berkeley. Held on February 18 in San Francisco's Mark Hopkins Art Institute, the gathering was attended by more than three thousand admirers of LeConte. After they were presented with a gold loving cup, the LeContes listened to tributes from President Martin Kellogg and Governor James E. Budd. Many other prominent people gathered "to do honor to this prince of evolutionists," as the Oakland *Tribune* called LeConte.[19]

The San Francisco *Examiner* said after the affair that "the University of California is not known abroad for its revenues, for the cost of its buildings, or the extent of its collections. It is known and honored for the work of such men as Professor LeConte." Others praised Joseph and Bessie in warm and adoring words, and the Berkeley *Gazette* reported that they "were visibly affected, and when the Professor rose to reply tears glistened his eyes, and it was some moments before his voice was steady." When he was finally able to speak, the snowy-haired idol of affection said, "I easily perceive two things which more than all others have contributed to the attainment by me of whatever is worthy to be remembered." These he identified as his marriage and his removal to California. His marriage to Bessie, he declared, had awakened and stimulated "all that was best within" him, while his coming

18 LeConte, *Autobiography*, 327–28; Joseph LeConte to Mrs. Asa Gray, November 3, 1896, in Asa Gray Letters, Gray Herbarium, Harvard University; Josiah Royce to Mary Gray Ward Dorr, November 5, 1896, in Clendenning (ed.), *Letters of Josiah Royce*, 348–49.

19 LeConte, *Autobiography*, 328–29; various newspaper clippings in McMillan Collection, including the Oakland *Tribune*, February 20, 1897, and the Berkeley *Advocate*, February 18, 1897, which reported that five thousand invitations had been issued. See also the *Berkeleyan*, IV (January 29, 1897), 1, and (February 23, 1897), 4.

to California had furnished "a suitable environment" for his "best activity." After gentle words of loving praise for Bessie, he spoke further
on his life in California. "In the South . . . we take life easy—much too
easy," he said. "On coming here I felt at once the quickening effect of
contact with a young-fresh-rapidly progressive community, the quickening effect of a new and rich field of scientific work, especially in my
own favorite department of Geology." He added, "Nearly all my work
had been done since I came to California when I was near 47." For
twenty-eight years he had labored in his adopted land, and despite
his occasional thoughts of leaving during the first decade, he had indeed "absorbed the inspiration of the new civilization," and he had
added to it his own strength. To his wife of fifty years, he gave much
of the credit for his success. Bessie had maintained a steady sense of
devotion to her husband and his career, constantly bolstering his
spirit by warm encouragement, faithful duty, and womanly companionship. For over fifty years Caroline Elizabeth Nisbet LeConte gave of
herself to make not only a successful but also a genuinely happy marriage.[20]

Meanwhile, LeConte was basking in the praise accorded the new editions of his *Elements of Geology* and *Sight*. On the latter, however, he
received more criticism than he had on the first edition, primarily
from psychologists. Since the original edition of *Sight* in 1881, LeConte
had virtually ceased to write on physiological optics. By the mid-
1890s, however, he had resumed his interest in the subject, and he
decided to revise the book. Unfortunately, he had not kept up with
advances in psychology, many of which dealt with visual perception.
Thus, when the revised edition came off the press early in 1897, it was
reviewed less favorably than the original had been.

20 Memorials to Joseph LeConte, including clippings from the San Francisco *Examiner*, February 19, 1897, and the Berkeley *Gazette*, February 20, 1897, in Joseph LeConte
Memorials, Archives Division, Bancroft Library, UCB; Golden Wedding Reception Address, February 18, 1897, in LeConte Family Papers, UCB. The depth of Bessie's love for
her devoted husband is intimately revealed in her very touching letters written to his
memory on January 14 and March [?], 1903, and January 14, 1904; see also her poem
"God Is Near," July, 1915, and her letters to Sallie LeConte Davis, October 28, 1903, and
March 29 [1914], all in LeConte Family Papers, UCB. See also LeConte, *Autobiography*,
329–31; *Proceedings of the Chit-Chat Club* (1894), 23; and LeConte, "The Cooperation of
Religion and Science in Uplifting Humanity," Unitarian Club Addresses, April 26, 1897, in
Joseph LeConte, Miscellaneous Writings, Bancroft Library Archives, UCB.

Popular Science Monthly praised LeConte for adding many new topics and for giving "fuller explanations" and "clearer statements" of other subjects that he had discussed in the original edition of *Sight*, and the *Archives of Ophthalmology* lauded it as the only treatise of its kind that gave "so clear an account of the fundamental phenomena of monocular and binocular vision." The *Archives* declared that experts would profit from reading the work, especially the section on binocular vision, which contained "acute observations" by the author. It further praised LeConte for his explanations of ocular motion and the horopter.[21]

The new edition of *Sight* was also reviewed at length by E. W. Scripture in the *Psychological Review*. An instructor of experimental psychology at Yale University, Scripture had made important contributions to the physiology and psychology of vision. His review of *Sight* was both laudatory and caustically critical. On the one hand, he hailed the "conspicuous merits" of the book, such as "the ingenuity of the illustrations, the clearness of the statements and the fascinating character of the experiments described," but on the other, he criticized it for presenting a "misunderstanding of the psychological principles involved in monocular vision."[22]

The psychologist J. McKeen Cattell was also generally pleased with the revised edition of *Sight*, and he reviewed it in his own journal *Science* (of which LeConte had served for a time on the editorial board). The book, Cattell said, was "extremely welcome," for, "as it was . . . when the first edition was published, . . . there is no other book in the English language covering this field." He also hailed LeConte for his practical approach, noting that he had again published the only nontechnical work on vision. "While it is surprising . . . ," said Cattell, "that we have only one book on the subject, it is fortunate that it is so excellent." Although Cattell criticized LeConte for devoting "two-thirds of his book to binocular vision," he nevertheless commended *Sight* for its "important original contributions." What bothered Cattell most, however, was LeConte's inadequate treatment of such important topics as "intensity of sensations, their time relations, the field of vision, illusions, the combinations of colors, etc." In addition, he took issue

21 *Popular Science Monthly*, LII (January, 1898), 423; *Archives of Ophthalmology*, XXVI (1897), 490–91.

22 E. W. Scripture, review of *Sight* (2nd ed.), *Psychological Review*, IV (September, 1897), 543–45.

with LeConte on three points, namely, "upright vision," the doubleness
of human sight, and the function of the fovea or central spot in the
eye. Cattell also correctly observed that LeConte did not understand
the psychological nature of retinal impressions.[23]

In a response to Cattell's criticisms of his statement on the fovea,
LeConte agreed that the editor of *Science* had grounds for objection
"if I had implied anything so absurd" as the notion that the central
spot was necessary to the development of the higher mental faculties
"in the *ontogeny*" of man. Admitting to "a possible ambiguity" of his
statement, LeConte protested that Cattell had misunderstood his
"meaning." He insisted that he had referred to "the phylogeny of man;
not in the education of the *individual*, but in the origin of the *race*." He
confessed that he "ought to have used the word *evolution* instead of
development." LeConte also addressed himself to Cattell's argument
on double vision. He agreed that "we learn to consciously see them
[*i.e.*, things] double. But if we see only what we consciously see, we
see comparatively little." The act of binocular vision, he said, is based
upon an immediate recall of things as we are used to seeing them;
hence, because they "lie so near the surface of consciousness," they
are immediately perceived as we *know* them. Thus, once again, Le-
Conte had resorted to the Neo-Lamarckian view of the accumulation
of residual experiences, believing that they are eventually inherited by
the race, or by mankind in general.[24]

Reserving the topic of upright vision until last, LeConte said that
Cattell misunderstood this phenomenon, for "seeing things upright is
not necessarily connected with an inverted image." Because the
"retina is *concave* instead of convex . . . the image is inverted, and,
therefore, [it] must be reinverted in the act of outward reference." Ca-
tell had asked the wrong question, LeConte declared. The important
point is that people see "things in their true places." Moreover, Le-
Conte believed that experiments with glasses that turn the viewer's
external images upside down, although "admirable," do not refute the
law of direction. Further, he noted, those experiments resulted in an
inversion of "the *whole external world*," and hence "all our movements

23 J. McKeen Cattell, review of *Sight* (2nd ed.), *Science*, n.s., VI (September 24, 1897),
491–92. LeConte had maintained (incorrectly) that "in passing down the animal scale,
the central spot is quickly lost. It exists only in man and the higher monkeys" (*Sight*, 2nd
ed., 78).

24 *Science*, n.s., VI (November 12, 1897), 737–39.

must be readjusted." In other words, these studies had shown only that "we see things in wrong places." They do not deal with *"the impressing forces"*; they deal only with "the relative places and positions of objects in the external world *as we know it.*" The law of direction, he countered, is not acquired through experience, except from *"ancestral experience"*; it is acquired or "inherited all along the line of evolution ever since eyes were formed, and finally embodied in brain structure." That, declared LeConte, is simply a fact of evolutionary development.[25]

LeConte's controversy with the psychologists did not end there, however, for E. W. Scripture introduced yet another subject on which LeConte could not keep silent. This final controversy had to do with a visual phenomenon called "cerebral light." In July, 1897, Scripture had written a letter to *Science*, explaining the "irregular forms of light in our visual field" when we close our eyes. He believed they were "cerebral rather than retinal" in their origin because: (1) only one rather than two illuminated fields appears; (2) the illuminated forms remain stationary and are unaffected by eye-movements; and (3) displacement of the eyes does not result in movement of the forms. He concluded that "cerebral light is located in those higher centers of the brain which are connected with visual memories and imaginations." LeConte quickly responded to Scripture's explanation. Noting that he had studied the phenomenon for a long time, he cited three brief experiments to demonstrate the validity of his own argument. Actually, both men misunderstood the photoreceptive nature of the retina and thus failed to grasp the true nature of those "afterimages" that appear when the eyes are closed after an intensity of exposure to brilliant objects.[26]

If Scripture read the reply, he remained oblivious to it, for two years later he again wrote to *Science* on the subject of cerebral light. In the letter he noted that a recent German reviewer had questioned the sufficiency of Scripture's proof, but he did not refer at all to LeConte's

25 *Ibid.* LeConte later aided his young colleague George Stratton with one of his experiments, being called upon by Stratton to confirm a point regarding visual distance. See G. M. Stratton, "A Mirror Pseudoscope and the Limit of Visible Depth," *Psychological Review*, V (November, 1898), 632–38. Stratton reported his experiments on "Vision Without Inversion of the Retina" in *Psychological Review*, III (November, 1896), 611–17, IV (July, 1897), 341–60, and IV (September, 1897), 463–81. Cattell's rejoinder to LeConte appeared in *Science*, n.s., VI (November 12, 1897), 739–40.

26 E. W. Scripture, "Cerebral Light," *Science*, n.s., VI (July 23, 1897), 138–39; Joseph LeConte, "Cerebral Light," *ibid.*, n.s., VI (August 13, 1897), 257–58.

criticism. Scripture then related another experiment and concluded once more that the phenomenon was cerebral. LeConte read Scripture's latest published letter, and after observing in the next issue of *Science* that no one had replied to Scripture, he wrote out his own criticism and sent it to Cattell. LeConte charged that Scripture's conclusion did not follow from the experiment reported. To satisfy himself, LeConte applied Scripture's experiment to afterimages and got the same results that Scripture achieved with "cerebral light," indicating clearly that Scripture was experiencing difficulty in clarifying the phenomenon with which he was concerned. Although LeConte never fully apprehended the nature of retinal images, he at least shunned such mystical conceptions as "cerebral light" and endeavored always to base his explanations of vision upon carefully controlled experiments.[27]

LeConte maintained his interest in vision to the end of his life, and only a few months before his death he was still corresponding with Cattell on the subject. Moreover, his *Sight* continued to serve as a fundamental textbook for physiologists and psychologists for several years after he died, although the second edition did not have the impact of the original work. By the end of the first decade of the twentieth century, the book had been replaced by other works that covered new advances in the field of physiological optics, and LeConte's contributions to the study of vision were largely forgotten. Nevertheless, the simplicity of the explanations, the clarity of the diagrams, and the character of the experiments in *Sight* all represented classical contributions to the study of human vision.[28]

In retrospect, it is now obvious that LeConte contributed to the development of the science of physiological optics in the United States. He was the first American to write extensively and systematically on the subject, and his book served as the basic English-language text for a generation of physiologists and psychologists. In the final analysis, LeConte's work on vision suffered from a number of defects, but at the same time it represented primary scientific research. Because he was not intimately familiar with the studies of German opticians, LeConte missed some of their important findings, and in some instances he

27 E. W. Scripture, "Cerebral Light: Further Observations," *ibid.*, n.s., IX (June 16, 1899), 850; Joseph LeConte, "Cerebral Light Again," *ibid.*, n.s., X (July 14, 1899), 58.
28 LeConte to Cattell, May 4, 1901, in J. McKeen Cattell Papers, Library of Congress.

unwittingly duplicated some of their experiments and observations. Yet his independent studies confirmed the conclusions of other physiologists, and they helped to shed light upon some of the most difficult problems in the field of binocular vision. Moreover, his concentration upon certain aspects of the physiology of sight caused him to neglect, or at least to downplay, other important visual phenomena. That was especially true of his study of color vision, and even though he gave more space to the topic in the second edition of *Sight* than he had in the original book, he never really mastered the subject. Finally, LeConte was handicapped by his too-rigid conformity to Neo-Lamarckian evolutionary views. His propensity to offer explanations on visual phenomena that were rooted in the theory of the accumulative effect of residual experiences thus created problems regarding the psychology of sensations.

Still, LeConte deserves credit as the father of physiological optics in the United States. His was the first textbook on the subject to be written in the Western Hemisphere, and by the consensus of contemporary critics and the extent of its use, *Sight* was an extremely influential work. Marked by lucid language, reduction of complex phenomena to easily comprehensible explanations, simple and distinct illustrations, and experimental verification of visual phenomena, it was a remarkably clear presentation of a complicated subject, and it paved the way for the scientific study of vision in the United States.

LeConte had remained very active throughout the year of 1897, but when the new year arrived, he found it impossible to do much writing. This was due primarily to a severe recurrence of Carrie's "nervous prostration." Early in the year, LeConte told his young colleague Andrew Lawson that "we are all in good health except Carrie who has been suffering for many months with nervous prostration & nervous dyspepsia." During the summer he accompanied Bessie and Carrie on a trip to Sonoma County and the Santa Cruz Mountains, hoping that a vacation would alleviate the anxieties suffered by his daughter. It seems to have done Carrie little good, but it rejuvenated LeConte. As he later noted, he went "swimming every day for a month or more," and he proudly claimed that his activity won "the admiration of all."[29]

Despite the attention given to Carrie, LeConte still found time to

29 LeConte, *Autobiography*, 330; LeConte, Autobiographical MS.; LeConte to Andrew Lawson, January 28, 1898, in Lawson Papers.

conduct his classes and write out a few articles. His students again honored him on his birthday by decorating his lectern with flowers and presenting him with a handsome bookcase. Moved by the gesture, LeConte told his students how much they helped to stimulate his mind and make him stay young at heart. In the course of the year, he also completed a revised edition of his *Compend of Geology*, gave the Charter-Day address on "The True Idea of a University," prepared and delivered a paper on William James's *The Will to Believe*, and presented a lecture on the physical geography and geology of Palestine. In the seventy-fifth year of his life, LeConte reveled in his good health. To a friend who asked him who he would like to be if he could come back in another life, LeConte replied that it would be as himself, for he got so much fun out of life as Joseph LeConte.[30]

30 *Daily Californian*, February 4, 25, and 28, August 29 and 31, and September 1, 1898, pp. 1, 2, 1, 3, 1, and 1, respectively; Joseph LeConte to Lester Frank Ward, February 26, 1898, in Ward Papers; LeConte to David Starr Jordan, February 14, 1898, in Jordan Papers.

──◆◆❃XV ❃◆◆──
Spokesman for the Contractionalist Theory

By 1898 LeConte had published nearly sixty articles on geological top-ics as well as four editions of his *Elements of Geology* and two of his *Compend of Geology*. Although this record of scholarly productivity did not equal that of many contemporary geologists, it is nevertheless impressive, especially since it was achieved mostly during the twenty-five-year period after LeConte had settled in California and began anew his efforts to contribute to the field of earth science. The record is all the more remarkable when one considers that LeConte pro-duced these works in the face of a heavy teaching load and a commit-ment to other scholarly endeavors, including numerous publications in various fields of thought. Given his broad range of interests, his lack of specialization in a single discipline, and his limited field experi-ence, LeConte attained a high degree of success as a geologist, and he was accordingly recognized by his peers for his contributions to geo-logical knowledge. True, these contributions were primarily theoreti-cal rather than specific factual discoveries, but they certainly pro-voked useful discussions and advanced the prospects of critical inquiry in the science of geology.

In the judgment of one of his contemporaries, LeConte was "a com-piler of information and a teacher rather than an original investigator" in geology. This accurate assessment was confirmed by another peer, who called LeConte "a noble example of the type of man, now rare, whose distinction and influence are due to breadth of knowledge in all fields of science and clearness of perception and exposition rather than to exclusive mastery of the minute details of some one branch of science so essential to the expert specialist." Perhaps the whole matter of LeConte's universalist approach to science was best summed up by

Herman L. Fairchild, a professor of geology at the University of Rochester, longtime secretary of the GSA, and reviser of *Elements* after LeConte's death. Fairchild aptly noted that LeConte was "perhaps the last distinguished representative of the general geologist as typified during the past century," and he placed him alongside James Dwight Dana as a transitional figure in the development of the discipline of geology. In Fairchild's judicious assessment, these two men were the "noblest American representatives of the passing type, for while they grew up under the influence of the older intellectual attitude, they grew out of it in spirit while they studied and guided the transition." Both men, he concluded, "were preeminently students of the accumulated data of the literature of the science, with generalization and philosophic inference as their dominant inspiration."[1]

As a synthesizer of information and a reconciler of conflicting theories, LeConte had few equals, and certainly he was one of the very last of the nineteenth-century universalist types. Yet he did contribute some specialized knowledge to the field of geology, and, even more, he clarified a number of concepts and developed many hypotheses that stimulated both his students and his peers to grapple with important problems in the discipline, especially with respect to coral reefs, glaciation, coal measures, ore deposits, lava floods, structural faults, historical geology, geological classifications, orogeny, and diastrophism. LeConte also wrote at length upon vulcanism, earthquakes, and geysers, but, in general, most of his comments on those topics merely constituted syntheses of the literature. But, all in all, his chief area of interest in geological phenomena lay with the question of the "great features" of the earth's surface, and it was on the special topic of orogeny, or mountain formation, that he gained the greatest recognition as a geologist. Eventually, of course, his theories were replaced by further discoveries, but, in the context of late nineteenth-century geology, LeConte contributed significantly to contemporary thought on the origins and processes of mountain formation.[2]

---

1 *American Journal of Science*, 4th ser., XII (October, 1901), 248; Fairchild, "Memoir of Joseph LeConte," 53–54.

2 Among his articles on these topics were: "Earthquakes," *University of California Echo*, II (April, May, June, and July, 1872), n.p.; "Note [Regarding Source of Volcanic Energy]," *American Journal of Science*, 3rd ser., V (February, 1873), 156; "On Some of the Ancient Glaciers of the Sierras," *ibid.*, 3rd ser., V (May, 1873), 325–42; "Astronomy and

One of the early hypotheses regarding the thermal origins of mountain making had been developed in the eighteenth century by James Hutton, who contended that the internal heat of the earth caused rocks to expand, forcing them at certain points to rise above sea level and become continents and mountains. Other European geologists followed the Huttonian hypothesis, including the French scientist Léonce Élie de Beaumont, whose three-volume study, *Notice sur les Systèmes de Montagnes* (1852), seems to have influenced LeConte's ideas on mountain formation. An advocate of the contractional theory—that is, that the cooling of the earth resulted in tectonic convulsions of the earth's crust—Élie de Beaumont argued that changes in the earth's surface occurred in "paroxysmal stages" and that mountains were therefore born in times of critical upheavals.[3]

Meanwhile, many American scientists had turned their attention to the topic of mountain formation. In 1842, for example, H. D. Rogers, who had studied the Appalachian ranges in great detail, proposed that "wave-like pulsations" of magma pushed upward at the thinnest points of the earth's crust. This theory was, of course, based upon the idea that the interior of the earth was molten. Dana objected to that explanation, however, and in 1846 and 1847 he published several papers, arguing that a cooling of the earth resulted in shrinkage and the consequent wrinkling of the earth's surface. Dana maintained that crustal blocks broke through the surface along "cleavage lines," or the

---

Geology," *Berkeleyan*, I (May, 1874), 1–4; "On Some of the Ancient Glaciers of the Sierra Nevada," *American Journal of Science*, 3rd ser., X (August, 1875), 126–39; "Hog Wallows or Prairie Mounds," *Nature*, XV (April 19, 1877), 530–31; "Geysers and How They Are Explained," *Popular Science Monthly*, XII (February, 1878), 407–17; "Geological Climate and Geological Time," *Nature*, XVIII (October 24, 1878), 668; "Rate of Denudation," *Geological Magazine*, n.s., IX (June, 1882), 288; "Earthquake-Shocks More Violent on the Surface Than in Mines," *Science*, VI (December 18, 1885), 540; "Determination of the Depth of Earthquakes," *ibid.*, X (July 8, 1887), 22–24; "The Flora of the Coast Islands of California in Relation to Recent Changes of Physical Geography," *Bulletin of the California Academy of Sciences*, VIII (September 19, 1887), 515–20; "Glacial Motion," *Philosophical Magazine*, 5th ser., XXV (May, 1888), 452; and comments in the "Symposium on Geological Classification and Nomenclature," *Journal of Geology*, VI (May–June, 1898), 337–38.

3 For an account of the history of orographic theories, see von Zittel, *History of Geology and Paleontology*, 186–323; George P. Merrill, "Contributions to the History of Geology," *Report of the United States National Museum, 1904* (Washington, D.C., 1906), 189–733; George P. Merrill, *The First One Hundred Years of Geology* (New Haven, 1924; reprint ed., New York, 1964), *passim*. For a concise and critical analysis of Hutton's work, see Bailey, *James Hutton: The Founder of Modern Geology*.

points of least resistance in the unequal strata of the earth's crust. Thus had continents and mountains been formed over the ages.[4]

Later, James Hall set forth the new theory that when the weight of steady accumulations of sediment caused sufficient subsidence, a gravitational effect resulted in the consequent upheaval of adjacent land masses. Essentially an extension of the ideas expounded by the Englishmen Charles Babbage and Sir John Herschel, who had suggested that sedimentary deposits expanded as a result of heat in their lowest depths, Hall's theory eventually became a focal point in the debate on the origin of mountains. From it came the concept of geosynclines (named by Dana), which are basically linear surface depressions extending over a region. Geosynclines are formed over a long period of time and contain accumulations of sedimentary and volcanic rock.[5]

Hall's development of the idea made no great impact at first, and for over a decade the subject of mountain formation stood in limbo, although Dana and others, like the competent but not-well-known Englishman, Osmond Fisher, continued to speculate occasionally upon the topic. Then, in 1872, LeConte published his "A Theory of the Formation of the Great Features of the Earth's Surface," which revived interest in the contractional theory of orogeny. As indicated by later writers, it is often difficult to separate the theories of LeConte and Dana. Certainly, the credit for initial work goes to Dana, but just as surely the credit for clarifying Dana's views belongs to LeConte. At any rate, the ideas set forth by these two able geologists served to make

4 William B. Rogers and Henry Darwin Rogers, "On the Physical Structure of the Appalachian Chain, as Exemplifying the Laws Which have Regulated the Elevation of Great Mountain Chains Generally," *Report of the First, Second, and Third Meetings of the Association of American Geologists and Naturalists at Philadelphia in 1840 and 1841 and at Boston in 1842* (Boston, 1843), 474–531; Merrill, *The First One Hundred Years, passim.* Dana's articles are: "A General Review of the Geological Effects of the Earth's Cooling from a State of Igneous Fusion," *American Journal of Science,* 2nd ser., IV (1847), 88–92; "Geological Results of the Earth's Contraction," *ibid.,* 2nd ser., III (1847), 176–88; "On the Origin of Continents," *ibid.,* 2nd ser., III (1847), 94–100; "On the Volcanoes of the Moon," *ibid.,* 2nd ser., II (1846), 335–55; and "Origin of the Grand Outline Features of the Earth," *ibid.,* 2nd ser., III (1847), 381–98.

5 James Hall, *Palaeontology of New York* (8 vols.; Albany, N.Y., 1859), III, pt. I, 68–85; Clarke, *James Hall of Albany,* 325–36; T. Sterry Hunt, "On Some Points in American Geology," *American Journal of Science,* 2nd ser., XXXI (May, 1861), 392–414; von Zittel, *History of Geology and Palaeontology,* 305.

the contractional theory the leading explanation of mountain formation in the 1870s and the 1880s.[6]

In his initial paper on deformation, LeConte sought first to refute
Rogers' theory and second to articulate his own tentative explanation.
Objecting to the view that the earth's interior was in a molten state
and that the earth's crust was relatively thin, LeConte clearly showed
the defects of Rogers' theory. If the earth's crust were merely "floating"
upon a vast sphere of molten magma, argued LeConte, then the
"greater inequalities of the surface cannot be produced by alternate
convex and concave bendings of the crust" because it would be impossible for a large "arch" or continent to sustain itself, nor indeed
could an "inverse arch" or ocean basin of such magnitude exist for
very long. In other words, a floating crust would inevitably assume a
state of equilibrium and therefore flatten out very quickly. At the same
time, LeConte rejected the idea that continents and mountains arose
from the unequal cooling of segments of the earth's crust. Thus, he
maintained, if the inequalities were due to varying rates of loss of heat
by conduction, and if they were floating upon a liquid interior, then
the "thicker parts" would rise and the "thinner parts," subside. But as
the process continued, continents and mountains would "rise higher
and higher" while the ocean beds would "sink deeper and deeper."[7]

LeConte likewise rejected his own earlier theory, propounded at the
1850 meeting of the AAAS, in which he subscribed to an "interior-flu-
idity" view of the earth. According to that theory, continents and
mountains were formed "by unequal crushing together by lateral

6 Joseph LeConte, "A Theory of the Formation of the Great Features of the Earth's
Surface," *American Journal of Science*, 3rd ser., IV (November, 1872), 345–55, 460–72; R. H.
Dott, Jr., "The Geosyncline—First Major Geological Concept 'Made in America,'" in Cecil
J. Schneer (ed.), *Two Hundred Years of Geology in America* (Hanover, N.H., 1979), 239–64.
The Reverend Osmond Fisher, who has received little recognition for his geological theo-
ries, published a number of insightful papers and, later, *Physics of the Earth's Crust*
(London, 1881). See Stephen G. Brush, "Nineteenth-Century Debates About the Inside of
the Earth," *Annals of Science*, XXXVI (May, 1979), 225–54.

7 LeConte, "Great Features of the Earth's Surface," 346–48. Perhaps LeConte's most
extensive efforts at reconciliation of geological theories were his papers on "critical periods" in the history of the earth, whereby he sought to harmonize the catastrophist and
uniformitarian views; these were given in detail in his articles "On Critical Periods in the
History of the Earth and Their Relation to Evolution," *American Journal of Science*, 3rd
ser., XIV (August, 1877), 99–114, and "Critical Periods in the History of the Earth," *Bulletin
of the Department of Geology, University of California*, I (1895), 313–36.

pressure" as a consequence of the "unequal cooling" (hence "unequal thickening") of a solid earth crust. Although this interpretation would account for the folding of strata, it failed to explain why the crust folded in the thicker parts rather than in the thinner ocean beds. By 1872 LeConte had come to realize that the concepts of a liquid interior and a floating crust encountered insuperable obstacles, and he abandoned them for the contractional theory of a solid earth.[8]

The contractional theory was based upon the idea of a "cooling earth ... composed of concentric isothermal shells, each cooling by conduction ... to the next outer and the outermost by radiation into space." Since the interior areas were cooling, they would contract more quickly than the shell, and the surface therefore "should be subjected to powerful horizontal pressure." Eventually, the surface would have to yield, and it was at those points of greatest stress that such yielding occurred, giving rise to mountain chains. Deformation was thus achieved by "a mashing or crushing together horizontally like dough or plastic clay, with foldings of the strata, and an upswelling and thickening of the whole squeezed mass." For LeConte, no other explanation was then available to account for "the complex foldings so universal in great mountain chains." More specifically, he concluded that if such horizontal pressure crushed and mashed sediments of 10,000-foot thickness in a range of 2.5 miles to 1 mile, the corresponding upheaval would result in a vertical thickness of 25,000 feet, of which 15,000 feet would be elevated above sea level. Thus, the earth's crust deformed by "unequal radial contraction."[9]

But LeConte also had to deal with the subsidence theory advocated by Hall and, in modified form, by J. D. Whitney. Although he praised Hall for his studies of the Appalachians and his ideas on the sedimentary nature of mountain chains, LeConte nevertheless found fault with his subsidence theory because it failed "to explain the actual process" of mountain formation and because it ignored "the immense horizontal crushing and plication of the strata." Hall and T. Sterry Hunt believed that the Appalachians were formed from "an immense convex

8 In a paper presented before the American Association for the Advancement of Science in 1859, LeConte maintained that the earth's interior was fluid (see T. Sterry Hunt's abstract of the paper in the *Canadian Naturalist*, IV [1859], 293–94). LeConte never published the paper because he "soon became dissatisfied with it," as he explained many years later in a letter to the editor of *Nature*, XXIX (January 3, 1884), 212–13.
9 LeConte, "Great Features of the Earth's Surface," 349–55.

mass of submarine sediment," which, through gradual subsidence and consequent horizontal pressure, was "thrown into folds." Then, by continental upheavals, the great convex mass became an enormous plateau. In subsequent ages, erosion created valleys and ridges, leaving the great mountains to stand as monuments in the denuded plateau. Whitney likewise subscribed to the subsidence theory, but he argued that plication (or folding) resulted from "the subsidence of a mountain axis, previously elevated by other agency." In response to those views, LeConte pointed out the inability of subsidence alone to account for massive folding, and he noted that the processes of sedimentation and subsidence occurred simultaneously, not in a simple cause-effect relationship. Moreover, he rejected Whitney's version because it was neither logically consistent nor clear on what other agencies operated to create a "previous elevation."[10]

LeConte also discussed the various mountain ranges of the United States, examining the variety of landforms in this country and elaborating upon the various geological agents at work in deformation of the earth's crust. He readily acknowledged that his theory was very incomplete and that it failed to explain "those great and widespread oscillations which have marked the great divisions of time, and have left their impress in the general unconformability of the strata." Although he could not fully account for those oscillative or vibrative "waves" that constitute disturbance of the earth's equilibrium (or a sort of geophysical pulsation), LeConte had nevertheless published a plausible hypothesis, which, in a nutshell, proposed that "mountain chains are formed by the mashing together and the upswelling of sea bottoms where immense thickness of sediments have accumulated; and as the greatest accumulations usually take place off the shores of continents, mountains are usually formed by the up-pressing of marginal sea bottoms."[11]

Although most of the responses to LeConte's article were highly favorable, at least one was not. T. Sterry Hunt, of the Canadian Geological Survey, reacted negatively, not only because he disagreed in part with the theory but also because he believed that LeConte had purloined some of his own ideas. In his reclamation, Hunt contended

10 *Ibid.*, 460–72; Hall, *Palaeontology of New York*, III, pt. 1, 68–86; Hunt, "On Some Points in American Geology," 392–414; Josiah D. Whitney, *Earthquakes, Volcanoes, and Mountain Building* (Cambridge, Mass., 1871), 90–104.
11 LeConte, "Great Features of the Earth's Surface," 463–72.

that he had been propounding "the theory of igneous agencies on the basis of a solid earth" since 1858, and that he alone had advocated the theory until LeConte appropriated it. In fact, LeConte was not even partially guilty of plagiarism, though as he admitted in his reply to Hunt's review, he might have avoided any pique to Hunt if he had used his name more frequently in the article.[12]

The plain fact of the matter was that Hunt, who had attained fame as a chemical geologist in the early 1850s, felt slighted because, by the 1870s, his reputation lay in question, due in large measure to his "erratic tendency, and a disregard for facts that in any manner conflicted with or failed to substantiate his views." As LeConte noted, Hunt seemed anxious "to press *yet once more* upon the attention of geologists his own labors." Moreover, stated LeConte, Hunt could not distinguish between "the *use of similar* materials for the *similar use* of materials." LeConte never claimed originality for his theory; what he did justifiably claim was the development of a more comprehensive explanation of mountain formation, to which were added his own hypothetical views. Indeed, as LeConte observed, the problem was that Hunt and others had failed to be precise in their definitions, often confusing the various types of geological deformation.[13]

Nevertheless, Hunt's criticism that LeConte had not adequately discussed "the chemical cause of the intense heat so often found in the lavas" caused LeConte to modify his theory. That criticism, coupled with a review of a recently published paper by the English seismologist Robert Mallet, led LeConte to the following revision of the formation of mountain chains: "Lines of thick sediments, rise of geo-isotherms and aqueo-igneous softening determine lines of yielding; then crushing together *horizontally* and swelling up *vertically* forms the chain; but once the yielding commences, then mechanical energy is changed into heat, which may thus be increased to any amount and produce true igneous fusion."[14]

12 T. Sterry Hunt, "On Some Points in Dynamical Geology," *American Journal of Science*, 3rd ser., V (April, 1873), 264–69; Joseph LeConte, "On the Formation of the Features of the Earth-Surface: Reply to Criticisms of T. Sterry Hunt," *ibid.*, 3rd ser., V (June, 1873), 448–53.

13 LeConte, "On the Formation of the Features of the Earth-Surface," 448–50; Merrill, *The First One Hundred Years*, 447; James D. Dana to Joseph LeConte, March 12 and April 16, 1873, both in LeConte Family Papers, UCB.

14 LeConte, "On the Formation of the Features of the Earth-Surface," 453; Robert Mallet, "Volcanic Energy: An Attempt to Develope Its True Origin and Cosmical Relations,"

The question of mountain formation now revived, Dana published a series of papers on the subject in the 1873 issues of his *American Journal of Science*, and he soon incorporated his views into the second edition of his *Manual of Geology*. Dana had first advocated the contractional theory in 1846, but he did not systematize the theory and give it credence until many years later. His success in doing so was in large measure due to LeConte's thoughts on the subject. The efforts of these two men eventually resulted in the widespread adoption of the theory, which prevailed as the dominant explanation until Clarence Dutton later introduced the notion of isostasy.[15]

Like LeConte, Dana took exception to Hall's views on the origin of mountains. Citing LeConte to enforce his criticism, he contended that Hall's theory ignored the question of mountain elevation and overplayed the role of subsidence. But at the same time Dana was not entirely satisfied with LeConte's view of contraction because it indicated that the geological activity "must have been confined to the underlying rocks," and therefore their composition would have resulted in an ascension of heat that "would have produced expansion instead of contraction." In other words, he argued that subsidence could not be explained solely on the basis of LeConte's hypothesis. Further, contended Dana, LeConte's notion that plication resulted from crushing was insufficient to account for the enormous amount of elevation. Nevertheless, he accepted "the truth" of "Professor LeConte's important principle" so long as it was applied only to "monogenetic mountains"—that is, to mountains that were formed in a single process.[16]

Unable to accept fully the idea of a solid globe, Dana commended

Proceedings of the Royal Society of London, XX (1871–72), 438–41. In a letter to the editor of the *American Journal of Science*, 3rd ser., V (February, 1873), 156, LeConte claimed prior credit, explaining that his paper had been written before Mallet's but that it had been delayed because of slow mail deliveries, his absence on an excursion into the High Sierra, and the inability of the editor to publish it immediately.

15 James D. Dana to Joseph LeConte, May 31, 1873, in LeConte Family Papers, UCB; James Dwight Dana, *Manual of Geology* (2nd ed.; New York, 1874), 735–54; George P. Merrill, "Dana the Geologist," *Bulletin of the Geological Society of America*, XXIV (March 24, 1913), 66; Edward B. Mathews, "Progress in Structural Geology," in *Fifty Years' Progress in Geology*, ed. Edward B. Mathews (Baltimore, 1927), 140; William N. Rice, "The Geology of James Dwight Dana," in *Problems of American Geology*, ed. William N. Rice (New Haven, 1915), 16; Joseph Barrell, "A Century of Geology," in *A Century of Science*, ed. Dana, 178–82.

16 James Dwight Dana, "On the Origin of Mountains," *American Journal of Science*, 3rd ser., V (May, 1873), 347–50.

LeConte for his hypothetical explanations, but he also stated that the California geologist failed, as LeConte had already admitted, to account for oscillation within the interior of continents. In part, this failure resulted from the limited perspective of mountain and continent formation in North America, where, unlike the intracontinental cordillera of Europe, the great mountain chains are basically marginal to the two great oceans. From a much broader view, Dana noted that neither LeConte nor Hunt had dealt with the problems of interior subsidence, which could not have been "due to contraction" and did not correlate with the geosynclinal phenomenon. They had not, in Dana's view, appreciated the work of the English physicist William Hopkins, who in 1839 and again in 1847 had concluded that pressure and corresponding temperature would necessarily make the center of the earth a solid and that surface cooling would inevitably form a crust. But Hopkins had likewise concluded that there was a "viscous layer" between the interior and the exterior that itself must contract over time. This was the view finally accepted by Dana—and much later by LeConte. Such a hypothesis, argued Dana, would explain the development of geosynclines and crustal weaknesses that would produce "a scene of catastrophe and mountain-making." To LeConte he gave credit for making "prominent" the principle whereby "the rise of isogeotherms" would weaken the earth's crust.[17]

Dana modified his earlier views of the contractional theory in the formation of continents, mountains, and ocean basins, declaring that he had done so "in consequence of some new considerations of my own" and as a result of LeConte's publications. In particular, he accepted LeConte's idea that elevation occurs as a result of plication that causes "shoving along fractures" and consequent crushing and uplifting, but he believed that these actions were limited to the second stage of the formation of monogenetic cordillera. He also adopted LeConte's hypothesis that subsidence due to sedimentation, which by weight and heat weakened geosynclines, produced catastrophic uplift in the adjacent area.[18]

Subsequently, Dana credited Hunt with the original conception of

17 James Dwight Dana, "On Some Results of the Earth's Contraction from Cooling, Including a Discussion of the Origin of Mountains and the Nature of the Earth's Interior," *American Journal of Science*, 3rd ser., V (June, 1873), 423–43, and VI (July, 1873), 6–13.

18 *Ibid.; American Journal of Science*, 3rd ser., VI (September, 1873), 170–72, (October, 1873), 304, and (November, 1873), 381–82.

the second idea but argued that it was LeConte's view of the role of "lateral pressure from the earth's contraction" that represented a significant breakthrough in the knowledge of mountain formation. Still later, he confessed that Hunt had even proposed that lateral pressure was a significant agent, although he persisted in pointing out that it was LeConte who had successfully demonstrated the deficiency of the Hunt-Hall theory of subsidence. But, aside from these considerations, it is obvious that LeConte had posited a theory that, despite its shortcomings, brought the contractional theory to the forefront as the major explanation of mountain formation.[19]

During the next twenty-five years, LeConte wrote a great deal more on mountain making and related areas of diastrophism. Although much of his writing continued to be theoretical and synthetic, it also represented findings from field explorations. In 1874, for example, LeConte published a paper on the great Cascade lava flood that during the Miocene epoch covered much of the region from upper California to British Columbia and eastward into Idaho. His comments were based largely upon his firsthand observations in the region. While his findings proved beneficial to other geological topics (e.g., landforms), they also bore directly upon the contractional theory of tectonics. Accepting Dana's contention that there was an "inverse ratio between the amount of mountain mashings and foldings ... and the amount of fissurings and fissure eruptions," he refuted his own earlier position and demonstrated the validity of Dana's hypothesis.[20]

19 Mathews, "Progress in Structural Geology," 140.

20 Some of the more important of LeConte's papers dealing with diastrophism and landforms were: "On the Great Lava-Flood of the West; and the Structure and Age of the Cascade Mountains," *American Journal of Science*, 3rd ser., VII (March, 1874), 167–80, and (April, 1874), 259–67; "On the Evidences of Horizontal Crushing in the Formation of the Coast Range of California," *ibid.*, 3rd ser., XI (April, 1876), 297–304; "On the Extinct Volcanoes About Lake Mono, and Their Relation to the Glacial Drift," *ibid.*, 3rd ser., XVIII (July, 1879), 35–44; "The Old River-Beds of California," *ibid.*, 3rd ser., XIX (March, 1880), 176–90; "A Post-Tertiary Elevation of the Sierra Nevada Shown by the River-Beds," *ibid.*, 3rd ser., XXXII (September, 1886), 167–81; "Tertiary and Post-Tertiary Changes of the Atlantic and Pacific Coasts," *Bulletin of the Geological Society of America*, II (March 16, 1891), 323–30; "The Origin of Transverse Mountain-Valleys and Some Glacial Phenomena in Those of the Sierra Nevada," *University Chronicle*, I (December, 1898), 479–97; and "The Ozarkian and Its Significance in Theoretical Geology," *Journal of Geology*, VII (September–October, 1899), 525–44. An assessment of LeConte's contributions to the study of landforms is given in Chorley, *et al.*, *The History of the Study of Landforms*, I, 591–92, 612–14. For a contemporary evaluation of LeConte's hypothesis on the lava flood, see Leo Lesquereux to J. Peter Lesley, May 18, 1874, in Lesley Papers.

The initial stage of the formation of mountain ranges, reiterated LeConte, resulted from "interior contraction of the earth," which then placed "irresistible horizontal pressure" on the crust. Where the surface yielded, mountain ranges were born; and this generally occurred "along marginal sea bottoms." As strata were mashed and folded, fissures developed, allowing "any sub-mountain liquid matter" to be "squeezed out by the same force which elevated the mountain range." This process led to a second stage, wherein "lines of successive elevation" occurred from lateral force. But LeConte noted a significant difference in the processes involved in the two stages. The first, because it involved the "yielding of the soft sea-bottom sediments," met with "little" resistance; hence, the amount of heat generated was comparatively small. In the second stage, however, the transformation was "paroxysmal," for the newly hardened surface encountered strong resistance and a corresponding quantity of heat. Strata thus metamorphosed in the first stage, but, in the second, they actually fused "with the fissuring of the strata and the out-squeezing of the sub-mountain liquid." Thus, concluded LeConte, "the Appalachians, the Sierra and the Coast Ranges were formed almost fully grown by the first method, while the Cascade Range was formed almost wholly by the second process." This argument, he held, would be true whether one applied his own theory or those of either Mallet or Dana.[21]

In 1878 LeConte was invited to address the National Academy of Sciences on the subject of mountain formation. In this paper, LeConte further clarified and elaborated the contractional theory. He did so because a number of objections had recently appeared in the literature. The objections he placed into two categories: (1) those "based upon structure," and (2) those based upon "mathematical formulae." But he considered only the first a "serious objection." It was raised by such competent structural geologists as G. K. Gilbert and J. W. Powell, both of whom had extensive field experience in the western region of the United States. Each of these geologists contended that LeConte's theory of horizontal pressure did not seem to apply to the Great Plateau and the Basin region. In general, LeConte accepted their criticism, but he argued that Gilbert and Powell expected the "universal applicability" of his theory. Instead, countered LeConte, they must realize that the Plateau and Basin regions were "not monogenetic ranges

21 LeConte, "On the Great Lava-Flood," 179–80.

at all, but only the displaced parts of one great monogenetic bulge."[22]

LeConte was much more adamant in his opposition to the position held by "mathematical physicists." In essence, his opposition stemmed from the Baconian idealism that had flourished so long in nineteenth-century American science. Although vaguely formulated and variable in its interpretation, the Baconian philosophy stressed the importance of "empiricism" or observation of individual facts until from them broader generalizations could be derived. Likewise, it tended to shun hypothesizing and theorizing as antiempirical in nature. Deeply steeped in this tradition, which he had imbibed principally from intensive reading in William Whewell's *History of the Inductive Sciences*, LeConte had long followed the basic proposition that the inductive method was the only valid scientific approach. While he had somewhat modified the notion in his own writings by engaging in hypothetical and theoretical explanations, he nevertheless embraced the basic position of Baconianism. Hence, LeConte was disturbed by the new school of scientific thought that promoted the role of hypothesis-testing as a valid form of scientific inquiry. It is little wonder, then, that he praised the "thorough structural geologists educated in the field" and criticized Sir William Thomson and Osmond Fisher for a "demonstrative certainty" of geological knowledge bolstered by "a bristling array of mathematical formulae."[23]

More specifically, he objected to the "mathematical-physicist" argument that interior contraction in a "solid" earth could not "concentrate its effects along certain lines (viz., mountain ranges) without a slipping or a shearing of the exterior shell upon the interior nucleus." That would, admitted LeConte, be a valid criticism only if he had assumed "the complete solidity of the earth," and he was clearly in favor of the notion of "a layer of semi-fused matter." In his judgment, however, this plastic layer did not extend universally as a "semi-liquid subcrust" but only locally, as in the ocean beds. To the "mathematical-physicist" contention that large foldings could not occur from contraction because the amount of heat-loss was too small to produce

22 LeConte, "On the Structure and Origin of Mountains, with Special Reference to the 'Contractional Theory,'" *American Journal of Science*, 3rd ser., XVI (August, 1878), 95–105. See also Stephen J. Pyne, *Grove Karl Gilbert: A Great Engine of Research* (Austin, Tex., 1980), *passim*.

23 *Ibid*. For a critical analysis of "Baconianism," see Daniels, *American Science in the Age of Jackson*, 63–137. See also Pyne, *Grove Karl Gilbert*, 126–34, for an example of the new school of "geophysics" that was developing.

them, LeConte countered that the argument was limited to a single diastrophic factor. Nowhere, as he indicated, had the contractional theory been limited solely to the factor of heat-loss. True, he confessed, heat-loss was "the most obvious cause of contraction," but it was only a necessary, not a sufficient, cause.[24]

Yet LeConte could not himself account for all of the causes of contraction, and he said that, until more scientific evidence became available, geologists must accept the reality that "the *fact* of contraction is one thing and the cause of contraction another and quite a different thing." In other words, correlation was not tantamount to a cause-effect relationship. As a geologist, he was more concerned with the "phenomena of structure" than with the causative factors of contraction, and he therefore declared that earth scientists must admit their ignorance while continuing their search for "still other causes yet unknown."[25]

After dispensing with those objections, LeConte turned to other criticisms raised by his old friend Clarence Dutton, whose extensive studies of the great plateaus of the West led him to challenge the validity of the contractional theory, especially with respect to the feasibility of the hypothesis of unequal radial contraction of the earth's surface. Dutton pointed out that the contractional theory seemed inapplicable in the cases of the Himalayas and the Alps, which were once sea bottoms but had rapidly elevated since the Tertiary. He thus questioned how the radii of those ranges had suddenly changed from

24 LeConte, "On the Structure and Origin of Mountains," 105–106. In his "Address at the Glasgow Meeting of the British Association," published in the *American Journal of Science*, 3rd ser., XII (November, 1876), 336–54, Thomson had sought to refute "the hypothesis of a perfectly rigid crust containing liquid" because it "violates physics," and he supported his argument by elaborate mathematical formulae. Osmond Fisher had previously raised similar objections in his *Physics of the Earth's Crust* (1881) and again in his modified explanation, "On the Amount of Elevations Attributable to Compression Through Contraction During Cooling of a Solid Earth," *Philosophical Magazine*, 5th ser., XXIII (February, 1887), 145–49. For an excellent discussion of the views of Thomson and Fisher, and indeed on the extended debate between the geologists and physicists on the subject, see Brush, "Nineteenth-Century Debates About the Inside of the Earth." A useful discussion regarding the relative strengths of geology and physics in the United States is given in Daniel J. Kevles, *The Physicists* (New York, 1978), 37–38. Also pertinent is Joe D. Burchfield, *Lord Kelvin and the Age of the Earth* (New York, 1975), which contains a helpful discussion on the disagreement between physicists and geologists on earth mechanics.

25 LeConte, "On the Structure and Origin of Mountains," 106–107.

a conductive to a nonconductive state—that is, why they had so quickly ceased to be affected by thermal activity. To that question LeConte replied that Dutton misunderstood the contractional theory, confusing orogeny with continent formation. He maintained that geologists "know the time of birth of mountains," but they possess no accurate knowledge of continental origins and development.[26]

LeConte acknowledged that some of the objections to the contractional theory were "well worthy of serious attention" and that the theory would undoubtedly undergo further modifications. He was especially aware of the deficiencies of the theory regarding its applicability to continent formation, and a few years later he would declare that the theory had been "pushed too far." As always, he cautioned against extreme positions, holding that "the truth" was to be found in reconciliation. For LeConte, that meant more than a mere medley of hypotheses. Instead, once more following the Hegelian model that served as his basis for synthesizing conflicting ideas, LeConte insisted upon a genuinely harmonious theory arrived at by serious criticism and finely honed explanations. Yet there would always be deficiencies in the theory, and as new geological evidence was uncovered the problem was so compounded that certain assumptions had to be abandoned. Unfortunately, LeConte could never quite bring himself to that point.[27]

For a decade LeConte devoted his attention to other geological topics, some of which represented his best practical work in the discipline. By the late 1880s, however, geologists were struggling to develop a workable hypothesis regarding the earth's interior. Thus, once more, LeConte took up the case for the contractional theory. In a paper on "The General Interior Condition of the Earth," published in 1889, he said: "The house of Science seems . . . to be divided against itself" on the question of the nature of the earth's interior and its crust. He reiterated his view that the core of the earth is solid as is its thick crust, but between those lies a "sub crust liquid layer." This reconciled view, he maintained, would "satisfy the physicists . . . and the geologists,"

26 *Ibid.*, 108–10. A brief summary of the argument over theories of mountain formation is given in Bailey Willis, "American Geology, 1850–1900," *Proceedings of the American Philosophical Society*, LXXXVI (1942), 40–42. Dutton's objection was first given in his "A Criticism upon the Contractional Hypothesis," *American Journal of Science*, 3rd ser., VIII (August, 1874), 113–23.

27 LeConte, "On the Structure and Origin of Mountains," 110–12.

because it eliminated the former's objection to the diffused effects of contraction in a solid earth while it accepted the latter's argument for the concentration of uplift in specific areas of the earth's crust.[28]

Although, strangely, LeConte said virtually nothing on the cosmological origins of the earth, his theoretical explanations clearly suggest a belief in the nebular hypothesis—the rather commonly held Laplacian view that when the centrifugal force of hot nebulous matter in the primeval solar system equaled the centripetal force of the mass, the body of material separated into more rapidly revolving rings that ultimately evolved into planets. He did at one point in this paper contend that originally the earth was an "incandescent liquid" that gradually cooled and solidified "at the centre, because the pressure was greatest there," and he argued that the "nucleus thus formed" continued to enlarge "until it nearly reached the surface." Moreover, he maintained that the "surface liquid" became "viscous" as "convective currents were impeded" and finally by radiation formed a "superficial solid crust." Eventually, he added, the continued cooling of both the interior and the crust would decrease the liquid stratum and, even in some places, solidify it completely. Where the viscous layer remained, however, crustal deformation was still in process.[29]

In a subsequent paper of the same year, LeConte applied the contractional theory to the faulting processes found in the Great Basin. This paper, based upon LeConte's field studies in the region, represented an important modification of his views. Whereas he had previously argued that "mountain ranges were formed by lateral pressure acting upon thick sediments folding and swelling up the mass along the line of yielding," LeConte now contended that large crustal blocks were "not formed by lateral pressure but by tension of lifting." From that modified hypothesis, he was persuaded to adopt a two-fold conception of mountain formation, viz., "those formed by lateral crushing and folding" and "those formed by adjustment of crust-blocks." The first produced "reverse faults," while the latter created "normal faults."[30]

28 Joseph LeConte, "The General Interior Condition of the Earth," *American Geologist*, IV (July, 1889), 38–44.

29 *Ibid.*

30 Joseph LeConte, "On the Origin of Normal Faults and of the Structure of the Basin Region," *American Journal of Science*, 3rd ser., XXXVIII (October, 1889), 257–63; James D. Dana to Joseph LeConte, September 6 and 13, 1890, both in LeConte Family Papers,

LeConte's most comprehensive paper on orogeny appeared in 1893, titled "Theories of the Origin of Mountain Ranges." The stated purpose of that paper, written over two decades after LeConte's initial venture into the theoretical bases of mountain making, was to give "the present condition of science on this subject" and to distinguish between the "region of comparative certainty" and the "region of uncertainty" in order to "clear the ground, narrow the field of discussion and direct the course of profitable investigation." In a remarkably concise summary, LeConte restated the contractional theory and the evidence to support it. He concluded with a discussion of the contemporary objections to the contractional theory and then offered a synopsis of three "alternate physical theories" that formed the basis of his last major defense of the theory that had such a profound impact upon geological speculation about the origin and development of mountains. Although the theory eventually gave way to others, it continued to have currency for another seventy-five years.[31]

A generally flexible thinker given to changing his views, however slowly, LeConte declared that he would "give it up" if a "better" theory came forth, but he confessed that he would be "reluctant to do so." "We all dearly love our intellectual children," he said, "especially if born of much labor and thought; but I am sure that I am willing . . . to sacrifice, if need be, this my fairest daughter on the sacred altar of Truth." Nevertheless, in spite of the objections that were coming "thick and fast from many directions," the grand old defender of contractionalism sketched out his last major reply to the critics of the Dana-LeConte theory. Briefly, the most "serious" objections were that: (1) the theory did not sufficiently account for "the amount of lateral thrust" necessary to produce massive plication; (2) following Dutton, interior contraction would produce "a shearing of the crust on the interior" rather than concentrate at certain points; (3) any action due to contraction would be haphazard rather than "definite" in direction, at least over a large expanse of the crust; and (4) the quantity of "circum-

---

UCB. See also Walter H. Bucher, *The Deformation of the Earth's Crust* (Princeton, 1933), 345.

31 Joseph LeConte, "Theories of the Origin of Mountain Ranges," *Journal of Geology*, I (September–October, 1893), 543–73. Overlooked by LeConte was the theory propounded by Albert Heim, *Untersuchungen über den Mechanismus der Gebirgsbildung im Anschluss an die geologische Monographie der Tödi-Windgällen-Gruppe* (2 vols.; Basel, 1878). Again, this reflected his general neglect of German-language publications.

ferential shortening" essential to massive folding "would disarrange the stability" of the earth's rotation. To each of these arguments Le-Conte gave his now-standard reply, and he was convinced that, while serious, they were not fatal to the theory.[32]

To a fifth objection, however, he admitted reservations and an inability to respond with any degree of certainty. That objection was raised by the English engineer T. Mellard Reade, who introduced the principle of "level of no strain." Working from the supposition that the interior layers could be affected by both radial and circumferential contraction, Reade demonstrated that the former would laterally crush every concentric layer while the latter would expand them. Thus, at certain depths the two actions would equalize and form a "level of no strain," offsetting thereby the crushing and tension-fissuring forces. LeConte declared that Reade's important hypothesis "must not hereafter be neglected," but he likewise demurred in accepting it as a refutation of the contractional theory because it posited loss of heat as the principal agent of contraction. The "loss of constituent water," he said, "would put an entirely different aspect on the subject." Again, he held reservations about purely "mathematical" and "physical" explanations of contraction, but clearly his strong defense of contractionalism had been shaken.[33]

Finally, LeConte considered and ultimately rejected three contemporary "alternative physical theories." The first of these was expounded by the same Reade, who had recently published his *Origin of Mountain Ranges*. Reade posited that mountains arose in the fol-

---

32 LeConte, "Theories of the Origin of Mountain Ranges," 562–66.

33 *Ibid.*, 566–68; T. Mellard Reade, *The Origin of Mountain Ranges Considered Experimentally, Structurally, Dynamically, and in Relation to Their Geological History* (London, 1886). Previously LeConte had written a letter in which he called Reade's work "a very suggestive paper," but he claimed that Reade's "view of Appalachian structure" was erroneous (in *Philosophical Magazine*, 5th ser., XXV [May, 1888], 450–51). As late as 1898 LeConte was still puzzled over the concept of "level of no strain," and he asked the able astronomer, physicist, and mathematician Simon Newcomb for help in clarifying the idea and applying it to the contractional theory (LeConte to Newcomb, September 7, 1898, in Simon Newcomb Papers, Library of Congress). LeConte seemed to be unfamiliar with Charles Davison's articles on the same subject in the *Proceedings of the Royal Society of London*, XLII (May 5, 1887), 325–28, and LV (February 15, 1894), 140. But, again, these he probably would have viewed as the work of "mathematical physicists" and therefore debatable. By his own admission, LeConte did not possess a thoroughly competent grasp of advanced mathematics.

lowing manner: sedimentary accumulations led to subsidence and the consequent rise of geoisotherms, which in turn caused an expansion of the sedimentary mass. As the crust would not yield laterally, plication and vertical uplift occurred at points of thickest sedimentation, followed by a surge of the underlying liquid mass in the direction of those points. Eventually, then, the mass would move upward through fissured strata, "folding these back on either side." To this theory LeConte objected on two grounds. First, he argued that, unlike the contractional theory, it supposed "local" rather than circumferential expansion and ignored the "enormous foldings of the mountains." Secondly, he maintained that Reade's theory failed to account completely for the "sum of heat in the earth." Since sedimentation alone was inadequate to generate sufficient expansive heat, "whence did it derive," queried LeConte. Thus, he concluded, "Reade's theory cannot be accepted as a substitute."[34]

The other "alternative" explanations of mountain building were Dutton's "isostatic theory" and the "gliding theory" of Eduard Reyer, a Viennese seismologist. As previously noted, Dutton had gradually developed the concept of isostasy and had offered it as a substitute for contractionalism. Arguing against the homogeneity of the earth's crust, Dutton contended that the crust was composed of heavy and light masses that continually move toward states of equilibrium. Thus, heavy masses subside while lighter masses ascend and buckle the crust. Through the steady processes of sedimentation and surface erosion, the center of a mountain-arch bears less weight and is therefore further elevated. To that theory LeConte took exception because, unless isostasy were complete, "the sediments quickly rise to sea-level and stop the process of sedimentation at that place." Moreover, he objected that the theory also failed to explain adequately the relationship between the isostatic phenomenon and "landward" sloping from the sedimentary mass. Reyer's theory, on the other hand, held that, as a mountain axis arose, it pushed up the surrounding strata until the slope became so steep that they slid down and crumpled into "complex folds." To Reyer's credit, the hypothesis was based on considerable experimentation with "plastic materials," but, as LeConte pointed out, the Reyer theory neither cited a cause for "granitic uplift" nor

34 LeConte, "Theories of the Origin of Mountain Ranges," 569–71.

explained the complex plication of ranges that contained no granitic axis.[35]

In conclusion, LeConte found those theories "untenable," and he reiterated his belief in the contractional theory, "not indeed with our old confidence, but with the conviction that it is even yet the best working hypothesis we have." His confidence in the theory was repeated in his presidential address before the Geological Society of America in 1896, but LeConte also indicated misgivings, especially because he was unable to account for the variety of "oscillations" in the earth's surface. Soon other geologists were proposing objections and other hypotheses that fractured the foundation of the contractional theory. Undoubtedly, the most serious of these was T. C. Chamberlin's planetesimal theory, which held that the earth was formed slowly by accretion of cold particles; thus, it was a solid from its origin, and its surface was always cold. Because it undermined the theory of an originally incandescent mass that gradually cooled and contracted, forming a wrinkled crust, the planetesimal hypothesis naturally vitiated many of the ideas associated with diastrophic movements. It did not necessarily negate every aspect of contractionalism, but it certainly brought many important points into question.[36]

The contractional theory lived on, however, having followers until as late as the 1950s, when it finally began to give way to the theories of convection currents, continental drift, and plate tectonics. It not only enjoyed a long life but also proved to be a major milestone in the development of geological theory. To Dana belongs the credit for developing the theory after it had first been proposed by older European geologists; to Hunt and Hall belongs the credit for developing important principles associated with the theory; and to LeConte belongs the credit for formalizing and articulating the theory so clearly that it be-

---

35 *Ibid.*, 571–72; Eduard Reyer, *Theoretische Geologie* (Stuttgart, 1888); Eduard Reyer, *Ursachen der Deformationen und der Gebirgsbildung* (Leipzig, 1892); Eduard Reyer, *Geologie und geographische Experimente* (3 vols.; Leipzig, 1892–94), Vol. I.

36 LeConte, "Theories of the Origin of Mountain Ranges," 573; Joseph LeConte, "Earth-Crust Movements and Their Causes," *Bulletin of the Geological Society of America*, VIII (February 15, 1897), 113–26; Joseph LeConte, "A Century of Geology," *Popular Science Monthly*, LVI (February, 1900), 431–43, and (March, 1900), 546–56; Stephen G. Brush, "A Geologist Among Astronomers," *Journal of the History of Astronomy*, IX (February, 1978), 1–41, and (June, 1978), 77–104.

came the paramount explanation of mountain formation before the turn of the century.[37]

In 1899 LeConte published yet another paper on geology. He considered his "The Ozarkian and Its Significance in Theoretical Geology" one of the "most important" papers he had published. It was an able argument in favor of attaching greater significance to the Ozarkian epoch as a "critical period" in geological evolution. In the paper, LeConte modified some of his earlier views of geological classification and attempted to demonstrate how the Ozarkian epoch was one of the so-called "lost intervals" in the "great cycles in the evolution of the earth." From that point he proceeded to plead, as he had previously done, for a new geological division, which he called the "Psychozoic," or the most recent period in the earth's history. The idea fitted nicely into his evolutionist scheme of human progress, but, in spite of the forcefulness of the argument, LeConte's proposal reflected more a desire than a valid reclassification. Once more, the aging geologist had relied too heavily upon theoretical speculations and not enough upon the evidential data needed to develop hard-and-fast proof in the branch of historical geology.[38]

Yet the mind of the seventy-six-year-old LeConte was as active as ever. Geology was still his "favorite department," and he continued almost to the end to contribute to the advancement of the field. When he published his last paper on geology in 1900, he sought to provide a brief history of the development of the discipline in the nineteenth century, and in that history he viewed himself as having played a small role. He was indeed correct in his humble estimate, though his role was more that of a great teacher, theoretician, and stimulator of ideas than an original researcher. Still, that role alone was no mean achievement.[39]

---

37 Stephen J. Pyne, "From the Grand Canyon to the Marianas Trench: The Earth Sciences After Darwin," in *The Sciences in the American Context*, ed. Reingold, 165–92; [Thomas C. Chamberlin], "Editorial," *Journal of Geology*, IX (July–August, 1901), 439–40.

38 LeConte, "The Ozarkian and Its Significance in Theoretical Geology," 525–44.

39 LeConte, "A Century of Geology."

"Among the Monuments of the Centuries": Final Years

Joseph LeConte's students once again saluted their beloved old teacher on his birthday in 1899. In a letter to Sallie he said: "They crowded my room—covered my desk with flowers—presented me with a pretty silver set of desk-furnishings ... and engraved with initials." He also told Sallie that, although Bessie was sick, he was himself "in excellent health—never better—although ... touched with Lumbago." Of Carrie, he reported that she was "greatly improved, in fact looks as well as usual—but can't stand yet any great excitement." But LeConte himself was feeling so good that he decided to resume work on a textbook in comparative physiology, a task he had set aside the year before in order to nurse Carrie. By mid-1899, he had reworked his physiology lectures into book form and sent them off to the publisher. Meanwhile, he occupied himself with correspondence, public lectures, and teaching.[1]

As the twentieth century dawned, LeConte's students planned another birthday celebration for their idol. To honor him as he entered upon his seventy-seventh year of life and his thirty-first year as a Californian, the students again decorated his lectern with flowers and presented him with an engraving of "The Prophets." The revered

1 Joseph LeConte to Sallie LeConte Davis, March 2, 1899, in Davis Papers; LeConte to Thomas Nelson Page, July 18, 1899, in Thomas Nelson Page Papers, Duke University Library, Durham, N.C.; *Daily Californian*, February 14, 1899, p. 3; *ibid.*, February 15, 1899, p. 3; *ibid.*, August 23, 1899, p. 1; *ibid.*, August 31, 1899, p. 1; *ibid.*, September 1, 1899, p. 1; *ibid.*, September 5, 1899, p. 3; *ibid.*, September 13, 1899, p. 3; *ibid.*, September 20, 1899, p. 4; *Bulletin of the Geological Society of America*, VII (1900), 587; Meeting of the Council, December 28, 1899, in Minutes of the Geological Society of America.

teacher responded as always with humble thanks and a brief horta-tory speech. On this occasion, he chose to relate his long-held belief in the dual necessity of teaching and research by the university's fac-ulty. But he cautioned against excessive emphasis upon research at the expense of teaching. The two functions, he firmly believed, must go hand-in-hand to strengthen any university, but, above all, teaching must be of the most worthy kind. "Teaching and research," he said, "are indissolubly married, and should never be divorced. Each vivifies the other, and saves it from becoming mere drudgery."[2]

Meanwhile, the *Outline of the Comparative Physiology and Morphol-ogy of Animals* had come off the press, and LeConte looked forward to reviews of his latest book. In the preface to the work, LeConte had explained that it was intended to complement other books that em-phasized "selected types." That approach had many advantages, he admitted, but, at least in his view, "the attempt to explore thoroughly a few small spots here and there" in recent biological textbooks re-sulted in a loss of perspective regarding the "general connection of all parts to one another." Yet he confessed that because of space limita-tions his book slighted the invertebrates, and he noted that it dwelt disproportionately on the physiological functions of the eyes and the liver. But he justified his predilections on the grounds that "a certain insistence on points best known to and most thoroughly investigated by the teacher . . . is necessary to give life and interest to any subject." Though remarkable in its comprehension and, as with LeConte's other books, lucid in its explanations, the *Outlines* was in many re-spects dated before the plates were struck by the printer. Unfortu-nately, LeConte did not recognize this. He was still the explainer and teacher of broad fields, and modern advances in highly specialized physiological and zoological inquiries were rapidly outdating his gen-eral knowledge.[3]

2 *Daily Californian*, February 27, 1900, p. 1.
3 LeConte, *Outline of the Comparative Physiology and Morphology of Animals*, pref-ace. His work on vision has already been discussed (see chapters VIII and XIV). His studies of the liver were: "Some Thoughts on the Glycogenic Function of the Liver," *American Journal of Science*, 3rd ser., XV (February, 1878), 99–107, and XIX (January, 1880), 25–29; "Glycogenic Function of the Liver," *American Naturalist*, XX, pt. 1 (May, 1886), 473–74; "Ptomaines and Leucomaines, and Their Relation to Disease," *Science*, XIV (November 8, 1889), 322–23. For a brief account and evaluation of LeConte's hypothesis on glycogen-esis, see John J. McNamara, "Joseph LeConte: Nineteenth-Century Physician, Twentieth-

From the popular magazine the *Nation* LeConte received a reasonably favorable review. "Professor LeConte," said the *Nation* reviewer,
"has a really wonderful capacity for selecting what is interesting in
science ... and the little book before us can safely be said to have
much more than the interest of the ordinary novel." The reviewer
praised the book as "not at all unnecessarily technical" and commended it for use in high schools and colleges. Admitting that some
statements "may be lacking in absolute accuracy," the reviewer par-
doned LeConte because "it is impossible in a bare outline to make
room for every important exception to general statements." In a review
of the work for *Science*, however, J. S. Kingsley of Tufts College was less
forgiving of the "too many inaccurate statements" of LeConte's book.
Kingsley proceeded to cite well over a dozen such inaccuracies. He
added that "the worst feature, physiologically, of the book is the recognition of a vital force." Above all, however, Kingsley charged that the
most "serious fault is the lack of a broader grasp of the results of recent morphological and physiological research."[4]

LeConte published a rejoinder to Kingsley's critical review, relying
mainly upon the argument that his work was "a bare *outline*" intended
merely as an introduction to comparative physiology. LeConte accused his critic of "pure mistakes or else misunderstandings of my
meaning" and argued that some of Kingsley's objections "concern
points still in doubt." But he admitted there were "some real mistakes"
in the book. It was to Kingsley's disdain for the concept of vital force
that LeConte objected most strenuously, however. This "really amused
me," said LeConte, and he cited his older paper on the correlation of
vital with chemical and physical forces in support of the idea. He also
added that the paper had been cited by a number of authorities. In
fact, it had, but most of those were works that had appeared almost
three decades before. Indeed, the *Outline* belonged to an earlier era,

Century Educator," *Southern Medicine*, LVI (June, 1973), 43–47. In the *Autobiography* (p.
334), LeConte stated, "My views on glycogeny, although not yet certain, have undoubtedly contributed to clearness of scientific thought on that important subject." Perhaps
he was correct in claiming a contribution toward clarity, but he was mistaken regarding
the relation of the liver to diabetes, and he was in error on the etiology of diseases
commonly associated with that organ.

    4 *Nation*, LXXI (October 25, 1900), 334; *Dial*, XXVIII (March 16, 1900), 208; *Science*, n.s.,
XI (April 27, 1900), 661–62.

and while it demonstrated the universalism of its author, it likewise indicated that his day had passed. The book quickly faded into obscurity.[5]

As the summer of 1900 arrived, Joseph and Bessie began to prepare for a long-planned journey to Europe, his request for a sabbatical having been readily approved by the regents several months previously. Their primary destination would be France, where LeConte was to serve as the university's representative to the Paris Exposition. He was eager to attend the various congresses on geology, mining, zoology, psychology, and education. But just before the scheduled departure, Carrie again fell into a state of nervous anxiety, and the trip had to be postponed. Meanwhile, Young Joe had planned a camping trip into the High Sierra, and after it became apparent that LeConte could not leave for Europe, he invited his father to go along. "I had thought my camping days were over, and had taken an affectionate leave of the grand scenes of the high Sierra," LeConte later wrote. "But from time to time the yearning for camp life comes upon me." Thus, when Young Joe urged his father to accompany him into the King's River Canyon, the devoted camper could not resist. Said LeConte, "I was now seventy-seven, but in good health and spirits, so determined to try it."[6]

Just before he left for the excursion, LeConte wrote to Niece Tillie, telling her that the European trip had been delayed because of "Carrie's health." He reported that Carrie's bodily functions were apparently sound, but he lamented that something was "not quite right in her nerve centres" and that she was so weary and weak that she spent most of her time in bed. "The case is a perfect mystery to me," he said to Tillie. To a grand nephew he also wrote that Carrie was in "feeble

5 LeConte to J. McKeen Cattell, May 24, 1900, in Cattell Papers; Joseph LeConte, "Reply to Professor Kingsley's Criticism," *Science*, n.s., XI (June 8, 1900), 909–11.

6 Minutes of the Board of Regents, University of California, Vol. XII, January 25, 1900; *Daily Californian*, March 13, 1900, p. 1; *ibid.*, March 15, 1900, p. 1; *ibid.*, March 28, 1900, p. 4; *ibid.*, April 4, 1900, p. 3; *ibid.*, April 13, 1900, p. 4; LeConte to George Davidson, May 26, 1900, in Davidson Papers; LeConte, *Autobiography*, 331–32; LeConte, "My Trip to King's River Cañon," *Sierra Club Bulletin*, IV (June, 1902), 88–99; Joseph N. LeConte, "Recollections"; Emma LeConte Furman, manuscript journal, 1900 (written in two separate log books, one of which is in the possession of Katherine Smith Adams, Carrboro, N.C., and the other in the possession of Anne Brown, Athens, Ga.).

health" and noted that he was "often buoyed up with hope & then she falls back again."[7]

But he was pleased that Emma had come for another visit, and he invited her to accompany them on their camping trip. Helen Gompertz (later to become Joe's wife) also joined the party, which set out on June 7 for a six-weeks' trip into King's River Canyon. At the end of the first day, LeConte wrote in his notebook that "the pain of starting this morn[ing] ... Carrie's cond[itio]n & dear Bessie's anxious tearful face as I folded her in my arms haunted me all day. But I must have some recreation." The group traveled by train to Sanger and then caught a stagecoach to Milwood, where they hired two horses and two packmules. Joseph and Emma would ride most of the journey while Joe and Helen walked and led the mules. As they passed through the Big Tree Grove, LeConte observed the careless devastation of the trees by lumberjacks, and he later wrote that "here we had a striking object lesson in the necessity of reservations." He reported that the original periphery of the grove "was a scene of horrible desolation, ruin, and ugliness.... It was literally a slaughter-pen." Such wanton destruction of thousand-year-old trees led him to comment that "like the prodigal son, the nation has wasted his substance in riotous living, careless of the future."[8]

But the party did not tarry long among the "grand sequoias." For four days they pushed on through heat and thunderstorms to their destination, arriving in the canyon on June 12. Joseph spent most of his time in camp while the others took trips into the surrounding areas. He had tried a longer walk but found it too difficult: "Oh my days of climbing are over forever—came back to camp, bathed my feet in the river & felt better." Still, he enjoyed the excursion with great gusto, sitting for hours to study rock formations or "merely musing, day-dreaming, or even dozing, but always taking in great health-giving draughts of pure air and bright sunshine."[9]

Writing home to Bessie, Joseph told her not to worry, because he felt fine. Yet he was still feeling sad about leaving his beloved wife and

7 Joseph LeConte to Matilda "Tillie" Harden Stevens, July 27, 1899, and June 6, 1900, both in LeConte Family Papers, APS; Joseph LeConte to James A. LeConte, April 27, 1900, in possession of Mrs. Joseph Nisbet LeConte.

· 8 Joseph LeConte, notebook labeled "King's River Cañon, 1900," in LeConte Family Papers, UCB; LeConte, "My Trip to King's River Cañon," 90–91.

9 LeConte, notebook, "King's River Cañon."

Carrie: "I dreamed about you and Carrie last night. I dreamed that Dr. Devecchi [Paolo de Vecchi] told you that massage would do Carrie good & that you must push it. But of course it was only a dream." A few days later he wrote to Bessie again to assure her that both he and Emma were well and enjoying themselves. Shortly thereafter he heard from Bessie, who reported that Carrie's depression had worsened. In reply, the perplexed father said, "I am quite convinced that there is nothing the matter with Carrie but a loss of interest in life, but this is a serious condition and she will come out of it slowly." Diversion he felt was the best recourse for Carrie, and he recommended that Bessie and Carrie "go off somewhere," no matter what it cost.[10]

After two weeks in camp, Joseph wrote to Bessie, "I have never felt better in my life than I do now. The climate here is admirable—much better than Yosemite." On the twenty-eighth of June, the party moved on to Bull Frog Lake, near Kearsarge Pass and about 10,500 feet above sea level. This "produced an exhilaration such as I have not felt before in 10–15 yrs.," said LeConte. "It is simply glorious." So the camping trip continued until the middle of July. Upon arriving back in Berkeley on the twentieth, Joseph wrote a brief account of the trip for the *Sunset*. Like all other camping trips, he stated, this excursion had been "a renewal of my life." Now he could resume his plans to go to Europe. Carrie seemed better upon his return, and again he secured steamship tickets to make the journey. In September, he and Bessie went to New York to prepare for their long-awaited departure.[11]

Once in New York, however, LeConte caught a severe cold and had to be confined to bed for several weeks. He claimed he was "never very sick" during this time, and by early November he was completely well. Nevertheless, he and Bessie determined to delay their trip abroad until the following spring, not wishing to be in Europe during the winter. Instead, they entrained for South Carolina and Georgia, where they would stay until conditions were right for going abroad. It would be a leisurely stay with their daughter, grandchildren, and great-grandchildren. LeConte also hoped he might have an opportunity to visit his old plantation. He had last been there in December, 1896, with

10 LeConte, "My Trip to King's River Cañon," 94; Joseph LeConte to Bessie LeConte, June 12, 15, and 21, 1900, all in LeConte Family Papers, UCB.
11 Joseph LeConte to Bessie LeConte, June 26 and 29, 1900, both in LeConte Family Papers, UCB; LeConte, notebook, "King's River Cañon"; Emma LeConte Furman's manuscript journal, 1900; LeConte, *Autobiography*, 332.

Young Joe, who on that occasion snapped a photograph of his father standing near one of the surviving great camellias.[12]

In early January, 1901, Joseph went to Midway to visit the old church in which he had worshiped as a young boy and to view the cemetery where lay the remains of his father, mother, and other relatives. If he got down to see his own property, he never said so. Probably he did not. There was little to see; nothing remained but an overgrown wilderness and slight vestiges of the once-famous botanical garden. Meanwhile, Bessie's health had declined, and the devoted mother expressed her strong desire to be back in Berkeley to recuperate and to look after Carrie. LeConte thought momentarily of going on to Europe by himself, but, as he wrote to Tillie, he and Bessie had "so grown together that separation is painful." Thus, he gave up the idea, and he and Bessie returned to California in early March.[13]

LeConte's students were waiting to greet him. In his absence they had again celebrated his birthday and purchased two paintings to present to him upon his return. Once more LeConte became active in university affairs, but clearly he could never pick up as vigorously as before. Now seventy-eight, his health was still good, but, as he had informed Tillie a few weeks before, his physical strength was declining. Thoughts of going to Europe were yet in his mind, but the trip must wait until Joe's wedding in June. Meanwhile, he continued to write his autobiography, begun a few months before at Carrie's urging. For many weeks he dutifully recorded the events of his life. As time passed, however, his interest in the project flagged, and toward the end of the account he ceased to record matters in detail, giving mainly a brief sketch of his later career, noting his principal papers, and offering a self-evaluation of his contributions to science and philosophy.[14]

12 LeConte, *Autobiography*, 332; Joseph LeConte to Matilda "Tillie" Harden Stevens, November 3, 1900, in LeConte Family Papers, APS; *Daily Californian*, October 18, 1900, p. 3; *ibid.*, October 31, 1900, p. 1.

13 LeConte, *Autobiography*, 332–33; Joseph LeConte to Matilda "Tillie" Harden Stevens, January 23, 1901, in LeConte Family Papers, APS; LeConte to University of California President Benjamin Ide Wheeler, February 7, 1901, in the President's Correspondence, Archives Division, Bancroft Library, University of California, Berkeley; Joseph N. LeConte, "Recollections"; Joseph LeConte, notebook labeled "Misc. 1894–to 1900," in LeConte Family Papers, UCB.

14 Telegram from "The Students of the University of California" to Joseph LeConte, February 26, 1901, in LeConte Family Papers, UCB; *Daily Californian*, February 14, 1901,

LeConte also managed to write two very brief articles in 1901, the latter of which he titled "What Is Life?" In it he repeated his long-held views on vital force, explaining that what constituted the difference between "actual life" and dead matter was "a difference in *molecular arrangement*—a difference in *allotropic condition.*" Whenever life is extinguished, he said, the molecular structure is simply rearranged; thus it does not necessarily die. For LeConte, life continued in some other form after its extinction. Perhaps it was merely an argument to refute scientific "conceptions of life" or even a reply to his critic J. S. Kingsley. In a large sense, however, it was a credo of immortality. Fear of personal death never seems to have plagued LeConte, but the loss of so many loved ones, especially during his formative years, impressed him thoroughly with a hope for the sustentation of life beyond material existence. The hope remained with him throughout his long years, and it proved strong psychological support in the face of the bewildering conclusions of scientific inquiry.[15]

Feeling financially secure, LeConte was sure he had provided sufficiently for Bessie and Carrie after his death. In 1900, he had recorded in his notebook that his savings were in excess of $32,000, and he was still receiving royalties from his books. Added to these were his property holdings in Berkeley (a lot and a house on Bancroft Way and a lot on Durant Street). He also owned approximately two thousand acres of land in Liberty County, and he carried an unknown amount of life insurance. Further, he noted in his will that after Bessie's death he wished the house and lot on Bancroft Way to go to Carrie. But these were mere cautions to be taken. LeConte still felt good, and when he heard that Sallie would come to Berkeley for her brother's wedding, the devoted father determined that he must show her the beauties of Yosemite. Sallie had lived in California from 1869 to 1876, of course, but she had never accompanied her father on one of his camping trips. What a grand idea, thought LeConte, to take Sallie and Bessie into the High Sierra for a pleasant vacation after Joe's marriage. Bessie was not overjoyed by the plan, however, and she tried to dissuade him from taking the trip, fearing the "fatigue and anxiety" would be too

---

p. 1; *ibid.*, February 21, 1901, p. 1; *ibid.*, February 27, 1901, p. 1; *ibid.*, March 6, 1901, p. 1; LeConte, *Autobiography*, preface and pp. 333–37.

15 Joseph LeConte, "What Is Life?" *Science*, n.s., XIII (June 21, 1901), 991–92; Joseph LeConte, "The Larynx as an Instrument of Music," *ibid.*, n.s., XIII (May 17, 1901), 790.

much for him. Joseph yearned to see the valley again, however, and he could not be talked out of going.[16]

After his marriage to Helen Gompertz on June 10, 1901, Joe and his bride set out for King's River Canyon for a camping-honeymoon trip. At the end of the month, Joseph and Sallie prepared to leave for Yosemite. Bessie decided to stay at home with Carrie. It would be a grand time, and a number of Sierra Club members, including Frank Soulé, would go along. Soulé found LeConte at the Oakland railway station, "as light hearted and expectant as when we had set out together from the same town, for the same destination, thirty-one years earlier." To his beloved friend, Soulé said, "We had a delightful time thirty-one years ago. Let us hope for as good a one now." LeConte replied that he hoped they would, and he said, "I am as eager and enthusiastic now as I was then." Stopping over at the hotel in Wawona for two days, Soulé "marvelled at his [LeConte's] wonderful memory." The party arrived in the valley at midday on July 4 and promptly settled in at Camp Curry. LeConte was "fatigued," but, as Soulé described him, "he was as joyous and enthusiastic as ever."[17]

Among the Sierra Club members and other camping friends who pitched camp near LeConte's own double tent were Dr. Edward R. Taylor, dean of the university's School of Law, William and Mary Keith, Dr. C. Hart Merriam, chief of the United States Biological Survey, John Muir, and many others, including a young woman, Bertha Chapman, who maintained a diary of the events surrounding those days. Joseph was delighted to be with this enthusiastic group of campers, and he hoped to accompany them on a few of their briefer sojourns. In fact, his camping outfit was already packed for an intended trip into the Tuolomne Meadows. On July 5, however, he and Sallie decided to remain at Camp Curry while a large party climbed to Glacier Point, feeling that it would be easier to travel after a day of rest. Thus, in the morning, along with Taylor, the Keiths, and another friend, Joseph and Sallie "engaged a coach and drove comfortably to Mirror Lake and the Happy Isles and to Yosemite Falls." Later in the afternoon, they

16 MS draft of final will (in Joseph LeConte's handwriting), and reports of Joseph LeConte's estate, both in Ellen J. McHenry's LeConte Album, Manuscripts Division, Bancroft Library, UCB. Superior Court Deeds, Book A–F, 1903, pp. 433–36, in Liberty County Records; Soulé, "Joseph LeConte in the Sierra," 9.

17 Soulé, "Joseph LeConte in the Sierra," 9–10; Joseph N. LeConte, "Recollections."

rode out to Bridalveil Fall, where Joseph "took off his hat, waved it, and fairly shouted at the sight." Afterwards, LeConte and the others made arrangements for horses to carry them to Glacier Point the next day to join the rest of the party.[18]

Around the camping grounds, LeConte greeted everyone in his usual warm way. The aging figure was the center of attraction, and Bertha Chapman's diary entries vividly pictured the gentle man in the words of an innocent youthful admirer, yet in the very spirit of all those who venerated their great friend and teacher:

I shall ever see him as he stood by the campfire in his rather shabby suit of black clothes. His tall lank figure bent forward with his delicate hands clasped behind his back. . . . His slender, refined face surrounded by the soft waving gray hair and his eyes lifted & allowed to rest on the hights [sic] above him. His gentle voice broken by a nervous . . . huskiness & irregular flow of words often delighted us, for he always explained the interesting things about us. . . . His whole manner was that of a child. I never dreamed such childlikeness possible—simply & entirely without self-pride of his greatness. [He was] always gentle & uncomplaining . . . , permitting no special courtesy, treating all alike, bearing all with the rest and showing a calmness and sweetness that awed us & taught us to love the man.[19]

Such was the adoration and esteem enjoyed by Joseph LeConte, and as night fell, several of his former students gathered round the campfire to reminisce with their old master and listen again to his enthusiastic praise of Nature's grandeur.

Joseph felt "unusually well and stayed up rather later than usual." Before leaving the little party for the evening, LeConte told Taylor that he looked forward to Sunday when he would visit "old South Dome and hear him preach." But, after retiring for the night, he began to complain of pains in his chest. The pains became more severe as dawn approached, and when Sallie arrived at 5:00 A. M. to arouse him for the trip, LeConte told her he must give up going because of he "slight pains in his heart." He insisted that Sallie and the others go on anyway, but she refused, saying that she "could not enjoy it without him." Sallie was not overly alarmed at this point, and she left for a few

18 Soulé, "Joseph LeConte in the Sierra," 10; Journal of Bertha L. Chapman, Yosemite National Park Museum, Yosemite, Calif.; Sallie LeConte Davis to Emma LeConte Furman, July 8, 1901, in Davis Papers; clipping from the San Francisco Call, July 8, 1901, in Mc-Henry LeConte Album.

19 Chapman Journal.

minutes to tell the group that her father was not feeling well. When she returned, however, she found him in a "violent spasm of pain."[20]

Frightened by her father's sudden turn for the worse, Sallie hastily summoned Dr. Taylor and another camper. When they looked in on Joseph, all three decided that they must quickly fetch the local doctor. It was much later, however, before the physician, Dr. Charles Cross, could be located. Dr. Cross checked on Joseph and told him that it was probably only "a slight attack of indigestion." But Cross seems to have known it was angina pectoris, and he gave LeConte an injection to ease the pain. LeConte apparently knew it was serious, and he told Cross that he was sure he was experiencing a heart seizure. The pain did not go away, and so Cross gave his patient another injection.[21]

The physician stayed until Joseph fell asleep, and, after assuring Sallie that her father "would get over his attack," he left to secure "some further remedies." LeConte continued to sleep until after ten o'clock. When he awoke, he told Sallie that he felt better and could use some light nourishment. Quickly, Sallie set out to get a "trained nurse" who was tending another ill patient in the vicinity, but she could not locate her. Upon returning to her father's tent, she asked him how he felt, and he replied that he was resting comfortably and turned over on his left side. His daughter insisted that he should not do that, but he replied that it did not matter. A few moments later, LeConte told Sallie that he "felt overpowered with sleep," and he closed his eyes. As the distraught daughter hovered over her beloved father, she noticed a "dreadful change come over his face," and she knew "he was dying." Flinging her arms around his frail form, she called to him, but "he was quite unconscious." Sallie burst from the tent and summoned two friends. But within minutes, Joseph gave "a few gasps," and "he was gone."[22]

"A greater shock could not have come to us all," wrote Miss Chapman. "A light mist actually settled above the granite heights, veiling the moment of his passing.... It seemed a fitting tribute from Nature to the giant man—passing [sic]." Dr. Cross embalmed the body as best he could, and Joseph's remains were placed in a "rough pine box ....

---

20 Sallie LeConte Davis to Emma LeConte Furman, July 8, 1901, in Davis Papers; clipping from the San Francisco *Call*, July 8, 1901, in McHenry LeConte Album.
21 Sallie LeConte Davis to Emma LeConte Furman, July 8, 1901, in Davis Papers.
22 *Ibid.*

made from the wood of the trees he loved so tenderly." The makeshift coffin was then placed upon the back seat of a coach and covered with fir boughs. Miss Chapman and a friend "made a great wreath" of black oak leaves and placed it atop the fir boughs. Sallie and Dean Taylor climbed aboard the coach in the front seat with the driver, and at five o'clock in the afternoon of July 6, the little group of mourners set out for Haywood, making the day-long trip in fourteen hours by means of relay horses. By train the body was carried on to San Francisco, where it was met by Joseph's nephew Julian and Little Joe's close friend, Clarence Cory.[23]

"Our one wish," said Bertha Chapman, "was that his body might rest here [in Yosemite] among the monuments of the centuries with the murmur of the trees to lull, and distant roar of the cateracts [sic] to sooth [sic] the sleep of the man who dwelt upon their mysteries to be sure, but who above all else knew how to love, to enter into their moods." But the Oakland Cemetery would be Joseph's final resting place, and Yosemite would come to him in the form of a large rough granite rock taken from his beloved valley to mark his grave. Yet in spirit LeConte would always be in Yosemite:

> Here in these consecrated haunts sublime,
>   Beloved through thy life's perennial youth,
> Kind Nature whispered of the parting time
>   And led thy spirit on its quest of truth
> Beyond these symbols into high domains
> Where goals that baffle now, the soul attains.[24]

Indeed, his spirit would live with all whom it had touched, for

> No man was cast in gentler mould,
>   Yet stronger none in firm command.
> His thought our lesser thought controlled,
>   Our hearts he held within his hand.[25]

23 Chapman Journal; Sallie LeConte Davis to Emma LeConte Furman, July 8, 1901, and Sallie to her husband, July 13, 1901, both in Davis Papers; Joseph N. LeConte, "Recollections"; clipping from the San Francisco Call, July 8, 1901, in McHenry LeConte Album.

24 Chapman Journal; Charles Keeler, "To Joseph LeConte," Sierra Club Bulletin, VI (January, 1908), 223–24. LeConte was buried in the academic gown he wore at the Princeton sesquicentennial celebration in 1896 (unidentified newspaper clipping in McHenry LeConte Album).

25 Ina Coolbrith, "Joseph LeConte," University of California Magazine, VII (September, 1901), 213. The entire issue of the Magazine consisted of tributes to the memory of

In 1904 a memorial lodge bearing LeConte's name was completed and dedicated in Yosemite Valley. The bronze memorial tablet, still present on the wall of the lodge, is simply inscribed. "Joseph LeConte, Scientist and Savant, Died In This Valley, July 6, 1901."[26]

LeConte. Numerous other magazines, professional journals, and newspapers likewise eulogized the deceased idol of the University of California. A memorial meeting for LeConte was held at the University of California on August 21, 1901 (see the *Daily Californian*, August 15–22, 1901, various pages). The tributes delivered at this meeting were published in the *University Chronicle*, IV (October, 1901), 241–64.

26 ["Memorial to Joseph LeConte"], *Sierra Club Bulletin*, IV (1903), Plates LI–LIV; William E. Colby, "The Completed LeConte Memorial Lodge," *ibid.*, V (1904), 66–67; "Secretary's Report," *ibid.*, V (1904), 134–35; Alexander G. Eells, "Address at Memorial Exercises," *ibid.*, V (1904), 176–77; ["Memorial Tablet for LeConte Lodge"], *ibid.*, VI (1906), Plate XII [p. ·24]; Willoughby Rodman, "In Memoriam: Joseph LeConte," *ibid.*, VI (1906), 25–31.

Sources

Manuscripts

Academy of Natural Sciences of Philadelphia.
  American Entomological Society Papers.
  LeConte, John Lawrence. Manuscripts.
  Leidy, Joseph. Papers.
American Philosophical Society Library, Philadelphia, Pennsylvania.
  Bowers, Stephen. Papers.
  LeConte Family. Papers.
  Lesley, J. Peter. Papers.
  Sharswood, William. Papers.
Auburn University Library, Auburn, Alabama.
  Broun, William LeRoy. Correspondence.
Brown University Library, Providence, Rhode Island.
  Ward, Lester Frank. Papers.
University of California, Berkeley.
  Board of Regents Minutes, 1868–1901 (Regents' Office).
  Davidson, George. Papers (Manuscripts Division, Bancroft Library).
  Hearst, Phoebe. Papers (Manuscripts Division, Bancroft Library).
  Lawson, Andrew. Papers (Manuscripts Division, Bancroft Library).
  LeConte, Caroline E. Letters (Files of the University of California Press).
  LeConte, Joseph. Miscellaneous Writings (Archives Division, Bancroft Library).
  LeConte, Joseph. Memorials (Archives Division, Bancroft Library).
  LeConte Family. Papers (Manuscripts Division, Bancroft Library).
  McHenry, Ellen J. LeConte Album (Manuscripts Division, Bancroft Library).
  Palache, Charles. Collection (Manuscripts Division, Bancroft Library).
  Regents' Correspondence File. (Archives Division, Bancroft Library).
  Sutro, Adolph. Papers (Manuscripts Division, Bancroft Library).

Duke University Library, Durham, North Carolina.
  Nisbet, E. A. Manuscripts.
  Page, Thomas Nelson. Letter.
  Simpson, W. D. Collection.
Geological Society of America, Boulder, Colorado.
  Minutes of the Society, 1888–95.
Georgia Historical Society, Savannah, Georgia.
  LeConte-Furman. Papers.
Harvard University, Cambridge, Massachusetts.
  Agassiz, Alexander. Papers (Museum of Comparative Zoology).
  Edwards, Henry. Papers (Museum of Comparative Zoology).
  Gray, Asa. Letters (Gray Herbarium Library).
  James, William. Letters (Houghton Library, Manuscripts).
Historical Society of Pennsylvania, Philadelphia, Pennsylvania.
  Youmans, E. L. Letter.
  Ward, William H. Letter (Dreer Collection).
Huntington Library, San Marino, California.
  Keeler, Charles. Correspondence.
Johns Hopkins University Library, Baltimore, Maryland.
  Gilman, Daniel. Papers.
Georgia State Archives, Atlanta, Georgia.
  Liberty County Records:
    Accounts of Executors, Administrators and Guardians, 1829–54.
    Liberty County, Georgia, Land Records, Book A–L, 1840.
    Superior Court Deeds, 1842–54, 1903.
    Tax Reports, 15th Militia District, 1847, 1848, 1861.
    Wills and Appraisements, 1789–1850.
Library of Congress, Washington, D.C.
  Bache, Alexander D. Papers.
  Cattell, J. McKeen. Papers.
  Gibbes, Lewis R. Papers.
  Newcomb, Simon. Papers.
Maryland Historical Society, Baltimore, Maryland.
  Hyatt, Alpheus. Correspondence. MS 1007.
Michigan Historical Collections, University of Michigan Library, Ann
Arbor, Michigan.
  Winchell, Alexander. Correspondence.
Minnesota Historical Society, St. Paul, Minnesota.
  LeConte, Joseph. "An Early Geological Excursion."
New York State Library, Albany, New York.
  Hall, James. Papers.

North Carolina State Department of Archives and History, Raleigh, North Carolina.
  Cowles, Calvin J. Letter.
Princeton University Library, Princeton, New Jersey.
  Mayer, Alfred M. Correspondence.
Private Collections of LeConte Manuscripts.
  Adams, Katherine. Carrboro, North Carolina.
  Anderson, Richard LeConte. Macon, Georgia.
  Brookshire, Katherine. Athens, Georgia.
  Brown, Anne. Athens, Georgia.
  LeConte, Helen Malcolm. San Francisco, California.
  LeConte, Mrs. Joseph Nisbet. Greenville, North Carolina.
  McMillan, Carolyn. St. Simons Island, Georgia.
  Smith, Joseph LeConte. Macon, Georgia.
  Smith, Nancy. In possession of Katherine Adams, Carrboro, N.C.
Smithsonian Institution Archives, Washington, D.C.
  Merrill, George P. Collection. Record Unit 7177.
  Office of the Registrar. Accession Records.
  Record Units 2, 26–29, 33, 53.
  Secretary. Correspondence.
  Casey, Thomas Lincoln. Papers.
South Caroliniana Library, University of South Carolina, Columbia, South Carolina.
  Charles Family. Papers.
  Davis, R. Means. Papers.
  South Carolina College Faculty Minute Books, 1856–69.
  LeConte, Joseph. Papers.
  Rivers, William J. Papers.
Southern Historical Collection. University of North Carolina, Chapel Hill, North Carolina.
  Alexander, E. P. Papers..
  Anderson, Clifford. Papers.
  LeConte-Furman. Papers.
  Talley, Elizabeth Furman. Papers.
Stanford University Library, Stanford, California.
  Jordan, David Starr. Papers.
UCLA Library, University of California at Los Angeles, Los Angeles, California.
  Spalding, W. A. Collection.
University of Georgia Library, Athens, Georgia.
  Cobb, Howell. Papers.

Coulter, E. Merton. Collection.
Faculty Minutes, 1838–56.
LeConte, Joseph. Letter.
LeConte Family. Papers.
LeConte Genealogical Collection.
Phi Kappa Literary Society Minutes, 1838–41.
University of Georgia Prudential Committee Minutes, 1855–56.
Minutes of the University of Georgia Trustees, 1838–56.
Yale University Library, New Haven, Connecticut.
Beecher Family. Papers.
Brush Family. Papers.
Marsh, O. C. Papers.
Whitney Family. Papers.
Yosemite National Park Museum, Yosemite, California.
Chapman, Bertha L. Journal [copy in possession of Mrs. Carol C. Montoya, Vidalia, Georgia].

Newspapers

Athens (Ga.) *Southern Banner*, 1845, 1852–56.
Athens (Ga.) *Southern Watchman*, December 4, 1856.
Atlanta *Constitution*, May 1, July 29, and July 31, 1888.
Augusta (Ga.) *Chronicle*, January 29, 1853.
Augusta (Ga.) *Weekly Chronicle & Sentinel*, December 12 and 19, 1855.
Berkeley *Advocate*, February 18, 1897.
Berkeley *Daily Californian*, 1897–1901.
Berkeley *Gazette*, February 20, 1897.
Berkeleyan, 1893–96.
Macon *Georgia Journal and Messenger*, April 11, 1849.
Macon *Georgia Telegraph*, August 17, 1841, February 9 and June 26, 1849.
Milledgeville (Ga.) *Federal Union*, October 9, 1855.
Milledgeville (Ga.) *Southern Recorder*, November 30, 1852.
New York *Evening Post*, May 16, 1891.
New York *Times*, February 14, 1874, October 21 and 23, 1896.
Oakland *Tribune*, February 20, 1897.
San Francisco *Call*, July 8, 1901.
San Francisco *Chronicle*, May 19–26, 1895.
San Francisco *Examiner*, June 15, 1869, May 26, 1887, May 19–26, 1895, and February 19, 1897.
San Francisco *Evening Bulletin*, March 2 and June 22, 1870.
Savannah (Ga.) *Republican*, October 23, 1851.

Books

Abele, Rudolph von. *Alexander H. Stephens: A Biography.* New York: Alfred A. Knopf, 1946.

Adicks, Richard, ed. *LeConte's Report on East Florida.* Orlando: University Presses of Florida, 1978.

Ames, Mary Lesley, ed. *Life and Letters of Peter and Susan Lesley.* Vol. I. New York: G. P. Putnam's Sons, 1909.

Anderson, Richard LeConte. *LeConte Family History and Genealogy.* 2 vols. Macon, Ga.: Privately printed, 1981.

Annual Report of the Superintendent of the Coast Survey During the Year Ending November, 1851. Senate Executive Documents, No. III, 32nd Congress, First Session. Appendix No. X, 145–60. Serial Set 616. Washington, D. C.: Robert Armstrong, 1852.

Avery, Myrta Lockett, ed. *Recollections of Alexander H. Stephens.* New York: Doubleday, 1910.

Badè, William Frederic. *The Life and Letters of John Muir.* 2 vols. Boston: Houghton Mifflin, 1924.

Bailey, Edward Battersby. *James Hutton: The Founder of Modern Geology.* London: Elsevier Publishing Co., 1967.

Billington, Ray Allen. *Frederick Jackson Turner: Historian, Scholar, Teacher.* New York: Oxford University Press, 1973.

Boller, Paul F., Jr. *American Thought in Transition: The Impact of Evolutionary Naturalism, 1865–1900.* Chicago: Rand McNally, 1969.

Blue and Gold, Being a Record of the College Year. Vol. XVIII. Berkeley: Associated Students of the University of California, 1892.

Bucher, Walter H. *The Deformation of the Earth's Crust.* Princeton: Princeton University Press, 1933.

Buranelli, Vincent. *Josiah Royce.* New York: Twayne Publishers, 1964.

Burchfield, Joe D. *Lord Kelvin and the Age of the Earth.* New York: Science History Publications, 1975.

Carroll, James Parsons. *Report of the Committee Appointed to Collect Testimony in Relation to the Destruction of Columbia, S. C., on the 17th of February, 1865.* Columbia: Bryan, 1893.

Cash, Wilbur J. *The Mind of the South.* New York: Alfred A. Knopf, 1941.

Catalogue of the Trustees, Officers, Alumni, and Matriculates of the University of Georgia from 1785 to 1906. Athens: E. D. Stone Press, 1906.

[Chambers, Robert]. *Vestiges of the Natural History of Creation.* London: John Churchill, 1844.

Chesnut, Mary Boykin. *A Diary from Dixie.* Ed. Ben Ames Williams. Boston: Houghton Mifflin, 1949.

Chorley, Richard, Anthony J. Dunn, and Robert P. Beckinsdale. *The History of the Study of Landforms*. Vol. I. London: Methuen, 1964.

Chugerman, Samuel. *Lester F. Ward: The American Aristotle*. Durham: Duke University Press, 1939.

Clarke, John M. *James Hall of Albany: Geologist and Palaeontologist*. Albany: n.p., 1923.

Clendenning, John, ed. *The Letters of Josiah Royce*. Chicago: University of Chicago Press, 1970.

The Colonial Records of Georgia. Vols. VIII, IX. Atlanta and Athens, Ga.: State Printer and University of Georgia Press, 1904–79.

Columbia University Officers and Alumni, 1754–1857. New York: Columbia University Press, 1936.

Congrès Geologique International, Compte Rendu de la 5me Session, Washington, 1891. In *The Miscellaneous Documents of the House of Representatives*. Second Session, 53rd Congress, 1893–94. Vol. XIII. Washington, D. C.: Government Printing Office, 1894.

Connelly, Thomas L. *Discovering the Appalachians*. Harrisburg, Pa.: Stackpole Books, 1968.

Cornelius, Brother. *Keith, Old Master of California*. Vol. I. New York: G. P. Putnam's Sons, 1942.

Coulter, E. Merton. *College Life in the Old South*. 2nd ed. Athens: University of Georgia Press, 1951.

Cremin, Lawrence A. *The Transformation of the School*. New York: Alfred A. Knopf, 1961.

Curti, Merle. *The Growth of American Thought*. 2nd ed. New York: Harper & Brothers, 1951.

Dana, Edward S., ed. *A Century of Science in America*. New Haven: Yale University Press, 1918.

Dana, James Dwight. *Manual of Geology*. 2nd ed. New York: Ivison, Blakeman, Taylor and Co., 1874.

Daniels, George H. *American Science in the Age of Jackson*. New York: Columbia University Press, 1968.

―――. *Darwinism Comes to America*. Waltham, Mass.: Blaisdell Publishing Company, 1968.

Darlington, William, comp. *Reliquiae Baldwinianae*. Philadelphia: Kimber and Sharpless, 1843.

Darwin, Bernard. *British Clubs*. London: W. Collins, 1943.

Dupree, A. Hunter. *Asa Gray, 1810–1888*. Cambridge: Belknap Press of Harvard University Press, 1959.

Eaton, Clement. *Freedom of Thought in the Old South*. Durham: Duke University Press, 1940.

Engberg, Robert, and Donald Wesling, eds. *John Muir: To Yosemite and*

Beyond. Madison: University of Wisconsin Press, 1980.

Farquhar, Francis P. *History of the Sierra Nevada.* Berkeley: University of California Press, 1965.

———. *Place Names in the High Sierra.* San Francisco: Sierra Club, 1926.

Fisher, Osmond. *Physics of the Earth's Crust.* Rev. ed. London: Macmillan and Company, 1881.

Franklin, Fabian, *et al. The Life of Daniel Coit Gilman.* New York: Dodd, Mead and Company, 1910.

Fredrickson, George M. *The Black Image in the White Mind.* New York: Harper & Row, 1971.

Gallarati-Scotti, Tommaso. *The Life of Antonio Fogazzaro.* Trans. Mary Prichard Agnetti. London: Hodder and Stoughton, n.d.

Gaustein, Jeanette E. *Thomas Nuttall, Naturalist.* Cambridge: Harvard University Press, 1967.

Geikie, Archibald. *Textbook of Geology.* Vol. I. 4th ed. London: Macmillan and Company, 1903.

Gill, William I. *Evolution and Progress.* New York: Authors' Publishing Company, 1875.

Gillispie, Charles Coulston. *Genesis and Geology.* Cambridge: Harvard University Press, 1951. Reprint ed., New York: Harper Torchbooks, 1959.

Glass, Bentley, Owsei Temkin, and William L. Strauss, Jr., eds. *Forerunners of Darwin: 1745–1859.* Baltimore: Johns Hopkins University Press, 1959.

Glick, Thomas F., ed. *The Comparative Reception of Darwinism.* Austin: University of Texas Press, 1974.

Goldman, Eric. *Rendezvous with Destiny.* New York: Alfred A. Knopf, 1952.

Gulick, Addison. *Evolutionist and Missionary: John Thomas Gulick.* Chicago: University of Chicago Press, 1932.

Gulick, John Thomas. *Evolution, Racial and Habitudinal.* Washington, D.C.: Carnegie Institution of Washington, 1905.

Guralnick, Stanley M. *Science and the Ante-Bellum American College. Memoirs of the American Philosophical Society,* Vol. CIX. Philadelphia: American Philosophical Society, 1975.

Hall, James. *Palaeontology of New York.* Vol. III, Pt. I. New York: C. Van Benthuysen, 1859.

Haller, John S. *Outcasts from Evolution.* Urbana: University of Illinois Press, 1971.

Harper, Ida Husted. *The Life and Work of Susan B. Anthony.* Vol. II. Indianapolis: Hollenbeck Press, 1898.

Hawkins, Hugh. *Pioneer: A History of the Johns Hopkins University, 1874–1889*. Ithaca: Cornell University Press, 1960.

Heim, Albert. *Untersuchungen über den Mechanismus der Gebirgsbildung im Anschluss an die geologische Monographie der Tödi-Windgällen-Gruppe*. 2 vols. Basel: B. Schwabe, 1878.

Hirschberg, Julius. *Geschichte der Augenheilkunde*. Vol. III. Leipzig: W. Engelmann, 1915.

Hofstadter, Richard. *Social Darwinism in American Thought*. Rev. ed. New York: George Braziller, 1955.

Hollis, Daniel Walker. *University of South Carolina*. 2 vols. Columbia: University of South Carolina Press, 1951.

Holt, Thomas. *Black over White: Negro Political Leadership in South Carolina During Reconstruction*. Urbana: University of Illinois Press, 1977.

Hovenkamp, Herbert. *Science and Religion in America, 1800–1860*. Philadelphia: University of Pennsylvania Press, 1978.

Hynds, Ernest C. *Antebellum Athens and Clarke County, Georgia*. Athens: University of Georgia Press, 1974.

International Congress of Geologists. *Circular of Information, No. V.* Washington, D.C.: The Congress, 1891.

———. *Members of the Committee of Organization and List of Members and Delegates Present, Fifth International Congress*. Washington, D.C.: Gibson Bros., Printers and Bookbinders, 1891. 13 pp.

Jenny, Hans. *E. W. Hilgard and the Birth of Modern Soil Science*. Pisa: Collana della Revista "Agrochimica," 1961.

Johnson, Thomas Cary, Jr. *Scientific Interests in the Old South*. New York: Appleton-Century, 1936. Reprint ed., Wilmington, Del.: Scholarly Resources, 1973.

Johnson's New Universal Encyclopedia. New York: A. J. Johnson & Son, 1877.

Jones, Holway R. *John Muir and the Sierra Club: The Battle for Yosemite*. San Francisco: Sierra Club, 1965.

Jordan, David Starr. *The Days of a Man*. Vol. I. Yonkers-on-Hudson, N.Y.: World Book Company, 1922.

Keeler, Charles. *A Light Through the Storm*. San Francisco: William Doxey, 1894.

Keppel, Frederick P. *Columbia*. New York: Oxford University Press, 1914.

Kevles, Daniel J. *The Physicists*. New York: Alfred A. Knopf, 1978.

Kohlstedt, Sally Gregory. *The Formation of the American Scientific Community*. Urbana: University of Illinois Press, 1976.

Kraditor, Aileen S. *The Ideas of the Woman Suffrage Movement, 1890–1920*. New York: Columbia University Press, 1965.

LaBorde, Maximilian. *History of the South Carolina College From Its Incorporation, Dec. 19, 1801, to Dec. 19, 1865; Including Sketches of Its Presidents and Professors, With an Appendix.* Charleston: Walker, Evans and Cogswell Co., 1874.

LeConte, Caroline E., *The Statue in the Air.* New York: Macmillan, 1897.

———. *Yosemite, 1878: Adventures of N. & C.* Introd. Susanna B. Dakin. San Francisco: Book Club of California, 1964.

LeConte, Emma. *When the World Ended: The Diary of Emma LeConte.* Ed. Earl Schenck Miers. New York: Oxford University Press, 1957.

Loudon, John C. *An Encyclopedia of Gardening.* Rev. ed. London: Longman, Brown, Green, and Longmans, 1850.

Lucas, Marion Brunson. *Sherman and the Burning of Columbia.* College Station, Tex.: Texas A & M University Press, 1976.

Lunt, George. *The Origin of the Late War.* New York: D. Appleton and Company, 1866.

Lurie, Edward. *Louis Agassiz: A Life in Science.* Chicago: University of Chicago Press, 1960.

Mathews, Edward B., ed. *Fifty Years' Progress in Geology.* Baltimore: Johns Hopkins University Press, 1927.

Merriam, John C. *Published Papers of John Campbell Merriam.* Washington, D.C.: Carnegie Institute of Washington, 1938.

Merrill, George P. *The First One Hundred Years of Geology.* New Haven: Yale University Press, 1924. Reprint ed., New York: Hafner, 1964.

Moore, James R. *The Post-Darwinian Controversies.* Cambridge: Cambridge University Press, 1979.

Myers, Robert Manson, ed. *The Children of Pride.* New Haven: Yale University Press, 1972.

Osborn, Henry Fairfield. *Cope: Master Naturalist.* Princeton: Princeton University Press, 1931.

Perry, Ralph Barton. *Annotated Bibliography of the Writings of William James.* New York: Longmans, Green, and Co., 1920.

Persons, Stow, ed. *Evolutionary Thought in America.* New Haven: Yale University Press, 1950.

Pond, Cornelia Jones. *Life on a Liberty County Plantation.* Ed. Josephine Bacon Martin. Darien, Ga.: Privately printed, 1974.

Prestwich, Joseph. *Geology: Chemical, Physical, and Stratigraphical.* Vol. II. Oxford: Clarendon Press, 1888.

Pyne, Stephen J. *Grove Karl Gilbert: A Great Engine of Research.* Austin: University of Texas Press, 1980.

Reade, T. Mellard. *The Origin of Mountain Ranges Considered Experimentally, Structurally, Dynamically, and in Relation to Their Geological History.* London: Taylor and Francis, 1886.

Reingold, Nathan, ed. *Science in Nineteenth-Century America: A Documentary History.* London: Macmillan, 1966.

———, ed. *The Sciences in the American Context: New Perspectives.* Washington, D.C.: Smithsonian Institution Press, 1979.

Report of Alonzo Church, President of Franklin College, University of Georgia, to the Senatus Academicus. Milledgeville, Ga.: Boughton, Nisbet, and Barnes, State Printers, 1858.

Report of the Proceedings of the Fifth International Congress of Geologists, August 27–31, 1891. Washington, D.C.: The Congress, 1891.

Reyer, Eduard. *Geologie und geographische Experimente.* 3 vols. Leipzig: W. Englemann, 1892–94.

———. *Theoretische Geologie.* Stuttgart: Schweizerbart, 1888.

———. *Ursachen der Deformationen und der Gebirgsbildung.* Leipzig: W. Englemann, 1892.

Rice, William N., ed. *Problems of American Geology.* New Haven: Yale University Press, 1915.

Romanes, George J. *Darwin and After Darwin.* 3 vols. Chicago: Open Court Publishing Company, 1901–1906.

———. *The Life and Letters of George John Romanes.* 4th ed. New York: Longmans, Green, 1898.

Rothstein, William G. *American Physicians in the Nineteenth Century.* Baltimore: Johns Hopkins University Press, 1972.

Royce, Josiah, *et al. The Conception of God.* New York: Macmillan, 1898.

Schneer, Cecil J., ed. *Two Hundred Years of Geology in America.* Hanover, N.H.: University Press of New England, 1979.

Shryock, Richard H. *Medicine and Society in America, 1660–1860.* New York: New York University Press, 1960.

Smithsonian Institution. *Annual Reports, 1856, 1857, and 1873.* Washington, D.C.: Cornelius Wendell, James B. Steedman, and Government Printing Office, 1857, 1858, and 1874, respectively.

Stadtman, Verne A. *The University of California, 1868–1968.* New York: McGraw-Hill, 1970.

Stampp, Kenneth M. *The Era of Reconstruction.* New York: Alfred A. Knopf, 1965.

Stanton, William. *The Leopard's Spots: Scientific Attitudes Toward Race in America, 1815–59.* Chicago: University of Chicago Press, 1960.

Stocking, George W., Jr. *Race, Culture, and Evolution.* New York: Free Press, 1968.

Tankersly, Allen P. *College Life at Old Oglethorpe.* Athens: University of Georgia Press, 1951.

Taylor, Francis Long. *Crawford W. Long and the Discovery of Ether Anesthesia.* New York: Paul B. Hober, 1928.

U.S. War Department. *The War of the Rebellion: A Compilation of the Official Records of the Union and Confederate Armies.* Series IV. Vol. III. Washington, D.C.: Government Printing Office, 1900.

Wade, John Donald. *Augustus Baldwin Longstreet: A Study of the Development of Culture in the South.* New York: Macmillan, 1924.

Wall, Joseph Frazier. *Andrew Carnegie.* New York: Oxford University Press, 1970.

Wallace, Alfred Russel. *My Life.* 2 vols. New York: Dodd, Mead & Co., 1906.

Ward, Humphrey. *History of the Athenaeum, 1824–1925.* London: The Club, 1926.

Ward, Lester Frank. *Outlines of Sociology.* New York: Macmillan, 1923.

———. *The Psychic Factors of Civilization.* Boston: Ginn and Company, 1901.

Waring, Joseph I. *A History of Medicine in South Carolina, 1825–1900.* Charleston: South Carolina Medical Association, 1967.

White, Andrew D. *A History of the Warfare of Science and Theology in Christendom.* 2 vols. New York: D. Appleton, 1896.

Whitney, Josiah D. *Earthquakes, Volcanoes, and Mountain Building.* Cambridge: Harvard University Press, 1871.

Williamson, Joel. *After Slavery: The Negro in South Carolina During Reconstruction, 1861–1877.* Chapel Hill: University of North Carolina Press, 1965.

Wolfe, Linnie Marsh. *Son of the Wilderness: The Life of John Muir.* New York: Alfred A. Knopf, 1945.

Woodrow, Marion W., ed. *Dr. James Woodrow: Character Sketches by His Former Pupils, Colleagues, and Associates.* Columbia, S.C.: R. L. Bryan, 1909.

Wylie, Lollie Bell, ed. *Memoirs of Judge Richard H. Clark.* Atlanta: Franklin Printing and Publishing Co., 1898.

Young, Ida, *et al. History of Macon, Georgia.* Macon: Lyon, Marshall & Brooks, 1950.

Zittel, Karl von. *History of Geology and Palaeontology.* London: Walter Scott, 1901. Reprint ed., New York: Hafner, 1962.

Articles

Adams, Ansel. "The Photography of Joseph N. LeConte." *Sierra Club Bulletin*, XXIX (October, 1944), 40–46.

Agassiz, Alexander. "Note on the Florida Reef." *American Journal of Science*, 3rd ser., XLIX (1895), 154–55.

Bozeman, Theodore Dwight. "Joseph LeConte: Organic Science and a 'Sociology for the South.'" *Journal of Southern History*, XXXIX (November, 1973), 565–82.

Brücke, Ernst. "Über die stereoskopischen Erscheinunger und Wheatstones Angriff auf die Lehre von den identischen Stellen der Netzhaute." *Archiv für Anatomie und Physiologie* (1841), 459–76.

Brush, Stephen G. "A Geologist Among Astronomers." *Journal of the History of Astronomy*, IX (February, 1978), 1–41, and IX (June, 1978), 77–104.

———. "Nineteenth-Century Debates About the Inside of the Earth." *Annals of Science*, XXXVI (May, 1979), 225–54.

Buckley, James B. "Mountains of North Carolina and Tennessee." *American Journal of Science*, 2nd ser., XXVII (May, 1859), 286–94.

Bull, Mrs. Ole. "The Cambridge Conferences." *Outlook*, LVI (August 7, 1897), 845–49.

Carroll, A. W. de la Cour. "The Ascent of Mt. LeConte." *Sierra Club Bulletin*, I (May, 1896), 325–26.

[Chamberlin, Thomas C.]. "Editorial." *Journal of Geology*, IX (July–August, 1901), 439–40.

Christy, S. B. "Biographical Notice of Joseph LeConte." *Transactions of the American Institute of Mining Engineers*, XXXI (November, 1902), 765–93.

"Cinque Lettere Inedite del Fogazzaro al Professor LeConte." *La Cultura*, VIII (1934), 82–84.

Claparède, Édouard. "Quelques mots sur la vision binoculaire et stéréoscopique et sur la question de l'horoptre." *Archives des Sciences Physiques et Naturelles*, Nouvelle Série, III (1858), 138–68.

Coolbrith, Ina. "Joseph LeConte," *University of California Magazine*, VII (September, 1901), 213.

Cope, Edward D. "The Relation of the Sexes to Government." *Popular Science Monthly*, XXXIII (October, 1888), 721–30.

Coulter, E. Merton. "Why John and Joseph LeConte Left the University of Georgia, 1855–1856." *Georgia Historical Quarterly*, LIII (September, 1969), 16–40.

Dana, James Dwight. "On the Origin of Mountains." *American Journal of Science*, 3rd ser., V (May, 1873), 347–50.

———. "On Some Results of the Earth's Contraction from Cooling, Including a Discussion of the Origin of Mountains and the Nature of the Earth's Interior." *American Journal of Science*, 3rd ser., V (June, 1873), 423–43; VI (July, 1873), 6–14; VI (September, 1873), 161–72; VI (October, 1873), 304; and VI (November, 1873), 381–82.

————. "A General Review of the Geological Effects of the Earth's Cooling From a State of Igneous Fusion." *American Journal of Science,* 2nd ser., IV (1847), 88–92.

————. "Geological Results of the Earth's Contraction." *American Journal of Science,* 2nd ser., III (1847), 176–88.

————. "On the Origin of Continents." *American Journal of Science,* 2nd ser., III (1847), 94–100.

————. "On the Volcanoes of the Moon." *American Journal of Science,* 2nd ser., II (1846), 335–55.

————. "Origin of the Grand Outline Features of the Earth." *American Journal of Science,* 2nd ser., III (1847), 381–98.

Donnelly, Ralph W. "The Bartow County Confederate Saltpetre Works." *Georgia Historical Quarterly,* LIV (Fall, 1970), 305–19.

Dove, Heinrich. "Über Stereoskopie." *Annalen der Physik,* 2nd ser., X (1860), 494–98.

Duffield, John T. "Evolutionism Respecting Man and the Bible." *Princeton Review,* 4th ser., I (January, 1878), 150–77.

Dutton, Clarence E. "A Criticism upon the Contractional Hypothesis." *American Journal of Science,* 3rd ser., VIII (August, 1874), 113–23.

Eaton, Clement. "Professor James Woodrow and the Freedom of Teaching in the South." *Journal of Southern History,* XXVIII (February, 1962), 3–17.

Fairchild, Herman L. "Memoir of Joseph LeConte." *Bulletin of the Geological Society of America,* XXVI (1915), 47–57.

Farquhar, Francis P. "Place Names in the High Sierra." *Sierra Club Bulletin,* XII (1924), 52–54.

Fink, Paul M. "Smoky Mountains History as Told in Place Names." *East Tennessee Historical Society's Publications,* VI (1934), 3–11.

Fink, Paul M., and Myron A. Avery. "The Nomenclature of the Great Smoky Mountains." *East Tennessee Historical Society's Publications,* IX (1937), 53–64.

Fisher, Osmond. "On the Amount of Elevations Attributable to Compression Through Contraction During Cooling of a Solid Earth." *Philosophical Magazine,* 5th ser., XXIII (February, 1887), 145–49.

Fogazzaro, Antonio. "For the Beauty of an Ideal." *Contemporary Review,* LXVII (May, 1895), 671–94.

Georgia University Magazine, IX (November, 1855), 28 and 61; XI (November, 1856), 92; and XI (February, 1857), 124.

Gilbert, G. K. ["Review of *Elements of Geology.*"] *Science,* New Series, IV (October 23, 1896), 620–21.

Gordon, Alexander. "Principal Nurseries and Private Gardens in the

United States of America." In J. C. Loudon, ed., *Gardener's Magazine*, VIII (1832), 286–89.

Gray, Asa. "Some North American Botanists: IV. John Eatton LeConte." *Botanical Gazette*, VIII (April, 1883), 197–99.

Guyot, Arnold. "On the Appalachian Mountain System." *American Journal of Science*, 2nd ser., XXXI (March, 1861), 157–87.

Harden, John M. B. "Cases of Diffuse Cellular Inflamation." *American Journal of the Medical Sciences*, XX (August, 1838), 396–402.

Helmholtz, Hermann von. "On the Normal Motions of the Human Eye in Relation to Binocular Vision." *Proceedings of the Royal Society of London, 1863–1864*, Vol. B. (April, 1864), 186–99.

Hilgard, Ernest. "Memoir of Joseph LeConte." *National Academy of Sciences Biographical Memoirs*, VI (1907), 147–218.

————. "An Estimate of the Life Work of Dr. Joseph LeConte." *University of California Magazine*, VII (September, 1901), 231–34.

Hunt, T. Sterry. "On Some Points in American Geology." *American Journal of Science*, 2nd ser., XXXI (May, 1861), 392–414.

————. "On Some Points in Dynamical Geology." *American Journal of Science*, 3rd ser., V (April, 1873), 264–69.

Hutson, Charles Woodward. "The South Carolina College in the late Fifties." *Sewanee Review*, XVIII (July, 1910), 333–43.

James, George Wharton. "Camping with LeConte." *Sunset*, XI (October, 1903), 563–66; and XVI (December, 1905), 197–201.

Jones, William Louis. "Joseph LeConte." *Transactions of the Medical Association of Georgia* (1902), 42–44.

"Joseph LeConte." *Engineering and Mining Journal*, LXXII (July 13, 1901), 34.

Keeler, Charles. "To Joseph LeConte." *Sierra Club Bulletin*, VI (January, 1908), 223–24.

Keep, Josiah. "Doubts Concerning Evolution." *Overland Monthly*, 2nd ser., XVIII (August, 1891), 190–98.

"LeConte Memorial Meeting." *University Chronicle*, IV (October, 1901), 241–64.

McNamara, John J. "Joseph LeConte: Nineteenth-Century Physician, Twentieth-Century Educator." *Southern Medicine*, LVI (June, 1973), 43–47.

Mallet, Robert. "Volcanic Energy: An Attempt to Develope Its True Origin and Cosmical Relations." *Proceedings of the Royal Society of London*, XX (1871–72), 438–41.

Marsh, O. C. "On the Supposed Human Footprints Recently Found in Nevada." *American Journal of Science*, 3rd ser., XXVI (August, 1883), 139–40.

["Memberships."] *Proceedings of the American Academy of Arts and Sciences*, XI (1875–76), 317, 322–23, 378.

["Memberships."] *Transactions of the New York Academy of Sciences*, III (1883), 132.

["Memberships."] *Proceedings of the American Association for the Advancement of Science*, III (1850), xiii.

["Memorial to Joseph LeConte."] *Sierra Club Bulletin*, IV (1903), Plates LI–LIV.

Merrill, George P. "Contributions to the History of Geology." *Report of the United States National Museum, for the Year Ending June 30, 1904* (1906), 189–733.

———. "Dana the Geologist." *Bulletin of the Geological Society of America*, XXIV (1913), 64–68.

Muir, John. "Reminiscences of Joseph LeConte." *University of California Magazine*, VII (September, 1901), 209–13.

"Obituary [of Joseph LeConte]." *American Journal of Science*, 4th ser., XII (October, 1901), 248.

"Obituary: Joseph LeConte." *Mines and Minerals*, XXII (August, 1901), 24–25.

Pfeifer, Edward J. "The Genesis of American Neo-Lamarckism." *Isis*, LVI (Summer, 1965), 156–67.

Pizer, Donald. "Evolutionary Ethical Dualism in Frank Norris' *Vandover and the Brute* and *McTeague*." *Proceedings of the Modern Language Association*, LXXVI (December, 1961), 552–60.

Prévost, Alexandre-Pierre. "Note sur la vision binoculaire." *Archives des Sciences Physiques et Naturelles*, Nouvelle Série, IV (1859), 105–11.

Rabun, James Z., ed. "Alexander H. Stephens's Diary." *Georgia Historical Quarterly*, XXXVI (March, 1952), 71–96; and XXXVI (June, 1952), 163–89.

Ramsey, Annie Sabra, ed. "Church Going at Midway, Georgia, as Remembered by Matilda Harden Stevens." *Georgia Historical Quarterly*, XXVIII (December, 1944), 270–80.

["Review of LeConte's *Elements of Geology*, 5th edition."] *National Geographic*, XIV (November, 1903), 425–26.

["Review of LeConte's *Religion and Science*."] *Methodist Quarterly Review*, 4th ser., XXVI (July, 1874), 520.

["Review of LeConte's *Religion and Science*."] *Nation*, XVIII (January 15, 1874), 46.

Ritter, William E. "Professor LeConte As Seen Through His Biological Work." *University of California Magazine*, VII (September, 1901), 218–25.

Rogers, William B., and Henry Darwin Rogers. "On the Physical Structure of the Appalachian Chain, as Exemplifying the Laws Which Have Regulated the Elevation of Great Mountain Chains Generally." *Report of the First, Second, and Third Meetings of the Association of American Geologists and Naturalists at Philadelphia in 1840 and 1841, and at Boston in 1842* (1843), 474–531.

Royce, Josiah. "Joseph LeConte." *International Monthly*, IV (1901), 327–29.

Schuchert, Charles. "Lower Devonic Aspect of the Lower Helderberg and Oriskany Formations." *Bulletin of the Geological Society of America*, XI (May 10, 1900), 241–332.

Scudder, Samuel H. "Memoir of John Lawrence LeConte, 1825–1883." *National Academy of Sciences Biographical Memoirs*, II (1886), 261–307.

Scripture, E. W. "Cerebral Light." *Science*, New Series, VI (July 23, 1897), 138–39.

———. "Cerebral Light: Further Observations." *Science*, New Series, IX (June 16, 1899), 850.

Soulé, Frank. "Joseph LeConte in the Sierra." *Sierra Club Bulletin*, IV (January , 1902), 1–10.

Stephens, Lester D. "Birdflight and Aviation: An Unheralded View." *American Aviation Historical Society Journal*, XXIII (Winter, 1978), 267–71.

———. "Evolution and Woman's Rights in the 1890s: The Views of Joseph LeConte." *Historian*, XXXVIII (February, 1976), 239–52.

———. "Farish Furman's Formula: Scientific Farming and the 'New South.'" *Agricultural History*, L (April, 1976), 377–90.

———. "Joseph LeConte and the Development of the Physiology and Psychology of Vision in the United States." *Annals of Science*, XXXVII (1980), 303–21.

———. "Joseph LeConte on Evolution, Education, and the Structure of Knowledge." *Journal of the History of the Behavioral Sciences*, XII (April, 1976), 103–19.

———. "Joseph LeConte's Contribution to American Ornithology." *Georgia Journal of Science*, XXXV (June, 1977), 170–81; corrigendum, *ibid.*, XXXVI (January, 1978), 48.

———. "Joseph LeConte's Evolutional Idealism: A Lamarckian View of Cultural History." *Journal of the History of Ideas*, XXXIX (July–September, 1978), 465–80.

———. "Of Mercury, Moses, and Medicine: Views of Dr. John M. B. Harden." *Georgia Historical Quarterly*, LIX (Winter, 1975), 402–15.

Stratton, G. M. "A Mirror Pseudoscope and the Limit of Visible Depth." *Psychological Review*, V (November, 1898), 623–38.

————. "Vision Without Inversion of the Retina." *Psychological Review*, III (November, 1896), 611–17; IV (July, 1897), 341–60; and IV (September, 1897), 463–81.

Thomson, Sir William. "Address at the Glasgow Meeting of the British Association," *American Journal of Science*, 3rd ser., XII (November, 1876), 336–54.

Towne, Joseph. "Contributions to the Physiology of Binocular Vision." *Guy's Hospital Reports*, 3rd ser., XII (1866), 285–301; XIV (1869), 54–83; and XV (1870), 180–212.

————. "The Stereoscope and Stereoscopic Results." *Guy's Hospital Reports*, 3rd ser., X (1864), 125–41; and XI (1865), 144–80.

University of California Magazine, VII (September, 1901), 207–46.

Willis, Bailey. "American Geology, 1850–1900." *Proceedings of the American Philosophical Society*, LXXXVI (1942), 34–44.

Unpublished Works

Bozeman, Theodore Dwight. "Joseph LeConte: Evolutionary Science and the Sociology of Miscegenation." Unpublished paper.

Brown, Timothy Odum. "Joseph LeConte, Prophet of Nature and Child of Religion." M.A. thesis, University of North Carolina, Chapel Hill, 1977.

Gurr, Charles Stephen. "Social Leadership and the Medical Profession in Antebellum Georgia." Ph.D. dissertation, University of Georgia, 1973.

Lupold, John Samuel. "From Physician to Physicist: The Scientific Career of John LeConte, 1818–1891." Ph.D. dissertation, University of South Carolina, 1970.

Sheftall, John McKay. "Eugenius A. Nisbet." A.B. honors thesis, University of Georgia, 1980.

Miscellaneous

United States Census Population Schedule. Bibb County, Georgia, 1850, and Richmond County, South Carolina, 1860.

A Chronological List of the Published Works of Joseph LeConte

"On the Science of Medicine and the Causes Which Have Retarded Its Progress." *Southern Medical and Surgical Journal*, New Series, VI (August, 1850), 456–74.

"Salt Beds, and Their Connection with Climate." *Georgia University Magazine*, V (April, 1853), 8–14.

"Sun Drawing Water." *Georgia University Magazine*, VII (August, 1854), 225–29.

"Utilitarian Spirit of the Age." *Georgia University Magazine*, VII (March, 1855), 353–62.

"Classics *vs.* Mathematics." *Georgia University Magazine*, X (June, 1856), 69–75.

"On the Agency of the Gulf-Stream in the Formation of the Peninsula of Florida." *Proceedings of the American Association for the Advancement of Science*, X, Pt. 2 (August, 1856), 103–19. Reprinted in *American Journal of Science*, 2nd ser., XXIII (May, 1857), 46–60.

"Lectures on Coal." *Annual Report of the Board of Regents of the Smithsonian Institution, 1857.* Washington, D.C., 1858. Pp. 119–68. Translated and republished as pamphlet *Sur les plantes des terrains carbonifères* [1857, n.p.].

"Inaugural Address." Delivered on December 8, 1857, and published in Columbia, S.C., 1858. 31 pp. [title later inscribed by LeConte, "On the Function of Geology in a Course of Education"].

"Morphology and Its Connection with Fine Art." *Southern Presbyterian Review*, XII (April, 1859), 83–114.

"The Principles of a Liberal Education." Address delivered on April 20, 1859, and published in Columbia, S.C.: R. W. Gibbes, 1859. Reprinted in *Southern Presbyterian Review*, XII (July, 1859), 310–35.

"The Correlation of Physical, Chemical, and Vital Force, and the Conservation of Force in Vital Phenomena." *Proceedings of the American Association for the Advancement of Science*, XIII (August, 1859), 187–203. Reprinted in *American Journal of Science*, 2nd ser., XXVIII (November, 1859), 305–19; *Philosophical Magazine*, XIX (February, 1860), 133–48. Abstracted in *Canadian Naturalist*, IV (August, 1859), 291–93. Revised and reprinted in *Popular Science Monthly*, IV (December, 1873), 156–70; and in Balfour Stewart, ed., *The Conservation of Energy.* New York: D. Appleton, 1874. Pp. 169–201.

"Theory of the Formation of Continents." *Canadian Naturalist and Geologist*, IV (August, 1859), 293. [Abstract of an address delivered before the AAAS in August, 1859.]

"The Relation of Organic Science to Sociology." *Southern Presbyterian Review*, XIII (April, 1860), 39–77. Revised and reprinted as "Scientific Relation of Sociology to Biology." *Popular Science Monthly*, XIV (January, 1879), 325–36.

"An Address [on Female Education], Delivered on Commencement

Day of the Laurensville Female College, June 28, 1860." Published by request of the Trustees. Laurensville, S.C.: J. Hollingsworth, Printer, 1860. 27 pp. Reprinted as "Female Education." *Southern Presbyterian Review*, XIII (April, 1861), 60–91.

"Natural History as a Branch of School Education; and the School, the College, and the University, in Relation to One Another and to Active Life." *Southern Presbyterian Review*, XIV (July, 1861), 185–226.

Instructions for the Manufacture of Saltpetre. Booklet published in Columbia, S.C.: C. P. Pelham, State Printer, 1862.

"On the Nature and Uses of Art." *Southern Presbyterian Review*, XV (January, 1863), 311–48, and (April, 1863), 515–48.

"'On the Law of the Sexes,' or the Production of the Sexes at Will." *Nashville Journal of Medicine and Surgery*, New Series, I (October, 1866), 296–99.

["Corrigendum and Response to 'On the Law of the Sexes.'"] *Nashville Journal of Medicine and Surgery*, New Series, II (April, 1867), 332–33.

"On Some Phenomena of Binocular Vision: Adjustments of the Eye." *American Journal of Science*, 2nd ser., XLVII (January, 1869), 68–77. Reprinted in *Philosophical Magazine*, XXXVII (January, 1869), 131–40.

"On Some Phenomena of Binocular Vision: Rotation of the Eye on the Optic Axis and the Horopter." *American Journal of Science*, 2nd ser., XLVII (March, 1869), 153–78. Reprinted in *Philosophical Magazine*, XXXVIII (September, 1869), 179–202.

"On Some Phenomena of Binocular Vision: The Mode of Representing the Position of Double Images." *American Journal of Science*, 3rd ser., I (January, 1871), 33–44.

"On an Optical Illusion." *Philosophical Magazine*, XLI (April, 1871), 266–69.

"On Some Phenomena of Binocular Vision: Stereoscopic Phenomena." *American Journal of Science*, 3rd ser., II (July, 1871), 1–10.

"On Some Phenomena of Binocular Vision: So-called 'Images of Illusion,' and the Theory of Binocular Relief." *American Journal of Science*, 3rd ser., II (November, 1871), 315–23. Reprinted as "Sur les Images d'Illusion et sur la Théorie du Relief Binoculaire." *Archives des Sciences Physiques et Naturelles*, Nouvelle Série, XLI (1871), 394–422.

"On Some Phenomena of Binocular Vision." *American Journal of Science*, 3rd ser., II (December, 1871), 417–26.

"Natural Law of Circulation." *Proceedings of the California State Teachers' Institute, September 13–16*. Sacramento: D. W. Gelwicks, 1871. Pp. 54–67.

"Earthquakes." *University* [of California] *Echo*, II (April–July, 1872), n.p.

"A Theory of the Formation of the Great Features of the Earth's Surface." *American Journal of Science*, 3rd ser., IV (November, 1872), 345–55, 460–72.

"Sur la Transparence des Images Doubles." *Archives des Sciences Physiques et Naturelles*, Nouvelle Série, XLV (1872), 229–32.

"Note [Regarding Source of Volcanic Energy]." *American Journal of Science*, 3rd ser., V (February, 1873), 156.

"On Some of the Ancient Glaciers of the Sierras." *American Journal of Science and Arts*, 3rd ser., V (May, 1873), 325–42. Also published in *Proceedings of the California Academy of Science*, IV (1873), 259–62.

"On the Formation of the Features of the Earth-Surface. Reply to Criticisms of T. Sterry Hunt." *American Journal of Science*, 3rd ser., V (June, 1873), 448–53.

"Flight of Birds." *Nature*, IX (November 6, 1873), 5.

Religion and Science. New York: D. Appleton, 1873. [Numerous reprints through 1898.]

["Agassiz Memorial Address."] *Proceedings of the California Academy of Sciences*, V (1874), 230–36.

"On the Great Lava-Flood of the West; and on the Structure and Age of the Cascade Mountains." *American Journal of Science*, 3rd ser., VII (March and April, 1874), 167–80, 259–367 [pagination error; should be pp. 259–67].

"Astronomy and Geology." *Berkeleyan*, I (May, 1874), 1–4.

"LeConte on the Origin of Western Mounds." *Popular Science Monthly*, V (August, 1874), 506–507 [a summary of a portion of his article "On the Great Lava-Flood." *American Journal of Science*, 3rd ser., VII (March and April, 1874), 167–80, 259–67].

"Correlation of Vital with Chemical and Physical Forces." In Balfour Stewart, ed., *The Conservation of Energy*. New York: D. Appleton, 1874. Pp. 169–201.

"In Binocular Vision the Law of Corresponding Points May Be Opposed to the Law of Direction." *American Journal of Science*, 3rd ser., IX (March, 1875), 164–68.

"Comparative Physiology of Binocular Vision." *American Journal of Science*, 3rd ser., IX (March, 1875), 168–71.

"On Some Phenomena of Binocular Vision: Position of the Eyes in Sleepiness." *American Journal of Science*, 3rd ser., IX (March, 1875), 159–64.

"Rate of Growth of Corals." *American Journal of Science*, 3rd ser., X (March, 1875), 34–36.

"On Some of the Ancient Glaciers of the Sierra Nevada." *American Journal of Science*, 3rd ser., X (August, 1875), 126–39. Reprinted in *Pro-

ceedings of the California Academy of Sciences, VI (1875), 38–48.
"Instinct and Intelligence." *Popular Science Monthly*, VII (October, 1875), 653–64.
A Journal of Ramblings Through the High Sierra of California by the "University Excursion Party." San Francisco: Francis & Valentine, 1875. Portions reprinted in *Sierra Club Bulletin*, III (1900), 1–107; later republished in its entirety by the Sierra Club, 1930 and 1960 (with slight variations of illustrations); portions also reprinted in *Overland Monthly*, 2nd ser., VI (October, November, and December, 1885), 414–23, 493–505, and 624–38, respectively.
"On the Evidences of Horizontal Crushing in the Formation of the Coast Range of California." *American Journal of Science*, 3rd ser., XI (April, 1876), 297–304.
"The True Idea of a University." *Berkeleyan*, August, 1876. 3 pp.
"Review of Alfred Russel Wallace's *The Geographical Distribution of Animals.*" San Francisco *Evening Bulletin*, January 20, 1877.
"Hog Wallows or Prairie Mounds." *Nature*, XV (April 19, 1877), 530–31.
"Note on the Binocular Phenomena Observed by Professor Nipher." *American Journal of Science*, 3rd ser., XV (April, 1877), 252–53.
"On Critical Periods in the History of the Earth and Their Relation to Evolution; and on the Quaternary as Such a Period." *American Journal of Science*, 3rd ser., XIV (August, 1877), 99–114. Reprinted in *American Naturalist*, XI (1877), 520–57; *Kansas City Review of Science and Industry*, I (1877), 478–83, 522–30.
"Some Peculiarities of Phantom Images Formed by Binocular Combination of Regular Figures." *American Journal of Science*, 3rd ser. XIV (August, 1877), 97–107.
"On Some Phenomena of Binocular Vision: The Structure of the Crystalline Lens and Its Relation to Periscopism." *American Journal of Science*, 3rd ser., XIV (September, 1877), 191–95.
["Evolution in Relation to Religion."] *Proceedings at the Annual Dinner of the Chit-Chat Club*, (1877), 1–12.
Elements of Geology. New York: D. Appleton, 1877. Rev. eds., 1882, 1891, 1896; rev. and partly rewritten by Herman LeRoy Fairchild, as 5th ed., 1903. [Some of the numerous reprintings contained new plates and illustrations but no substantial textual revisions.]
"On Binocular Vision." *Philosophical Magazine*, 5th ser., V (January, 1878), 27–29.
"Some Thoughts on the Glycogenic Function of the Liver and Its Relation to Vital Force and Vital Heat." *American Journal of Science*, 3rd ser., XV (February, 1878), 99–107.
"Geysers and How They Are Explained." *Popular Science Monthly*, XII

(February, 1878), 407–17. [Extracted from *Elements of Geology*.]

"Science and Mental Improvement." *Popular Science Monthly*, XIII (May, 1878), 96–101.

"On the Structure and Origin of Mountains, with Special Reference to the 'Contractional Theory.'" *American Journal of Science*, 3rd ser., XVI (August, 1878), 95–112.

"Geological Climate and Geological Time." *Nature*, XVIII (October 24, 1878), 668.

"Man's Place in Nature." *Princeton Review*, 4th ser., II (November, 1878), 776–803.

"Scientific Relation of Sociology to Biology." *Popular Science Monthly*, XIV (January, 1879), 325–36, and (February, 1879), 425–34. Revision of "The Relation of Organic Science to Sociology." *Southern Presbyterian Review*, XIII (April, 1860), 39–77.

"On the Extinct Volcanoes About Lake Mono, and Their Relation to the Glacial Drift." *American Journal of Science*, 3rd ser., XVIII (July, 1879), 35–44.

"The Genesis of Sex." *Popular Science Monthly*, XVI (December, 1879), 167–79. Reprinted as "L'Origine des Sexes." *La Revue Scientifique*, XVI (February 7, 1880), 770–76.

"Some Thoughts on the Glycogenic Function of the Liver: Disposal of Waste." *American Journal of Science*, 3rd ser., XIX (January, 1880), 25–29.

"The Old River-Beds of California." *American Journal of Science*, 3rd ser., XIX (March, 1880), 176–90.

"The Essential Characteristics and Mutual Relations of the School, the College, and the University." *Princeton Review*, 4th ser., V (March, 1880), 177–204. Part 3 reprinted in condensed form in *California Teacher*, II (October, 1883), 153–62.

"The Effect of Mixture of Races on Human Progress." *Berkeley Quarterly*, I (April, 1880), 81–104.

"On Some Phenomena of Binocular Vision: Laws of Ocular Motion." *American Journal of Science*, 3rd ser., XX (August, 1880), 83–93.

"Coral Reefs and Islands." *Nature*, XXII (October 14, 1880), 558–59.

"Evolution in Relation to Materialism." *Princeton Review*, 4th ser., VII (March, 1881), 149–74. Abridged and reprinted in *California Teacher*, II (February, 1884), 520–29.

"Comte's Classification of the Sciences." *Berkeley Quarterly*, II (April, 1881), 97–117.

"Mr. Wallace's 'Island Life.'" *Californian*, III (June, 1881), 485–88.

"To Prevent Drowning." *Nature*, XXIV (July 21, 1881), 260.

"Devilution." *Occident*, I (November 3, 1881), 2.

"Illustrations of a Law of Evolution of Thought." *Princeton Review*, 4th ser., VIII (November, 1881), 373–93.

The Higher Utilities of Science. Published pamphlet of address delivered before the Chit-Chat Club, San Francisco, November 14, 1881, 15 pp. Reprinted in *Californian*, V (April, 1882), 306–11.

Sight: An Exposition of the Principles of Monocular and Binocular Vision. New York: D. Appleton, 1881; rev. ed., 1897.

"Brief Sketch of the Geology of California." In Alonzo Phelps, ed., *Contemporary Biography of California, with Contributions from Distinguished Scholars and Scientists.* Vol. I. San Francisco: A. L. Bancroft, 1881. Pp. 290–304.

"Review of John W. Judd's *Volcanoes. What They Are and What They Teach.*" *Californian*, V (January, 1882), 85–86.

"Rate of Denudation." *Geological Magazine*, New Series, IX (June, 1882), 289.

"The Phenomena of Metalliferous Vein-formation Now in Progress at Sulphur Bank, California." *American Journal of Science*, 3rd ser., XXIV (July, 1882), 23–33. (Coauthored by W. B. Rising.)

"On Certain Remarkable Tracks, Found in the Rocks of Carson Quarry." *Proceedings of the California Academy of Sciences* (August 27, 1882), 1–10 + plates [not published until 1883]. Also published in *Independent*, XXXIV (September 28, 1882), 7–9.

"Mutual Relations of Intellectual and Moral Culture." *Overland Monthly*, 2nd ser., I (January, 1883), 10–16.

"Science and Literature." *Overland Monthly*, 2nd ser., I (February, 1883), suppl., pp. 8–10.

"Movement of the Arms in Walking." *Science*, I (March 30, 1883), 220–21.

"Domestic Ducks That Fly Abroad Like Pigeons." *Science*, I (April 6, 1883), 249.

"Carson Footprints." *Nature*, XXVIII (May 31, 1883), 101–102.

"Review of Wright's *Studies in Science and Religion.*" *Science*, I (June 15, 1883), 543–45.

"On Mineral Vein Formation Now in Progress at Steamboat Springs Compared with the Same at Sulphur Bank." *American Journal of Science*, 3rd ser., XXV (June, 1883), 424–28.

"On the Genesis of Metalliferous Veins." *American Journal of Science*, 3rd ser., XXVI (July, 1883), 1–19.

"Continent Formation." *Geological Magazine*, New Series, X (November, 1883), 523–24.

"The Reefs, Keys, and Peninsula of Florida." *Science*, II (December 14, 1883), 764.

Die Lehre vom Sehen. Leipzig: F. A. Brockhaus, 1883. [Translation of *Sight: An Exposition of the Principles of Monocular and Binocular Vision*, with a special introduction by the publisher.]

"Elevation and Subsidence." *Nature*, XXIX (January 3, 1884), 212–13.

"Carrying-Power of Fluid Currents." *Popular Science Monthly*, XXIV (February, 1884), 555–56.

"Review of Fiske's *Excursions of an Evolutionist.*" *Overland Monthly*, 2nd ser., III (March, 1884), 329–31.

"Rightsideness: Letter on Deflection in Walking." *Nature*, XXIX (March 13, 1884), 452; and XXX (May 22, 1884), 76–77.

"A Singular Optical Phenomenon." *Science*, III (April 4, 1884), 404.

"Review of Guyot's *Creation.*" *Science*, III (May 16, 1884), 599–601.

"The Psychical Relation of Man to Animals." *Princeton Review*, 4th ser., XIII (May, 1884), 236–61. Revised and reprinted as "From Animal to Man." *Monist*, VI (April, 1896), 356–81.

"Review of *Annual Reports of the United-States Geological Survey*, Vols. II & III." *Science*, IV (July 18, 1884), 62–71.

"Pressure of Currents." *Mining Record*, XVI (November 29, 1884), 341–42.

Compend of Geology. New York: D. Appleton, 1884; rev. ed., New York: American Book Company, 1898.

"Relation of Berkeley to Certain Phases of Modern Thought." *Berkeleyan*, XIX (February 17, 1885), 27–28.

"The General Principles of Art and Their Application to the 'Novel.'" *Overland Monthly*, 2nd ser., V (April, 1885), 337–47.

"Review of Warren's *Paradise Found.*" *Science*, V (May 15, 1885), 406–407.

"Review of Royce's *The Religious Aspect of Philosophy.*" *Overland Monthly*, 2nd ser., V (May, 1885), 542–44.

"Immortality in Modern Thought." *Science*, VI (August 14, 1885), 126–27.

"The Result of Shad Propagation on the Atlantic Coast." *Science*, VI (December 11, 1885), 520.

"Earthquake-Shocks More Violent on the Surface Than in Mines." *Science*, VI (December 18, 1885), 540.

"Review of *Louis Agassiz: His Life and Correspondence.*" *Overland Monthly*, 2nd ser., VII (January, 1886), 103–105.

"Review of Fiske's *Darwinism and Other Essays.*" *Overland Monthly*, 2nd ser., VII (March, 1886), 334.

"On the Permanence of Continents and Ocean-Basins, with Special Reference to the Formation and Development of the North American Continent." *Geological Magazine*, III (March, 1886), 97–101; corrigendum, *ibid.*, III (June, 1886), 288.

"Glycogenic Function of the Liver." *American Naturalist*, XX, Pt. 1 (May, 1886), 473–74.

"Pharyngeal Respiratory Movements of Adult Amphibia Under Water." *Science*, VII (May 21, 1886), 462.

"Double Vision." *Science*, VII (June 4, 1886), 506.

"The Development of the North American Continent." *Geological Magazine*, III (June, 1886), 287–88.

"Germ of Hydrophobia." *Science*, VIII (July 30, 1886), 102.

"A Case of Inherited Polydactylism." *Science*, VIII (August 20, 1886), 166.

"A Post-Tertiary Elevation of the Sierra Nevada Shown by the River Beds." *American Journal of Science*, 3rd ser., XXXII (September, 1886), 167–81.

"Star Rays." *Science*, IX (January 7, 1887), 14.

"Relation of Biology to Sociology." *Berkeleyan*, XXIII (May, 1887), 123–31.

"Determination of the Depth of Earthquakes." *Science*, X (July 8, 1887), 22–24.

"On Some Phenomena of Binocular Vision: Some Peculiarities of the Phantom Images Formed by Binocular Combination of Regular Figures." *American Journal of Science*, 3rd ser., XXXIV (August, 1887), 97–107.

"The Flora of the Coast Islands of California in Relation to Recent Changes of Physical Geography." *Bulletin of the California Academy of Sciences*, VIII (September, 1887), 515–20. Reprinted in *American Journal of Science*, 3rd ser., XXXIV (December, 1887), 457–60; *American Geologist*, I (February, 1888), 76–81.

"What is Evolution?" *Popular Science Monthly*, XXXI (October, 1887), 721–35. [Reprinted from *Evolution and Its Relation to Religious Thought*.]

"Agassiz and Evolution." *Popular Science Monthly*, XXXII (November, 1887) 17–26. [Reprinted from *Evolution and Its Relation to Religious Thought*.]

"Sound-Blindness." *Science*, X (December, 1887), 312.

The Relation of Evolution to Religious Thought. Pamphlet published for the Pacific Coast Conference of Unitarian and Other Christian Churches. San Francisco: C. A. Murdock & Co., [1887]. 15 pp.

"Evolution and Religious Thought." *Popular Science Monthly*, XXXII (January, 1888), 311–15. [Reprinted from *Evolution and Its Relation to Religious Thought*.]

"A Difficulty Regarding Evolution." *Popular Science Monthly*, XXXII

(February, 1888), 554, and (May, 1888), 125.

"Monocular vs. Binocular Vision." *Science*, XI (March 9, 1888), 119.

"Sense-Training and Hand-Training in the Public Schools." *Pacific Educational Journal*, III (March, 1888), 41–52.

"Mountain Formation." *Philosophical Magazine*, 5th ser., XXV (May, 1888), 450–51.

"Significance of Sex." *Science*, XI (May, 1888), 229–30.

"Glacial Motion." *Philosophical Magazine*, 5th ser., XXV (May, 1888), 452.

"The Relation of Evolution to Materialism." *Popular Science Monthly*, XXXIII (May, 1888), 79–86. [Reprinted from *Evolution and Its Relation to Religious Thought*.]

"Experiments in Vision Again." *Science*, XI (May 14, 1888), 252.

"Responsive Address in Behalf of the Board of Regents." In *The Formal Transfer of the Lick Observatory to the Board of Regents of the University, Berkeley, June 27, 1888*. Sacramento: State Office, 1888. Pp. 14–24.

"Human Beings as Pack-Animals." *Science*, XI (June, 1888), 290.

["Comment on Hark's *The Unity of Truth*."] *Popular Science Monthly*, XXXIII (September, 1888), 699.

"The Problem of a Flying-Machine." *Popular Science Monthly*, XXXIV (November, 1888), 69–76.

Evolution and Its Relation to Religious Thought. New York: D. Appleton, 1888, and London: Chapman and Hall, 1888; rev. ed., *Evolution: Its Nature, Its Evidences, and Its Relation to Religious Thought*, 1891.

"Grounds of My Belief in Immortality." *Christian Register*, April 18, 1889, pp. 247–48.

"The General Interior Condition of the Earth." *American Geologist*, IV (July, 1889), 38–44.

"The South Revisited." *Overland Monthly*, 2nd ser., XIV (July, 1889), 22–31.

"Ptomaines and Leucomaines, and Their Relation to Disease." *Pacific Medical Journal*, XXXII (September, 1889), 529–32. Reprinted in *Science*, XIV (November 8, 1889), 322–23.

"On the Origin of Normal Faults and of the Structure of the Basin Region." *American Journal of Science*, 3rd ser., XXXVIII (October, 1889), 257–63.

"The Natural Grounds of Belief in a Personal Immortality." *Andover Review*, XIV (July, 1890), 1–13.

"Review of Rev. Howard MacQueary's *The Evolution of Man and Christianity*." *Overland Monthly*, 2nd ser., XVI (July, 1890), 110–11.

"On a Curious Visual Phenomenon." *American Journal of Psychology*, III (September, 1890), 364–66.

"Tertiary and Post-Tertiary Changes of the Atlantic and Pacific Coasts; with a Note on the Relation Between Land-Elevation and Ice-Accumulation During the Quaternary Period." *Bulletin of the Geological Society of America*, II (March 16, 1891), 323–30.

"The Factors of Evolution." *Monist*, I (April, 1891), 321–35.

"Evolution and Human Progress." *Open Court*, V (April 23, 1891), 2279–83.

"The Relation of the Church to Modern Scientific Thought." *Andover Review*, XVI (July, 1891), 1–11.

"Origin of Organic Forms—Is It Natural or Supernatural Process?" *Overland Monthly*, 2nd ser., XVIII (August, 1891), 198–203.

"Plato's Doctrine of the Soul and Argument for Immortality, in Comparison with the Doctrine and Argument Derived from the Study of Nature." *University of California Philosophical Union Bulletin*, VIII (1891), 1–19.

"The Relation of Philosophy to Psychology and to Physiology." *Bulletin of the Philosophical Society of Washington*, XII (February, 1892), 19–38.

"New Worlds in Time." *Proceedings of the Chit-Chat Club* (1892), 34–39.

"The Race Problem in the South." In *Man and the State* [Brooklyn Ethical Association]. New York: D. Appleton, 1892. Pp. 349–82. Reprinted, Miami, Florida: Mnemosyne Publishers, 1961.

"The Relation of Organic Evolution to Human Progress." *Pacific Coast Teacher*, II (February, 1893), 131–39.

"Address on Behalf of the Faculties." *Addresses at the Inauguration of Martin Kellogg, LL.D., President of the University of California*. Berkeley: University of California, March 23, 1893. Pp. 18–25.

"Estimated Distance of Phantoms." *Science*, XXI (June 16, 1893), 333–35.

"Theories of the Origin of Mountain Ranges." *Proceedings of the American Association for the Advancement of Science*, XLII (August, 1893), 1–27. Reprinted in *Journal of Geology*, I (September–October, 1893), 543–73; *Nature*, XLVIII (October 5, 1893), 551–54.

Evolutional Idealism. Pamphlet published by the University of California Philosophical Union, 1893. 4 pp.

[*Work of Agassiz*.] Pamphlet published by the Harvard Club of San Francisco, [1893 or 1894]. Pp. 10–17.

"Evolu[t]ion, An Element in Fundamental Religious Thought." *Jewish Progress*, March 23, [1894]. 4 pp.

"New Lights on the Problem of Flying." *Popular Science Monthly*, XLIV (April, 1894), 744–57.

"Criticisms of Posepny's *Genesis of Ore Deposits*." *Transactions of the*

American Institute of Mining Engineers, XXIV (1894), 996–1006.

["Conservation of Energy."] *Proceedings of the Chit-Chat Club* (1894), 18–22.

"Review of Geikie's *Memoir of Sir Andrew Crombie Ramsay.*" *Science*, New Series, I (May 3, 1895), 490–93.

"Review of Dana's *Manual of Geology.*" *Science*, New Series, I (May 17, 1895), 548–50.

"The Theory of Evolution and Social Progress." *Monist*, V (July, 1895), 481–500.

"Causes of the Gulf Stream." *Science*, New Series, II (August 16, 1895), 188–89.

"The Nature of Vowels." *Science*, New Series, II (August 16, 1895), 189.

"Are Consequences Ever a Test of Truth?" *Science*, New Series, II (September 20, 1895), 379–80.

"Erect Vision and Single Vision." *Science*, New Series, II (November 8, 1895), 629–30.

"Inverted Image Once More." *Science*, New Series, II (November 15, 1895), 667–68.

"A Last Word on Erect Vision." *Science*, New Series, II (December 20, 1895), 850–51.

"The Effect of the Theory of Evolution on Education." *Proceedings of the National Education Association* (1895), 149–61. Reprinted in *Educational Review*, X (September, 1895), 121–36.

"Critical Periods in the History of the Earth." University of California, *Bulletin of the Department of Geology*, I, No. 11 [1895], 313–36.

"Memoir of John LeConte, 1818–1891." *National Academy of Science Biographical Memoirs*, III (1895), 369–93.

"From Animal to Man." *Monist*, VI (April, 1896), 356–81. Revision of "The Psychical Relation of Man to Animal." *Princeton Review*, XIII (May, 1884), 236–61.

"Memoir of James Dwight Dana." *Bulletin of the Geological Society of America*, VII (April 21, 1896), 461–79. Reprinted in Daniel C. Gilman. *The Life of James Dwight Dana*. New York: Harper & Brothers, 1899. Pp. 248–60.

"Honors to James Hall at Buffalo." *Science*, New Series, IV (November 13, 1896), 698–99.

The Religion of a Man of Science. Pamphlet published in New York: Lenox Avenue Unitarian Church [1896]. 12 pp.

"The Relation of Biology to Philosophy." *Arena*, XVII (March, 1897), 459–67.

"The Cöoperation of Religion and Science in Uplifting Humanity." Unitarian Club Address, April 26, 1897.

"Cerebral Light." *Science*, New Series, VI (August 13, 1897), 257–58.

"Professor Cattell's Review of *Sight*." *Science*, New Series, VI (November 12, 1897), 737–39.

"Earth-Crust Movements and Their Causes." In *Annual Report of the Board of Regents of the Smithsonian Institution, 1896*. Washington, D.C.: 1897. Pp. 233–44. Reprinted in *Bulletin of the Geological Society of America*, VIII (February 15, 1897), 113–26; *Science*, New Series, V (February 26, 1897), 321–30.

"God, and Connected Problems, in the Light of Evolution." In Josiah Royce, Joseph LeConte, G. H. Howison, and Sidney Edward Mezes, *The Conception of God*. New York: Macmillan, 1898. Pp. 65–78. Originally delivered in August, 1895, and printed in the Philosophical Union of the University of California *Bulletin*, XV (1895), 43–50.

"The True Idea of a University." *University Chronicle*, I (February, 1898), 3–19.

"Main Characteristics of University Education." *Berkeleyan*, I (March 31, 1898), 110–13.

["Symposium on Geological Classification and Nomenclature."] *Journal of Geology*, VI (May–June, 1898), 337–38.

["Optical Illusion of a Rotating Fan."] *Science*, New Series, VIII (October 7, 1898), 480–81.

"The Origin of Transverse Mountain-Valleys and Some Glacial Phenomena in Those of the Sierra Nevada." *University Chronicle*, I (December, 1898), 479–97.

"Don's Work on Auriferous Veins." *Transactions of the American Institute of Mining Engineers*, XXVII (1898), 993.

"Cerebral Light Again." *Science*, New Series, X (July 14, 1899), 58.

"The Ozarkian and Its Significance in Theoretical Geology." *Journal of Geology*, VII (September–October, 1899), 525–44.

"Address [on James Dwight Dana]." In Daniel C. Gilman, *The Life of James Dwight Dana*. New York: Harper & Brothers, 1899. Pp. 248–60.

"A Note on the Religious Significance of Science." *Monist*, X (January, 1900), 161–66.

"Shadow Images on the Retina." *Psychological Review*, VII (January, 1900), 18–28.

"An Early Geological Excursion." *Science*, New Series, XI (February 9, 1900), 221.

"A Century of Geology." *Popular Science Monthly*, LVI (February, 1900), 431–43, and (March, 1900), 546–56. Reprinted in *Annual Report of the Regents of the Smithsonian Institution, 1900*. Washington, D.C., 1901. Pp. 265–87.

"Reply to Professor Kingsley's Criticism." *Science*, New Series, XI (June

8, 1900), 909–11.

"My Trip to King's River Cañon." *Sunset*, V (October, 1900), 275–85. Reprinted in *Sierra Club Bulletin*, IV (June, 1902), 88–99.

Outlines of the Comparative Physiology and Morphology of Animals. New York: D. Appleton, 1900.

"The Larynx as an Instrument of Music." *Science*, New Series, XIII (May 17, 1901), 790.

"What Is Life?" *Science*, New Series, XIII (June 21, 1901), 991–92.

The Autobiography of Joseph LeConte. New York: D. Appleton, 1903.

'*Ware Sherman: A Journal of Three Months' Personal Experience in the Last Days of the Confederacy.* Introduction by Caroline LeConte. Berkeley: University of California Press, 1937.

Index

333

0